Lead Compounds from Medicinal Plants for the Treatment of Cancer

A volume in the *Pharmaceutical Leads from Medicinal Plants* series

*'The only limit to our realization of tomorrow will be our doubts of today.
Let us move forward with strong and active faith.'*

Franklin D. Roosevelt
President of the United States of America (1933–1945)

Lead Compounds from Medicinal Plants for the Treatment of Cancer

Christophe Wiart, PharmD, PhD

Ethnopharmacologist

AMSTERDAM • BOSTON • HEIDELBERG • LONDON • NEW YORK • OXFORD
PARIS • SAN DIEGO • SAN FRANCISCO • SINGAPORE • SYDNEY • TOKYO
Academic Press is an imprint of Elsevier

Academic Press is an imprint of Elsevier

32 Jamestown Road, London NW1 7BY, UK

225 Wyman Street, Waltham, MA 02451, USA

525 B Street, Suite 1800, San Diego, CA 92101-4495, USA

First edition 2013

Notice
No responsibility is assumed by the publisher for any injury and/or damage to persons or property as
a matter of products liability, negligence or otherwise, or from any use or operation of any methods,
products, instructions or ideas contained in the material herein. Because of rapid advances in the
medical sciences, in particular, independent verification of diagnoses and drug dosages should be made.

British Library Cataloguing-in-Publication Data
A catalogue record for this book is available from the British Library

Library of Congress Cataloging-in-Publication Data
A catalog record for this book is available from the Library of Congress

ISBN: 978-0-12-398371-8

Front cover image: The illustration of *Piper longum* is used courtesy of Christophe Wiart.

For information on all Academic Press publications
visit our website at elsevierdirect.com

Typeset by MPS Limited, Chennai, India
www.adi-mps.com

Contents

Foreword

By Gordon M. Cragg

For many millennia, humans have relied on Nature as a source of medicines for the treatment of a wide spectrum of diseases. Plants, in particular, have formed the basis of sophisticated traditional medicine systems, which, according to the World Health Organization (WHO) estimates, continue to play an essential role in the primary healthcare of approximately 65% of the world's population. This traditional use as phytomedicines has guided chemists and pharmacologists in the selection of plants for investigation as a source of novel, bioactive natural products. Consequently, plants have historically been at the forefront of natural product drug discovery, and, in the anticancer area, plant-derived compounds have been an important source of several clinically useful chemotherapeutic agents. These include vinblastine, vincristine, the camptothecin derivatives, topotecan and irinotecan, etoposide, derived from epipodophyllotoxin, and paclitaxel (Taxol®).

With the discovery of antibiotics and the development of effective techniques for the collection of marine organisms, the study of microbes and marine invertebrates as a source of potential novel agents for the treatment of cancer and other serious diseases has greatly expanded the scope of natural product drug discovery. Analyses of the sources of new drugs reported in the past three to four decades have clearly demonstrated the continuing and valuable contributions of Nature as a source, not only of potential chemotherapeutic agents, but also of lead compounds which have provided the basis and inspiration for the semi- or total synthesis of effective new drugs. Thus, Nature, or in modern terminology, 'Nature's Combinatorial Library', has contributed to the discovery and development of over 60% of the anticancer drugs currently in regular use, or undergoing clinical study.

The potential of Nature in this regard, however, has barely been tapped, and plants still remain a relatively unexplored area for anticancer drug discovery. I am, therefore, delighted to have been invited to write a foreword for this volume. The author has carefully selected several medicinal plants from around the world based on reports of significant pharmacological evidence for their potential to treat one or more forms of cancer. For each plant, information concerning the relevant botanical and ethnobotanical aspects, medicinal uses, phytochemistry and phytopharmacology is given, followed by a proposal as to which particular class of chemical constituents merits further cancer-related study. A rationale for the selection of the particular chemical class to be studied is provided, including detailed antitumor screening data, related mechanistic data, and discussion of the structural features which affect the level of activity. The coverage of each plant and its constituents is extensive and well referenced to recent literature, and is accompanied by numerous chemical structures.

The probability of a directly isolated natural product, such as taxol® in the anticancer area, becoming the actual drug used for the treatment of a given disease is relatively low, but as the title of this volume implies, these natural molecules can serve as lead compounds for the development of analogues having the pharmacological properties necessary for advanced pre-clinical and clinical development. Recent advances in synthetic methodology and strategy are overcoming the barriers posed by the structural complexity of most natural products. In addition, natural products have been evolutionarily selected to bind to biological macromolecules, and they thus represent 'privileged structures' which are excellent templates for the synthesis of novel, bioactive, natural product-like molecules through the application of combinatorial and medicinal chemical methodology. Of course, suitable biological assays for evaluation of the structure–activity relationships (SAR) of the products of the optimization process are required for all these approaches, and thus a multidisciplinary, collaborative approach, often conducted at an international level, is required for effective drug discovery and development.

The wealth of information on an extensive array of bioactive lead molecules presented in this volume will provide the chemical, pharmacological and medical research communities with valuable leads for collaborative development of potentially more effective agents for the treatment of the many forms of cancer which afflict our global community. I feel that the author is to be congratulated on the preparation of a well-documented and comprehensive coverage of a topic which should be of considerable value to researchers involved in anticancer drug discovery and development.

Gordon M. Cragg
NIH Special Volunteer
Natural Products Branch
National Cancer Institute
Frederick, Maryland, USA

Gordon Cragg obtained his undergraduate training in chemistry at Rhodes University, South Africa, and his D. Phil. (organic chemistry) from Oxford University in 1963. After two years of post-doctoral research at the University of California, Los Angeles, he returned to South Africa to join the Council for Scientific and Industrial Research. In 1966, he joined the Chemistry Department at the University of South Africa, and transferred to the University of Cape Town in 1972. In 1979, he returned to the US to join the Cancer Research Institute at Arizona State University working with Professor G. R. Pettit. In 1985, he moved to the National Cancer Institute (NCI), National Institutes of Health (NIH) in Bethesda, Maryland, and was appointed Chief of the NCI Natural Products Branch in 1989. He retired in December 2004, and is currently serving as an NIH Special Volunteer. His major interests lie in the discovery of novel natural product agents for the treatment of cancer and AIDS, with an emphasis on multidisciplinary and international collaboration. He has given over 100 invited talks at conferences in many countries worldwide, and has been awarded NIH Merit Awards for his contributions to the development of the anticancer drug, Taxol® (1991), his leadership in establishing international collaborative research in biodiversity and natural products drug discovery (2004), his contributions to developing and teaching NIH technology transfer courses (2004), and his dedicated service to the NCI in developing and maintaining evidence-based PDQ cancer information summaries for health professionals and the public (2010). In 1998–1999 he was President of the American Society of Pharmacognosy and was elected to Honorary Membership of the Society in 2003; he was named as a Fellow of the Society in 2008. In November 2006, he was awarded the 'William L. Brown Award for Plant Genetic Resources' by Missouri Botanical Garden, which also named a recently discovered Madagascar plant in his honor, *Ludia craggiana*. In April 2010, he was awarded an Honorary Doctorate of Science by his South African alma mater, Rhodes University. He has established collaborations between the NCI and organizations in many countries, promoting drug discovery from their natural resources. He has published over 150 chapters and papers related to these interests.

Foreword

By David J. Newman

This book project, covering lead compounds from medicinal plants that may have utility in cancer treatment, is a massive operation that when published in its entirety will serve as a 'must go to' source for natural product scientists, particularly those who still practice pharmacognosy in schools of pharmacy and in other related departments.

The ability to find well-assessed details of secondary metabolites from nominally medicinal plants that have been reported in many ways to have potential as leads to cancer treatment, is something that cannot be understated in terms of scientific significance.

The major reason for this statement is that there are countless claims in both the lay and scientific literature about the utility of a given plant to produce 'a cure for cancer'. What is usually forgotten, sometimes conveniently by a correspondent, are two inconvenient facts: cancer is predominately a disease of the middle-aged to elderly (55 plus depending upon the geographic area) and cannot be diagnosed in the vast majority of cases without access to a well-equipped medical facility. As a result, any claim of activity from ethnobotanical sources in the multiplicity of diseases known as cancer needs to be very carefully examined by investigators.

The author has taken this admonition to heart, and has very carefully investigated the plant secondary metabolites isolated from the medicinal plants covered. The types of molecules range from alkaloids (amides, indoles, isoquinolines and terpenoid based) in Chapter 1, terpenes in many guises in Chapter 2, and the third chapter covers the manifold ways in which oxygen heterocycles appear in plants. In all cases, details are given of the chemistry, pharmacology and, where known, the biochemical mechanisms that might be involved in the quoted pharmacological actions.

As a result of the care with which the project was put together and then executed, the data sets given form an excellent basis for the derivation of biological and chemical hypotheses related to the pharmacological actions of the plant-derived agents covered. Though this is not a textbook in the normal sense of the term, this series has, as mentioned in the first paragraph, the potential to become the first place that one goes to in order to determine what is known and who reported it in the peer-reviewed literature with respect to medicinal plants and potential treatments for cancer.

I commend Dr. Wiart for this essential undertaking.

David J. Newman
Natural Products Branch
National Cancer Institute
Frederick, Maryland, USA

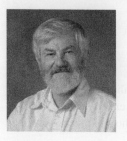

David Newman is the current chief of the Natural Products Branch (NPB) in the Developmental Therapeutics Program at the National Cancer Institute in Frederick, MD. Born in the UK he received an M.Sc. in synthetic organic chemistry and a D. Phil. for work in microbial chemistry.

He came to the USA in 1968 as a postdoc at the University of Georgia and in 1970 joined SK&F as a biological chemist. In 1985 he left SK&F and following work in marine and microbial discovery programs at various companies he joined the NPB in 1991 with responsibilities for marine and microbial collection programs. Following Gordon Cragg's retirement, he was acting chief from January 2005 until appointed chief in late 2006.

Research interests are in natural product structures as drugs and leads. He is the author or coauthor of over 150 papers, reviews and book chapters, and holds 21 patents, mainly on microbial products.

Preface

The twenty-first century has the promise to be the theater of one of the most extraordinary events in medical history: it is my hope that we once and for all defeat cancer. In 2012, cancer is still a tough enemy because it touches many organs, mutates permanently, propagates, and frequently defies surgery and chemotherapy. Of course a number of cancers are now treatable but, as yet, most malignancies are not. Indeed, I hope we or our children will see the day when being diagnosed with cancer is not such a big deal. The rationale for this hope is based on the fact that we are steadily identifying the manifold proteins and pathways involved in cancer. Indeed, the very process of treating the disease will change. In the future, oncologists will need to identify the proteins involved in each individual malignancy and then prescribe mixtures of specific anticancer agents to target them. Today, clinical anticancer agents derived from taxol, vinblastine, vincristine and camptothecin provide the evidence that medicinal plants are of immense oncological value. There is therefore a critical need to provide researchers with evidence and concepts to guide and facilitate the discovery of further chemotherapeutic agents from medicinal plants; and that is precisely the purpose of this book. In fact, this is the first of a series of volumes which will deal with specific diseases. I would like to express my gratitude to Kristine Jones and Andy Albrecht from Elsevier for their professionalism.

Christophe Wiart PharmD, PhD

About the Author

Dr. Christophe Wiart was born August 12, 1967 in Saint Malo, France. He obtained a Doctorate of Pharmacy from the University of Rennes in 1996 and was a pupil of the pharmacognosist Professor Loic Girre and the botanist Lucile Allorge from the Botanical section of the Museum of Natural History in Paris. Dr. Wiart has been studying the medicinal plants of India, Southeast Asia, China, Korea, Japan, Australia and the Pacific Islands for the last 20 years. He has collected, identified, classified and made botanical plates of about 2000 medicinal plants. Dr. Wiart is regarded as the most prominent living authority in the field of Asian ethnopharmacology, chemotaxonomy and ethnobotany. His research team currently works on the identification and pharmacological evaluation of Asian medicinal plants at the University of Nottingham. He has authored numerous bestselling books devoted to the medicinal plants of Asia and their pharmacological and cosmetological potentials.

Alkaloids

■ INTRODUCTION

Caspases are proteolytic cysteine proteases that mediate apoptosis. One such cysteine protease is caspase 3, which is involved with the apoptosis induced by alkaloids such as brucine from *Strychnos nux-vomica* L. (1.2.5), tetrandrine from *Stephania tetrandra* S. Moore (1.3.4), anonaine from *Michelia alba* DC. (1.3.7) and sanguinarine from *Macleaya microcarpa* (Maxim.) Fedde (1.3.8). During apoptosis caspase 3 is activated by cytochrome c, which is released in the cytoplasm by insulted mitochondria via either the 'intrinsic' and/or 'extrinsic' pathway. Several alkaloids induce apoptosis via the intrinsic pathway, which involves the activation of 'wild-type' pro-apoptotic protein (p53), resulting in the stimulation of pro-apoptotic Bcl-2-associated X protein (Bax)-induced mitochondrial insult and consequent leakage of cytochrome c. Berberine from *Corydalis ternata* (Nakai) Nakai (1.3.5), tetrandrine from *Stephania tetrandra* S. Moore (1.3.4) and liriodenine from *Goniothalamus amuyon* (Blco.) Merr (1.3.7) activate pro-apoptotic protein (p53). The pro-apoptotic protein (p53) is controlled by a protein kinase (Akt), which has an immense influence on cell survival. Interestingly, alkaloids such as sauristolactam from *Saururus chinensis* (Lour) Baill. (1.3.3) or tetrandrine (1.3.4) inhibit Akt. The extrinsic pathway involves the stimulation of death receptors such as TNF-α receptor and Fas by extracellular messengers. Activation of death receptors is followed by the activation of caspase 8, which incurs mitochondrial insult via the cleavage of Bid (BH3 interacting-domain death agonist) and activates caspase 3. The alkaloids 6-hydroxythiobinupharidine, 6,6′-hydroxythiobinupharidine and 6-hydroxythionuphlutine B activate caspase 8 (1.4.2).

Other cellular mechanisms that induce apoptosis by targeting nuclear factor kappa-light-chain-enhancer of activated B cells (NF-κB) involve cyclins, topoisomerase II, reactive oxygen species (ROS) and Ca^{2+} homeostasis, and microtubules. Note that piperine from *Piper nigrum* L. (1.1.3), 6-hydroxythiobinupharidine and 6-hydroxythiobinuphlutine B (1.4.2) and tylophorinidine from *Tylophora atrofolliculata* F.P. Metcalf (1.5.2) inhibits NF-κB and abrogates the survival of malignant cells. Berberine (1.3.1)

Christophe Wiart: Lead Compounds from Medicinal Plants for the Treatment of Cancer
DOI: http://dx.doi.org/10.1016/B978-0-12-398371-8.00001-5

inhibits cyclin B and cyclin-dependent kinase 1 (CDK1), the aristolactam SCH 546909 from *Aristolochia manshuriensis* Kom. (1.3.3) inhibits cyclin-dependent kinase 2 (CDK2) and tylophorine from *Tylophora indica* (Burm. f.) Merr. (1.5.2) inhibits cyclin A2, hence blockage of mitosis. Planar alkaloids such as ellipticine from *Ochrosia elliptica* Labill. (1.2.2), olivacine from *Aspidosperma australe* Müll. Arg. (1.2.2) and liriodenine (1.3.7) have the tendency to intercalate into DNA and inhibit the enzymatic activity of topoisomerase II, leading to DNA damage. Sanguinarine from *Sanguinaria canadensis* L. (1.3.8), jatrorrhizine (1.3.1) and tetrandrine (1.3.4) boost the cytoplasmic levels of ROS, whereas geissoschizine methyl ether and hirsutin from members of the genus *Uncaria* (1.2.3.) are Ca^{2+} blockers, and liriodenine (1.3.7) increases the cytoplasmic levels of nitric oxide (NO). Finally, alkaloids like colchicine, monoterpenoid indoles (1.2.1) and daphniglaucine C isolated from *Daphniphyllum glaucescens* Blume (1.4.1) inhibit the polymerization of tubulin and therefore abrogate the division of cancer cells.

In this chapter 1-[(9*E*)-10-(3,4-methylenedioxyphenyl)-9-decenoyl] pyrrolidine, *N-cis*-feruloyl tyramine, piplartine, guineensine, alstoyunine F, heptaphylline, vallesiachotamine, naucleaoral A, 12-methoxyechitamidine, alangiumkaloid A, SCH 546909, isotrilobine, demethylcorydalmine,racemosidineA,liriodenine,6-methoxydihydrosanguinarine, daphniglaucine C, 6-hydroxythionuphlutine B, epipachysamine E, 2-(2-bromophenyl)-5,6,7-trihydroxy-8-((2-hydroxy-4-methylpiperazin-1-yl) methyl)- 4H-chromen-4-one and phenanthroindolizidine and their derivatives are identified as lead alkaloids for the treatment of cancer.

Topic **1.1**

Amide Alkaloids

1.1.1 *Piper Boehmeriifolium* (Miq.) Wall. ex C. DC.

History The plant was first described by Nathaniel Wallich in *Prodromus Systematis Naturalis Regni Vegetabilis*, published in 1869.

Synonym *Chavica boehmeriifolia* Miquel

Family Piperaceae Giseke, 1792

Common Names False-nettle-leaved pepper, chavya, (Sanskrit), zhu ye ju (Chinese)

Habitat and Description This shrub grows wild in China, Bhutan, India, Malaysia, Myanmar, Thailand and Vietnam. The plant grows to 5 m high. The stems are terete smooth and glabrous. The leaves are simple. The petiole is 0.2–1 cm long. The blade is oblong-lanceolate, 10–25 cm×4–8 cm, membranaceous, asymmetrical at base oblique, acuminate at apex and exhibits 6–10 pairs of secondary nerves. The inflorescence is a spike which is 10–20 cm long and opposite to the leaves. The fruits are drupaceous, globose and 0.3 cm in diameter (Figure 1.1).

Medicinal Uses In China the whole plant is used to alleviate pain and for the treatment of rheumatism and arthritic conditions.

■ **FIGURE 1.1** *Piper boehmeriifolium* (Miq.) Wall. ex C. DC.

Phytopharmacology Phytochemical investigations of *Piper boehmeriifolium* (Miq.) Wall. ex C. DC. resulted in the isolation of the amide alkaloid piperine,[1] the aristolactam alkaloid piperolactam D[2,3] and aporphine alkaloids.[3] The anti-inflammatory and analgesic properties could be due to the isoquinoline contents of the plant since isoquinoline alkaloids suppress the receptor-mediated phospholipase A2 activation through uncoupling of a GTP binding protein from the enzyme.[4]

Proposed Research Pharmacological study of 1-[(9*E*)-10-(3,4-methylenedioxyphenyl)-9-decenoyl]pyrrolidine for the treatment of cancer.

Rationale In a recent study, Tang et al.[5] isolated a series of amide alkaloids, one of which 1-[(9*E*)-10-(3,4-methylenedioxyphenyl)-9-decenoyl]pyrrolidine (CS 1.1) abrogated the survival of human epitheloid cervix carcinoma (Hela) cells with an IC_{50} value of 2.7 μg/mL.[5] The mode of action of this alkaloid is to date unknown but one could think of apoptosis. In fact, irniine (CS 1.2), a pyrrolidine alkaloid extracted from the tubers of *Arisarum vulgare* Targ. Tozz. (family Araceae Juss) induced important alterations of the nuclei, induced DNA and mitochondrial damage and oxidative stress that led to cell death by necrosis and/or by apoptosis in rat liver epithelial cell line (RLEC).[6] Note that PHA-739358 (CS 1.3) is an aurora kinase inhibitor that shares some chemical features (1-(pyrrolidin-1-yl) ethanone and amide) with 1-[(9*E*)-10-(3,4-methylenedioxyphenyl)-9-decenoyl]pyrrolidine. Aurora kinases are pivotal for the successful execution of cell division and have been considered very attractive targets for the development of antitumor drugs.[7] Inhibitors of aurora kinases were shown to induce apoptosis via upregulation of p53, breakdown of the mitochondrial membrane potential and activation of caspase 3 and polyploidization in human colorectal carcinoma HCT-116 cells (wild-type p53).[8]

■ **CS 1.1** 1-[(9*E*)-10-(3,4-Methylenedioxyphenyl)-9-decenoyl]pyrrolidine

■ **CS 1.2** Irniine

1.1.2 *Piper Caninum* **Blume**

History The plant was first described by Carl Ludwig von Blume in *Verhandelingen van het Bataviaasch Genootschap van Kunsten en Wetenschappen*, published in 1826.

Synonym *Piper banksii* Miquel

Family Piperaceae Giseke, 1792

Common Names Cabai hutan, lada hantu, lada anjing, akar kalong (Malay)

Habitat and Description *Piper caninum* Blume is a climber that dwells in the rainforests of Southeast Asia and Australia. The leaves are simple. The petiole is 1–1.2 cm long, papery, glaucous beneath, 10–11 cm × 5–6 cm, ovate, wedge-shaped at base, cuspidate at apex and with two pairs of secondary nerves arising from the base. The inflorescence is a spike which is 4 cm long and opposite the leaves. The fruits are drupaceous, globose and 0.2 cm in diameter (Figure 1.2).

Medicinal Uses The leaves are chewed by Malays to allay sore throats.

Phytopharmacology Phytochemical study of the leaves revealed the presence of amide alkaloids.[9] The medicinal property is probably due to the hydroxycinnamic esters bornyl *p*-coumarate and bornyl caffeate, which inhibited the growth of *Bacillus cereus*, *Staphylococcus aureus* and *Streptococcus pneumoniae* cultured in vitro.[10] The plant is known to

■ **FIGURE 1.2** *Piper caninum* Blume

■ **CS 1.4** Cepharadione A

contain the isoquinoline alkaloid cepharadione A (CS 1.4); this potently inhibited the growth of a yeast strain lacking *RAD52* (IC_{50}: 50.2 nM), which is associated with the repair of double-strand DNA breaks.[11]

Since the discovery of the bleomycin antitumor antibiotics that work at the level of nucleic acid strand scission considerable effort has been made to identify natural products with such an activity.[12] The mechanism of action of bleomycin and congeners involves oxygen, iron or copper ions.[13] Cepharadione A inhibited the growth of human breast adenocarcinoma (MCF-7) cells, human glioblastoma (SF268), human large-cell lung cancer (NCI-H460), human colon adenocarcinoma (KM20L2) and human prostate carcinoma (DU-145) cells with IC_{50} values of 6.3 μg/mL, 2.9 μg/mL, 2.5 μg/mL, 17.3 μg/mL and 4.3 μg/mL, respectively.[14]

Proposed Research Pharmacological study of *N-cis*-feruloyl tyramine and derivatives for the treatment of cancer.

■ **CS 1.5** *N-cis*-Feruloyl tyramine

■ **CS 1.6** *N*-(4-Methoxyphenethyl)-*N*-methylbenzamide

Rationale *N-cis*-feruloyl tyramine (CS 1.5) isolated from *Piper caninum* Blume (family Piperaceae Giseke) in the presence of $20\,\mu M$ Cu^{2+}, exhibited potent strand scission activity in an in vitro assay that monitored the relaxation of supercoiled plasmid DNA.[12] In addition, it is interesting to note that the hydroxycinnamic acid amide caffeoylserotonin isolated from *Piper nigrum* L. (family Piperaceae Giseke) reduced the phosphorylation of H2AX in human keratinocytes HaCaT human hepatocellular liver carcinoma (HepG2) cells exposed to H_2O_2.[15,16] In normal physiological conditions, exposure of cells to ROS results in DNA damage and double-strand DNA breaks, which in turn induce the γ-phosphorylation of histone H2AX at Ser-139.[16] Thus, hydroxycinnamic acid amides may prevent double-strand DNA breaks or modulate the DNA damage signaling pathway through prevention of radical production or activation of signaling against oxidative stress-induced damage. One such compound is *N*-(4-methoxyphenethyl)-*N*-methylbenzamide (CS 1.6) isolated from *Zanthoxylum ailanthoides* Sieb. & Zucc. (family Rutaceae Juss.), which inhibited the generation of superoxide radical anions by human neutrophils in response to fMet-Leu-Phe/Cytochalasin.[17] Another compelling feature of hydroxycinnamic acid amides is their ability to inhibit the enzymatic activity of tyrosinase. *N*-feruloylserotonin (CS 1.7) isolated from *Carthamus tinctorius* L. (family Asteraceae Bercht. & J. Presl) inhibited the enzymatic activity of tyrosinase, with an IC_{50} value equal to $0.023\,\mu M$.[18] Tyrosinase is produced in massive amounts by malignant tumors and has the ability to react with tyrosine, one of the major constituents of proteins such as interferons.[19,20] In fact, tyrosinase catabolises tyrosine into series of genotoxic and mutagenic quinones, semiquinones and ROS, which may result in the generation of cancer cells.[21–23]

■ **CS 1.7** *N*-Feruloylserotonin

1.1.3 *Piper Chaba* Hunter

History The plant was first described by Alexander Hunter in *Asiatic Researches*, published in 1809.

Synonyms *Chavica officinarum* Miquel, *P. officinarum* (Miquel) C. de Candolle, *Piper retrofractum* Vahl

Family Piperaceae Giseke, 1792

Common Names Chavana (Sanskrit), jia bi ba (Chinese)

Habitat and Description *Piper chaba* Hunter is a climber that grows wild in India, Indonesia, Malaysia, the Philippines, Thailand and Vietnam. The stems are terete. The leaves are simple. The petiole is 0.5–1 cm and sheathed at base. The blade is elliptic, 8–15 cm × 3–8 cm, papery, the base cordate and asymmetrical, the apex acute and with 9–12 pairs of secondary nerves. The inflorescence is a conical spike which is 5 cm long and opposite the leaves. The fruits are drupaceous (Figure 1.3).

Medicinal Uses In India, the plant is used to treat asthma, bronchitis, piles and to expel worms from the intestines. In Vietnam, the plant is used to break fever and to treat jaundice. In Malaysia, the plant is used to promote digestion and to invigorate. In Indonesia, the plant is used to alleviate toothache. In the Philippines, the plant is used to promote digestion.

Phytopharmacology The plant is known to produce series of amide alkaloids: piperine, sylvatine, piperlonguminine,[24] piperamine

■ **FIGURE 1.3** *Piper chaba* Hunter

2,4-decadienoic acid piperidide, and pellitorine,[25] chabamide,[26] piper-chabamides A-D, piperanine, pipernonaline, dehydropipernonaline, ret-rofractamide B, guineensine, *N*-isobutyl-(2E,4E)-octadecadienamide, *N*-isobutyl-(2E,4E,14Z)-eicosatrienamide,[27] chabamide F and G (CS 1.8) and brachystamide B.[28] The plant has a carminative property, most probably owing to a synergistic effect of piperine and congeners and an essential oil predominantly containing caryophyllene oxide.[29] The analgesic and anti-inflammatory properties have been substantiated in rodents[30] and attributed to piperine, pipernonaline, dehydropipernona-line, retrofractamide B, *N*-isobutyl-(2E,4E)- octadecadienamide and *N*-isobutyl-(2E,4E,14Z)-eicosatrienamide.[27]

Proposed Research Pharmacological study of piplartine for the treatment of cancer.

Rationale Rao et al. (2011) isolated a series of amide alkaloids from the plant, including chabamide G, which abrogated the growth of human epitheloid cervix carcinoma (Hela), human breast adenocarcinoma (MCF-7),

■ **CS 1.8** Chabamide G

■ **CS 1.9** Piplartine

HepG2 and human colon adenocarcinoma (CoLo 205) cells, with IC_{50} values equal to 46.3 µg/mL, 27.9 µg/mL, 58.6 µg/mL and 0.018 µg/mL, respectively.[31] Another potently cytotoxic amide alkaloid isolated by Rao et al. (2011) is piplartine (CS 1.9), which was studied for cytotoxic activity by Jyothi et al. (2009). Murine macrophage-like (P388D1) cells treated with 10 µM and 15 µM doses of piplartine for 24 h showed a significant increase in percentage of cell death with values of 87% and 97%, respectively.[32] In addition, exposure of (BC-8) cells to 2 µM and 5 µM concentrations of piplartine for 18 h resulted in 13% and 78% cell deaths, respectively. Piplartine induced G1-arrest, inhibited Raf-1 activation, decreased the kinase levels of CDK2 and the protein levels of cyclin D1, and inhibited DNA synthesis.[32,33] After 72 h incubation with piplartine at a dose of 10 µg/mL, the number of viable human promyelocytic leukemia (HL-60), human erythromyeloblastoid leukemia (K562), Jurkat human T cell lymphoblast-like and human malignant T-lymphoblastic (MOLT-4) cells was reduced by 99%, 88%, 98% and 100%, respectively.[33]

In addition, piplartine inhibited the incorporation of the nucleotide BrdU by 50%, 88% and 100% in human promyelocytic leukemia (HL-60) cells at the concentrations of 2.5 µg/mL, 5 µg/mL and 10 µg/mL, respectively, while in human erythromyeloblastoid leukemia (K562) cells the inhibition was 48%, 37% and 100% at concentrations of 2.5 µg/mL, 5 µg/mL and 10 µg/mL, respectively,[33] suggesting DNA synthesis inhibition. Furthermore, human promyelocytic leukemia (HL-60) and human erythromyeloblastoid leukemia (K562) treated with piplartine at a concentration of 2.5 µg/mL showed chromatin condensation and DNA fragmentation, which are hallmarks of apoptosis.[33] The apoptotic theory was further substantiated by Kong et al. (2008) who demonstrated that 30 µM of piplartine induced the cleavage of the DNA repairing enzyme poly (ADP-ribose) polymerase by the effector caspase 3 in human prostate cancer (PC-3) cells.[34]

1.1.4 *Piper Longum* L.

History The plant was first described by Carl von Linnaeus in *Species Plantarum*, published in 1753.

Synonyms *Chavica roxburghii* Miquel

Family Piperaceae Giseke, 1792

Common Names Long pepper, chapala (Sanskrit), argadi (Tamil), peikchin (Burmese), bi ba (Chinese)

Habitat and Description *Piper longum* L. is a climber that grows wild in India, Malaysia, Nepal, Sri Lanka and Vietnam. The stems are flexuous, pubescent and terete. The leaves are simple. The petiole is grooved, of variable length and can reach 10 cm long. The blade is ovate, 6–12 cm × 3–12 cm, papery, the base cordate and asymmetrical, the apex acuminate and with two or three pairs of secondary nerves arising from the base. The inflorescence is a cylindrical and slightly curved spike which is 1.5–5 cm long and opposite the leaves. The fruits are drupaceous, globose and 0.2 cm in diameter (Figure 1.4).

■ **FIGURE 1.4** *Piper longum* L.

■ **CS 1.10** Guineensine

Medicinal Uses In India, the plant is used to treat tumors, depression, bronchitis, jaundice, impotency and rheumatism. In China, the plant is used to promote fertility in women and digestion. In Malaysia, the plant is used to treat coughs and promote digestion.

Phytopharmacology Phytochemical study of *Piper longum* L. resulted in the isolation of amide alkaloids such as piperine, piperlongumine and piperlonguminine,[35,36] pipernonaline, piperundecalidine,[37] lignans,[38] aristolactams and isoquinolines.[39]

The plant probably owes its hepatoprotective properties to the amide alkaloid piperine, which has been shown to protect hepatocytes against tert-butyl hydroperoxide and carbon tetrachloride poisoning both in vitro and in vivo.[40] Moreover, piperine at a dose of $25\,\mu M$ reduced the level of nitrite in lipopolysaccharide-stimulated Balb/C mice and may account for the anti-inflammatory property of the plant.[41] Piperine may also account for the antidepressant qualities of the plant since it inhibited the enzymatic activity of monoamine oxidases A and B, with IC_{50} values equal to $20.9\,\mu M$ and $7\,\mu M$, respectively.[42]

Proposed Research Pharmacological study of guineensine (CS 1.10) for the treatment of brain cancer.

Rationale Mishra et al. (2011) isolated a series of amide alkaloids from the plant, including guineensine, which abrogated the survival of human promyelocytic leukemia (HL-60) cells cultured in vitro with an IC_{50} value equal to $5\,\mu g/mL$.[43] The cytotoxic mode of action of guineensine is probably similar to the closely related amide alkaloid piperidine. Note that piperine at a concentration of $2.5\,\mu g/mL$, $5\,\mu g/mL$ and $10\,\mu g/mL$ inhibited collagen matrix invasion by murine melanoma (B16F10) cells by 69.03%, 82.36% and 86.5%, respectively, via inhibition of matrix metalloproteinase, and inhibited NF-κB.[44] Matrix metalloproteinases play a relevant role in tumor dissemination through their action of degrading the extracellular matrix component, which upregulates invasion and metastasis. Activation of NF-κB contributes to cancer development and progression via self sufficiency in growth signals, insensitivity to growth inhibitors, evasion of apoptosis, limitless replicative potential, tissue invasion and metastasis, and sustained angiogenesis.[45] Another compelling feature of guineensine is its ability to

■ **CS 1.11** Avasimibe

inhibit the enzymatic activity of cholesterol acyltransferase with an IC_{50} value of 3.12 µM in vitro.[46] The amide avasimibe (CS 1.11) is an inhibitor of cholesterol acyltransferase and has been seen to restrict the growth of human glioblastoma U87 cells by inhibiting cholesteryl esters synthesis and inducing apoptosis as a result of caspase 8 and caspase 3 activation.[47]

REFERENCES

[1] Mahanta PK, Ghanim A, Gopinath KW. Chemical constituents of *Piper sylvaticum* (Roxb) and *Piper boehmerifolium* (Wall). J Pharm Sci 1974;63(7):1160–1.

[2] Desai SJ, Chaturvedi R, Mulchandani NB. Piperolactam D, a new aristolactam from Indian piper species. J Nat Prod 1989;53(2):496–7.

[3] Desai SJ, Chaturvedi RN, Badheka LP, Mulchandani NB. Aristolactams and 4,5 dioxoaporphines from Indian piper species. Ind J Chem 1989;28:775–7.

[4] Akiba S, Kato E, Sato T, Fujii T. Biscoclaurine alkaloids inhibit receptor-mediated phospholipase A2 activation probably through uncoupling of a GTP-binding protein from the enzyme in rat peritoneal mast cells. Biochem Pharmacol 1992;44(1):45–50.

[5] Tang GH, Chen DM, Qiu BY, Sheng L, Wang YH, Hu GW, et al. Cytotoxic amide alkaloids from *Piper boehmeriaefolium*. J Nat Prod 2011;74(1):45–9.

[6] Rakba N, Melhaoui A, Rissel M, Morel I, Loyer P, Lescoat G. Irniine, a pyrrolidine alkaloid, isolated from *Arisarum vulgare* can induce apoptosis and/or necrosis in rat hepatocyte cultures. Toxicon 2000;38(10):1389–402.

[7] Pérez de Castro I, de Cárcer G, Montoya G, Malumbres M. Emerging cancer therapeutic opportunities by inhibiting mitotic kinases. Curr Opin Pharmacol 2008;8(4):375–83.

[8] Li M, Jung A, Ganswindt U, Marini P, Friedl A, Daniel PT, et al. Aurora kinase inhibitor ZM447439 induces apoptosis via mitochondrial pathways. Biochem Pharmacol 2010;79(2):122–9.

[9] Ahmad F, Bakar SA, Ibrahim AZ, Read RW. Constituents of the leaves of *Piper caninum*. Planta Med 1997;63(2):193–4.

[10] Setzer WN, Setzer MC, Bates RB, Nakkiew P, Jackes BR, Chen L, et al. Antibacterial hydroxycinnamic esters from *Piper caninum* from Paluma, north Queensland, Australia. The crystal and molecular structure of (+)-bornyl coumarate. Planta Med 1999;65(8):747–9.

[11] Ma J, Jones SH, Marshall R, Johnson RK, Hecht SM. A DNA-damaging oxoapor-phine alkaloid from *Piper caninum*. J Nat Prod 2004;67(7):1162–4.

[12] Ma J, Jones SH, Hecht SM. Phenolic acid amides: a new type of DNA strand scis-sion agent from *Piper caninum*. Bioorg Med Chem 2004;12(14):3885–9.

[13] Hecht SM. DNA strand scission by activated bleomycin group antibiotics. Fed Proc 1986;45(12):2784–91.

[14] Elban MA, Chapuis JC, Li M, Hecht SM. Synthesis and biological evaluation of cepharadiones A and B and related dioxoaporphines. Bioorg Med Chem 2007;15(18):6119–25.

[15] Kang K, Lee K, Ishihara A, Park S, Kim Y, Back K. Induced synthesis of caf-feoylserotonin in pepper fruits upon infection by the anthracnose fungus *Colletotrichum gloeosporioides*. Sci Hortic 2010;124:290–3.

[16] Choi JY, Kim H, Choi YJ, Ishihara A, Back K, Lee SG. Cytoprotective activities of hydroxycinnamic acid amides of serotonin against oxidative stress-induced damage in HepG2 and HaCaT cells. Fitoterapia 2010;81(8):1134–41.

[17] Chen JJ, Chung CY, Hwang TL, Chen JF. Amides and benzenoids from *Zanthoxylum ailanthoides* with inhibitory activity on superoxide generation and elastase release by neutrophils. J Nat Prod 2009;72(1):107–11.

[18] Roh JS, Han JY, Kim JH, Hwang JK. Inhibitory effects of active compounds isolated from safflower (*Carthamus tinctorius* L.) seeds for melanogenesis. Biol Pharm Bull 2004;27(12):1976–8.

[19] Priestman TJ. Interferon: an anti-cancer agent? Cancer Treatment Rev 1979;6(4):223–37.

[20] Chen YM, Lim BT, Chavin W. Serum tyrosinase in malignant disease, its activ-ity, and the electrophoretic patterns of the enzyme as carried by immunoglobulins. Cancer Res 1979;39(9):3485–90.

[21] Bonfigli A, Zarivi O, Colafarina S, Cimini AM, Ragnelli AM, Aimola P, et al. Human glioblastoma ADF cells express tyrosinase, L-tyrosine hydroxylase and melanosomes and are sensitive to L-tyrosine and phenylthiourea. J Cell Physiol 2006;207(3):675–82.

[22] Miranda M, Amicarelli F, Bonfigli A, Poma A, Zarivi O, Arcadi A. Mutagenicity test for unstable compounds, such as 5,6-dihydroxyindole, using an *Escherichia coli* HB101/pBR322 transfection system. Mutagenesis 1990;5(3):251–5.

[23] Stokes AH, Brown BG, Lee CK, Doolittle DJ, Vrana KE. Tyrosinase enhances the covalent modification of DNA by dopamine. Mol Brain Res 1996;42(1):167–70.

[24] Patra A, Ghosh A. Amides of *Piper chaba*. Phytochem 1974;13(12):2889–90.

[25] Connolly JD, Deans R, Haque ME. Constituents of *Piper chaba*. Fitoterapia 1995;66(2):188.

[26] Rukachaisirikul T, Prabpai S, Champung P, Suksamrarn A. Chabamide, a novel piperine dimer from stems of *Piper chaba*. Planta Med 2002;68(9):853–5.

[27] Morikawa T, Matsuda H, Yamaguchi I, Pongpiriyadacha Y, Yoshikawa M. New amides and gastroprotective constituents from the fruit of *Piper chaba*. Planta Med 2004;70(2):152–9.

[28] Rao VRS, Kumar GS, Sarma VUM, Raju SS, Babu KH, Babu KS, et al. Chabamides F and G, two novel dimeric alkaloids from the roots of *Piper chaba* Hunter. Tetrahedron Lett 2009;50(23):2774–7.

[29] Tewtrakul S, Hase K, Kadota S, Namba T, Komatsu K, Tanaka K. Fruit oil com-position of *Piper chaba* Hunt., *P. longum* L. and *P. nigrum* L. J Essential Oil Res 2000;12(5):603–8.

[30] Taufiq-Ur-Rahman Md, Shilpi JA, Ahmed M, Hossain CF. Preliminary pharmacological studies on *Piper chaba* stem bark. J Ethnopharmacol 2005;99(2):203–9.

[31] Rao VRS, Suresh G, Babu KS, Raju SS, Vishnu Vardhan MVPS, Ramakrishna S, et al. Novel dimeric amide alkaloids from *Piper chaba* Hunter: isolation, cytotoxic activitys, and their biomimetic synthesis. Tetrahedron 2011;67(10): 1885–92.

[32] Jyothi D, Vanathi P, Mangala Gowri P, Rama Subba Rao V, Madhusudana Rao J, Sreedhar AS. Diferuloylmethane augments the cytotoxic effects of piplartine isolated from *Piper chaba*. Toxicol in vitro 2009;23(6):1085–91.

[33] Bezerra DP, Militao GC, de Castro FO, Pessoa C, de Moraes MO, Silveira ER, et al. Piplartine induces inhibition of leukemia cell proliferation triggering both apoptosis and necrosis pathways. Toxicol. In vitro 2007;21:1–8.

[34] Kong EH, Kim YJ, Kim YJ, Cho HJ, Yu SN, Kim KY, et al. Piplartine induces caspase-mediated apoptosis in PC-3 human prostate cancer cells. Oncology Reports 2008;20(4):785–92.

[35] Chatterjee A, Dutta CP. Alkaloids of *Piper longum* Linn-I. Structure and synthesis of piperlongumine and piperlonguminine. Tetrahedron 1967;23(4):1769–81.

[36] Das B, Kashinatham A, Srinivas KVNS. Alkamides and other constituents of *Piper longum*. Planta Med 1996;62(6):582.

[37] Tabuneng W, Bando H, Amiya T. Studies on the constituents of the crude drug 'Piperis longi fructus'. On the alkaloids of fruits of *Piper longum* L. Chem Pharm Bull 1983;31(10):3562–5.

[38] Dutta CP, Banerjee N, Roy DN. Lignans in the seeds of *Piper longum*. Phytochem 1975;14(9):2090–1.

[39] Desai SJ, Prabhu BR, Mulchandani NB. Aristolactams and 4,5-dioxoaporphines from *Piper longum*. Phytochem 1988;27(5):1511–5.

[40] Koul IB, Kapil A. Evaluation of the liver protective potential of piperine, an active principle of black and long peppers. Planta Med 1993;59(5):413–7.

[41] Pradeep CR, Kuttan G. Effect of piperine on the inhibition of nitric oxide (NO) and TNF-α production. Immunopharmacol Immunotoxicol 2003;25(3):337–46.

[42] Lee SA, Hong SS, Han XH, Hwang JS, Oh GJ, Lee KS, et al. Piperine from the fruits of *Piper longum* with inhibitory effect on monoamine oxidase and antidepressant-like activity. Chem Pharm Bull 2005;53(7):832–5.

[43] Mishra P, Sinha S, Guru SK, Bhushan S, Vishwakarma RA, Ghosal S. Two new amides with cytotoxic activity from the fruits of *Piper longum*. J Asian Nat Prod Res 2011;13(2):143–8.

[44] Pradeep CR, Kuttan G. Piperine is a potent inhibitor of nuclear factor-κB (NF-κB), c-Fos, CREB, ATF-2 and proinflammatory cytokine gene expression in B16F-10 melanoma cells. Int Immunopharmacol 2004;4:1795–803.

[45] Naugler WE, Karin M. NF-κB and cancer – identifying targets and mechanisms. Curr Opin Genet Dev 2008;18(1):19–26.

[46] Seung WL, Rho MC, Jung YN, Eun HL, Oh EK, Young HK, et al. Guineensine, an Acyl-CoA: cholesterol acyltransferase inhibitor, from the fruits of *Piper longum*. Planta Med 2004;70(7):678–9.

[47] Bemlih S, Poirier MD, El Andaloussi A. Acyl-coenzyme A: cholesterol acyltransferase inhibitor Avasimibe affect survival and proliferation of glioma tumor cell lines. Cancer Biol Ther 2010;9(12):1025–32.

Topic **1.2**

Indole Alkaloids and Derivatives

1.2.1 *Alstonia Yunnanensis* **Diels**

History The plant was first described by Friedrich Ludwig Emil Diels in *Notes from the Royal Botanic Garden*, published in 1912.

Synonyms *Acronychia esquirolii* H. Lév., *Alstonia esquirolii* H. Lév.

Family Apocynaceae Juss., 1792

Common Names Ji gu chang shan (Chinese)

Habitat and Description This shrub grows wild to 3 m tall in Guangxi, Guizhou and Yunnan. The plant is laticiferous. The stems are lenticelled and pubescent when young. The leaves are simple, sessile and arranged in whorls of three to five. The blade is cuneate, 6–20 cm × 1–5 cm, membranaceous, pubescent, acuminate at apex, tapering at base and with 15–35 pairs of conspicuous secondary nerves. The inflorescence is a terminal cyme which is pubescent. The flowers are salver-shaped and pink. The calyx is pubescent and includes five minute lanceolate lobes. The corolla tube is 1–1.3 cm long and develops five lobes which are oblong, 0.2–0.6 cm and contorted. The fruits consist of pairs of follicles which are linear and 3–5 cm long. The seeds are oblong and ciliate at apex (Figure 1.5).

Medicinal Uses In China, the plant is used to treat fevers, headaches, inflammation, hypertension, hemostasis and fractures. An oil is produced from the seeds.

Phytopharmacology Phytochemical analysis of the plant revealed the presence of a series of indole alkaloids, including 11-methoxy-3-oxotabersonine, (–)-lochnerinine, 19(R)-hydroxy-11-methoxy-tabersonine, alstoyunines A–H, perakine, vinorine, tabersonine, raucaffrinoline, 11-methoxytabersonine, 19-acetoxy-11-methoxytabersonine, (–)-echitoveniline, echitoserpidine, vellosimine, vellosiminol, picrinine, picraline, 19(Z)-burnamine-17-*O*-3′,4′,5′-trimethoxybenzoate, compactinervine,

■ **FIGURE 1.5** *Alstonia yunnanensis* Diels

■ **CS 1.12** Alstoyunine F

19-*epi*-ajmalicine and alloyohimbine.[48–51] The anti-inflammatory effects of the plant are most probably conveyed by alstoyunines C, E and F (CS 1.12), which inhibited the enzymatic activity of cyclo-oxygenase 2 by 94.8%, 93.9% and 77.1%, respectively, at a dose of 100 μM.[51]

Proposed Research Pharmacological study of alstoyunine F for the treatment of colon cancer.

Rationale A compelling feature of monoterpenoid indole alkaloids is the vast array of cellular mechanisms they interfere with to bring about the death of cancer cells. Monoterpenoid indole alkaloids have the special ability to bind to β-tubulin, thus hampering the formation of microtubules. Microtubules are crucial for the formation and disappearance of the mitotic spindle, which, in turn, is responsible for the separation of duplicated chromosomes during cell division.[52] The commercial monoterpenoid indole alkaloids vincristine and vinblastine from *Catharanthus roseus* (L.) G. Don (family Apocynaceae Juss.) bind to β-tubulin, block microtubule polymerization, arresting mitosis, and lead to cancer cell death.[53]

Monoterpenoid indole alkaloids also induce DNA strand scission. DNA strand scission naturally occurs through the oxidation of the deoxyribose sugar ring, alkylation or oxidation of the aromatic nucleobase, or by hydrolysis of the phosphodiester backbone, and results in cell death. Natural products capable of mediating DNA strand scission are therefore cytotoxic and hold tremendous interest as potential anticancer agents. For instance, the monoterpenoid indole alkaloids turpiniside, 11-methoxyjavaniside, vincosamide, 3(R)-pumiloside, and paratunamide C from *Turpinia arguta* (Lindl.) Seem. (family Staphyleaceae Martinov) induced relaxation and strand scission of supercoiled pBR322 plasmid DNA in the presence of Cu^{2+}.[54] Another example of a DNA strand scission monoterpenoid alkaloid is javaniside; isolated from *Alangium javanicum* (Blume) Wangerin (family Cornaceae Bercht. ex J. Presl), this induced DNA relaxation and strand scission of the supercoiled pBR322 plasmid DNA at a dose of 200 μM in the presence of Cu^{2+}.[55]

Monoterpenoid indole alkaloids inhibit the enzymatic activity of topoisomerase II and the monoterpene indole alkaloid camptothecin isolated from *Camptotheca acuminata* Decne. (family Cornaceae Bercht. ex J. Presl), and its commercial analogues are routine anti cancer agents.[56] In addition, monoterpenoid indole alkaloids are apoptosis inducers. Tabernaemontanine and vobasine from *Tabernaemontana elegans* Stapf (family Apocynaceae Juss.) condensed chromatin, and brought about nuclear fragmentation and caspase activation in human hepatocellular carcinoma (HuH-7) cells at a dose of 20 μM[57] via a possible induction of c-jun.[58] The monoterpenoid indole alkaloid alstoyunine F isolated from *Alstonia yunnanensis* Diels (family Apocynaceae Juss.) inhibited the growth of human promyelocytic leukemia (HL-60) and human hepatoma (SMMC-7721) cells cultured in vitro with IC_{50} values equal to 3.8 μM and 21.7 μM, respectively, and inhibited the enzymatic activity of cyclooxygenase.[51] Note that the commercial anti-inflammatory cyclooxygenase inhibitors such as sulindac (CS 1.13) inhibited cell proliferation and induced apoptosis in cultured colon cancer cell lines[59] via inhibition of

■ **CS 1.13** Sulindac

prostaglandin E2 vascular endothelial growth factor (VEGF) induction,[60] cyclin D1, cytosolic beta-catenin, matrix metalloproteinases 2 and 9,[61] and reduction of oxidative processes.[62]

1.2.2 *Clausena Harmandiana* (Pierre) Guillaumin

History The plant was first described by André Guillaumin, in *Notulae Systematicae*, published in 1910.

Synonym *Glycosmis harmandiana* Pierre

Family Rutaceae Juss., 1789

Common Names Kasai (Malay), song faa (Thai)

Habitat and Description This treelet grows to 5 m high in the rainforests of Cambodia, Laos, Vietnam, Thailand, Malaysia and Indonesia. The leaves are pinnate and alternate. The rachis is 15–55 cm long. The folioles are alternate, 5–7, largest toward the apex of the leaf, ovate, 4–8 cm × 2–7 cm, cuneate at base, acute at apex, crenulate at margin and covered with translucent oil cells. The inflorescence is a terminal panicle. The flowers are small. The calyx produces five minute lobes. The corolla has five petals which are ovate, 0.3 cm long and whitish green. The androecium includes 10 stamens. The ovary is slightly raised on an hourglass-shaped stalk, rounded, and covered with oil cells. The fruit is a berry which is ovoid, 1.5 cm long, reddish and contains one to three seeds (Figure 1.6).

Medicinal Uses In Thailand the plant is used to break malarial fever, to promote digestion and health and to treat bronchitis, food poisoning and headaches.

Phytopharmacology Phytochemical analysis of the plant resulted in the identification of coumarins such as clausarin, dentatin, osthol, xanthoxyletin, nordentatin (1), and carbazole alkaloids including heptaphylline,[63] 2-hydroxy-3-formyl-7-methoxycarbazole and 7-methoxyheptaphylline,[64] 7-hydroxyheptaphylline, claurailas A–D, girinimbrine, clausines E, K and O, 3-formyl-1-hydroxy-7-methoxycarbazole, *O*-demethylmurrayanine, murrayanine, 7-methoxymurrayanine and lancine.[65]

The antimalarial property of the plant is most probably due to heptaphylline, dentatin and clausarin, which exhibited antiplasmodial activity against

■ **FIGURE 1.6** *Clausena harmandiana* (Pierre) Guillaumin

Plasmodium falciparum in vitro.[66] 7-Methoxyheptaphylline abrogated the survival of human epidermoid carcinoma (9-KB), human lung adenocarcinoma epithelial (A549) and human colon cancer (HT-29) cells with IC_{50} values equal to 3 μg/mL, 2.1 μg/mL and 1.3 μg/mL, respectively.[64]

Proposed Research Pharmacological study of heptaphylline and derivatives for the treatment of cancer.

Rationale Carbazole alkaloids are interesting because their size, planarity and the arc-shape configurations of these molecules are remarkably suited for DNA intercalation. It has been effectively recognized that one such alkaloid, ellipticine (CS 1.14) from *Ochrosia elliptica* Labill. (family Apocynaceae Juss.),[67] could intercalate in DNA.[68,69] Ellipticine intercalation into DNA is followed by DNA strand breaks and stimulation of topoisomerase II-mediated DNA cleavage.[70] The stabilized complex formed between topoisomerase II and DNA results in DNA damage and

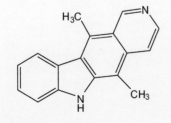

■ **CS 1.14** Ellipticine

cellular death.[71] Topoisomerase II is a key enzyme, the main function of which is to resolve DNA topological issues.[72] Ellipticine has been shown to inhibit pro-apoptotic protein (p53) phosphorylation through kinase inhibition in several wild-type p53 cancer cell lines,[73] uncouple mitochondrial oxidative phosphorylation,[74] and induce endoplasmic reticulum stress.[75] Another carbazole alkaloid that inhibits topoisomerase II is olivacine (CS 1.15) isolated from *Aspidosperma australe* Müll. Arg. (family Apocynaceae Juss.).[76] Several derivatives have been developed from olivacine and ellipticine, which have been further investigated in oncological studies.[77,78] The carbazole heptaphylline (CS 1.16) isolated from *Clausena harmandiana* (Pierre) Guillaumin (family Rutaceae, Juss.) showed strong cytotoxicity against human small cell lung cancer (NCI-H187) and human nasopharyngeal carcinoma KB cells with IC_{50} values equal to 1.3 µg/mL and 1.6 µg/mL, respectively.[65] An oxime derivative of heptaphylline (CS 1.17) displayed significant activity against both NCI-H187 and human nasopharyngeal carcinoma KB cells with IC_{50} values equal to 0.02 µM and 0.1 µM, and was inactive towards healthy African green monkey kidney (Vero) cells.[79]

1.2.3 *Melodinus Henryi* **Craib**

History The plant was first described by William Grant Craib, in *Bulletin of Miscellaneous Information Kew*, published in 1911.

Synonyms *Melodinus cochinchinensis* (Lour.) Merr., *Oncinus cochinchinensis* Lour.

Family Apocynaceae Juss., 1792

Common Name Si mao shan chen (Chinese)

Habitat and Description This stout climber grows to a length of 10 m in the montane forests of China, Burma, Thailand and Vietnam. The leaves are simple and opposite. The petiole is 0.6–1 cm long. The blade is narrowly elliptic, 6–19 cm × 2–7 cm, papery, cuneate at base, acuminate at apex and with numerous conspicuous secondary nerves. The inflorescence is a terminal cyme of white flowers. The calyx comprises five orbicular sepals which are 0.2 cm long and ciliate. The corolla is salver-shaped and presents a cylindrical tube which is 0.6 cm long, pilose, and dilated at staminal insertion. The five petals are ovate, contorted and 0.4 cm long. The corona scales are conspicuous and erect. The ovary is glabrous and develops a style which is 0.3 cm long and dilated at apex. The fruit is a

■ **CS 1.15** Olivacine

■ **CS 1.16** Heptaphylline

■ **CS 1.17** Oxime heptaphylline derivative

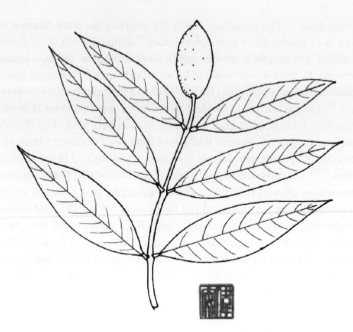

■ **FIGURE 1.7** *Melodinus henryi* Craib

berry which is elliptic and $9\,cm \times 5\,cm$. The seeds are oblong and $1.3\,cm$ (Figure 1.7).

Medicinal Uses The fruit are used to treat meningitis and bone injuries in Burma, Thailand and China.

Phytopharmacology *Melodinus henryi* Craib elaborates a broad array of indole alkaloids such as henrycinols A and B and monoterpenoid indole alkaloids including (+)-Δ14-vincamine, (+)-16-epi- Δ14-vincamine, and (+)-isoeburnamine,[80] melohenine A[81] and melodinines A–G, *O*-methylepivincanol,(–)-eburnamenine, vallesiachotamine, decarbomethoxydihydrogambirtannine, eburenine, rhazinilam, epivincanol, 19-(*R*)-methoxytubotaiwine, stemmadenine, stemmadenine-*N*-oxide, isositsirikine,[82] (+)-eburnamine, 14-epieburnamine, condylocarpine, isocondylocarpine, vincamenine, akuammicine, norfluorocurarine and 10,22-dioxokopsane.[83] The antibacterial and anti-inflammatory properties of the plant are most probably due to monoterpenoid indole alkaloids. Stemmadenine isolated from *Rhazya stricta* Decsne. (family Apocynaceae Juss.) exhibited antibacterial properties against Gramnegative *Pseudomonas aeruginosa* and *Escherichia coli*, Gram-positive *Staphylococcus aureus*, and the yeast *Candida albicans* with minimum inhibitory concentrations equal to 7.5 μg/mL, 1.2 μg/mL, 5 μg/mL and 37.5 μg/mL, respectively.[84] It could be interesting to evaluate the antibacterial activity of stemmadenine against the Gram-negative cocci *Neisseria meningitidis*. In addition, the monoterpenoid indole alkaloid strictamine isolated from *Alstonia scholaris* (L.) R. Br. (family Apocynaceae Juss.) inhibited the enzymatic activity of cyclooxygenases 1 and 2 by 47% and 95.6%, respectively, at a dose of 100 μM.[85]

Proposed Research Pharmacological study of vallesiachotamine and derivatives for the treatment of hematological cancers.

Rationale In a random screening of natural product derivatives, Wehner et al. (2008) identified a series of indoloquinolizidine as being strongly cytotoxic and pro-apoptotic. The indoloquinolizidines 6 (CS 1.18), 7 and 1 reduced the proliferation of human hepatocellular liver carcinoma (HepG2) cells with IC_{50} values equal to 6.6 μmol/L and 7.5 μmol/L and 8.3 μmol/L, respectively, via apoptosis induction and arrest in the G_2/M phase.[86] An interesting feature of indoloquinolizidines is their ability to block Ca^{2+} channels and thereby relax vascular smooth muscles. Geissoschizine methyl ether isolated from a member of the genus *Uncaria* Schreb. (family Rubiaceae Juss.) is a dependent Ca^{2+} blocker which relaxed Bay K8644-induced vasoconstrictions at a dose of 3×10^{-5} M.[87] Hirsutine from a member of the same genus exhibited Ca^{2+}-channel blocking activity.[88] The commercial indoquinolizidine alkaloid yohimbine used to treat erectile dysfunction and isolated from *Pausinystalia yohimbe* K. Schum. (family Rubiaceae Juss.) is a selective α_2-adrenoceptor antagonist and a Ca^{2+}-channel blocking agent.[89,90] Finally the monoterpenoid indole alkaloid vincamine isolated from *Vinca minor* L. (family Apocynaceae Juss.) is a commercial vasorelaxant used to treat poor cerebral perfusion and has been shown to exhibit Ca^{2+}-channel blocking effects.[91] The exact role of cytoplasmic Ca^{2+} levels in cell division and apoptosis is not yet fully understood but some evidence suggests that they

■ **CS 1.19** Flunarizine

■ **CS 1.20** Vallesiachotamine

play critical roles in the proliferation of certain types of cancer cell.[92,93] The commercial anti-migraine drug flunarizine (CS 1.19) is a Ca^{2+}-channel blocker which induced caspases 3 and 10 activation, poly (ADP-ribose) polymerase cleavage and laddering of DNA in Jurkat human T cell lymphoblast-like cells.[94] The Ca^{2+}-channel blocker carboxyamidotriazole reduced the growth, angiogenesis and metastasis in some solid tumors in ovarian cancer and has been under clinical trial as an anticancer agent.[95]

The indoloquinolizidine vallesiachotamine (CS 1.20) isolated from *Melodinus henryi* Craib (family Apocynaceae Juss.) showed cytotoxicity against human promyelocytic leukemia (HL-60), human hepatoma (SMMC-7721), human lung adenocarcinoma epithelial (A549), and human breast cancer (SK-BR-3) cells, with IC_{50} values of $2\,\mu M$, $16.8\,\mu M$, $26\,\mu M$, and $24.7\,\mu M$, respectively.[82]

1.2.4 *Nauclea Orientalis* Willd.

History The plant was first described by Carl Ludwig von Willdenow, in *Species Plantarum*, published in 1798.

Synonyms *Cephalanthus orientalis* L., *Nauclea orientalis* (L.) L.

Family Rubiaceae Juss., 1789

Common Names Leichhardt tree, cheesewood, bangkal (the Philippines), khan leuang (Laos), taniwatari no ki (Japanese)

Habitat and Description This is a timber tree which grows in moist soils in the forests of Sri Lanka, Burma, Laos, Cambodia, Vietnam, Malaysia, Japan (Kyuchu), Indonesia and Australia. The trunk is 20 m high. The bark is dark brown. The stems are lenticelled. The leaves are simple, and opposite. The stipule is interpetiolate, ovate, 2–4 cm × 1–2 cm. The petiole is stout, keeled and 1–4 cm long. The blade is glossy, ovate, acute at base and apex, 15–30 cm × 8–15 cm and presents five to eight pairs of conspicuous secondary nerves. The inflorescence is a terminal or axillary flowering head which is 3 cm in diameter. The flowers are minute and yellowish. The calyx produces five spathulate lobes which are 0.3 cm long. The corolla tube is 0.6–0.9 cm long and develops five lobes which are 0.5 cm long. The style produces a conspicuous spindle-shaped stigma which is 1.2–1.5 cm long and white. The fruit is a globose and woody syncarp which is 2.5–3 cm in diameter (Figure 1.8).

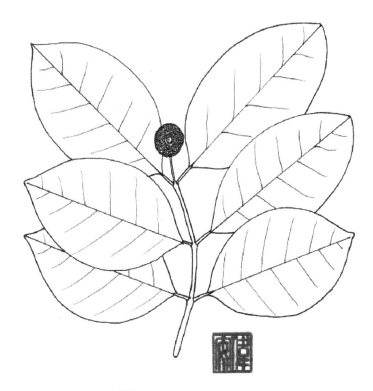

■ **FIGURE 1.8** *Nauclea orientalis* Willd.

Medicinal Uses In Papua New Guinea, the plant is used to alleviate abdominal pains and to treat animal bites and wounds. In Australia, the aborigines use the plant to alleviate pain and as a fish poison.

Phytopharmacology Phytochemical study of the plant resulted in the identification of chromones, triterpenoid saponins,[96] monoterpenoid indole alkaloids such as nauclealines A and B, naucleosides A and B, strictosamide, 10-hydroxystrictosamide, vincosamide, pumiloside, naucleaorals A and B, naucleaorine, epimethoxynaucleaorine and strictosidine lactam,[97–102] and triterpenes such as 3α-hydroxyurs-12-en-28-oic acid methyl ester (5), 3α,23-dihydroxyurs-12-en-28-oic acid (6), 3α,19α,23-trihydroxyurs-12-en-28-oic acid methyl ester and oleanolic acid.[102]

The analgesic property of the plant has not been validated yet but one can reasonably draw an inference that it is mediated by monoterpenoid alkaloids. In fact, the corynanthe-type monoterpenoid indole alkaloid mitragynine isolated from *Mitragyna speciosa* (Korth.) Havil. (family Rubiaceae Juss.) is a well-known central opioid analgesic.[103]

Proposed Research Pharmacological study of naucleaoral A and derivatives for the treatment of lung cancer.

Rationale Corynanthe-type monoterpenoid alkaloids such as corynanthine (CS 1.21) are interesting because they have the ability to bind and antagonize α-adrenergic receptors[104] and therefore to relax vascular smooth muscles. One such alkaloid is ajmalicine from *Rauvolfia serpentina* (L.) Benth. ex Kurz (family Apocynaceae Juss.), which abrogated phenylephrine-induced hypertension in rats.[105] Hirsutine from the genus *Uncaria* Schreb. (family Rubiaceae Juss.) at a dose of 3×10^{-5} M relaxed isolated rat aorta that had been contracted by norepinephrine (noradrenaline) and K^+.[88] Intravenous injections of 0.5 mg–5 mg/kg of dihydrocorynantheine from *Uncaria calophylla* Blume ex Korth. (family Rubiaceae Juss.) caused the mean arterial blood pressure of anesthetized normotensive rats to fall immediately by 34–49 mmHg.[106]

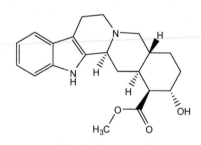

■ **CS 1.21** Corynanthine

Several lines of evidence suggest that α-adrenergic receptors may be somewhat linked to pro-apoptotic mechanisms. Bruzzone et al. (2011) observed that the promotion of growth of mouse mammary tumors induced by the commercial α_2-adrenergic agonist clonidine was reversed by the monoterpenoid alkaloid α_2-adrenergic antagonist corynanthidine.[107] In addition, quinazoline α_1-adrenoceptor antagonists, which are used to treat the symptoms associated with benign prostatic hyperplasia, induce apoptosis in prostate epithelial and endothelial cells, TGF-beta signaling and IκBα

■ **CS 1.22** Naucleaoral A

induction.[108] The pyridine derivative labedipinedilol-A induced apoptosis via a mechanism involving α_1-adrenoceptors in human prostate adenocarcinoma (LNCaP) and human prostate cancer (PC-3) cells, with a decrease in protein levels of cyclin D1/2, cyclin E, CDK2 and CDK4, CDK6 and increased expression of the cyclin-dependent kinase inhibitor proteins Cip/p21 and Kip/p27.[109] In fact, norepinephrine itself induced apoptosis in human lung adenocarcinoma epithelial (A549) cells with an IC_{50} value of 200 nM via the activation of α-adrenergic receptors and the autocrine generation of angiotensin-converting enzyme II.[110] Note that corynanthine inhibited the proliferation and DNA synthesis of human lymphocytes exposed to the mitogen concanavalin A.[111] The corynanthe-type monoterpenoid naucleaoral A (CS 1.22) isolated from *Nauclea orientalis* Willd. (family Rubiaceae Juss.) showed significant cytotoxicity to human epitheloid cervix carcinoma (Hela) cells with an IC_{50} value of 4 μg/mL.[53]

1.2.5 *Winchia Calophylla* **A. DC.**

History The plant was first described by Alphonse Louis Pierre Pyramus de Candolle, in *Prodromus Systematis Naturalis Regni Vegetabilis*, published in 1844.

Synonyms *Alstonia glaucescens* (Wallich ex G. Don) Monachino, *Alstonia rostrata* C.E.C. Fisch., *Alstonia pachycarpa* Merrill & Chun, *Alyxia glaucescens* Wallich ex G. Don, *Winchia glaucescens* (Wallich ex G. Don) K. Schum.

Family Apocynaceae Juss., 1789

Common Name Pen jia shu (Chinese)

Habitat and Description *Winchia calophylla* A. DC. is a timber tree that grows in the rainforests of China, India, Vietnam, Indonesia, Malaysia,

Burma and Thailand. The trunk is 30 m tall. The stems are angular at apex. The leaves are simple and arranged in whorls of three or four. The petiole is 1–2 cm long. The blade is narrowly elliptic, glossy above, glaucous beneath, 7–20 cm×2–4.5 cm, coriaceous, acuminate at apex, acute at base and shows numerous pairs of secondary nerves which are conspicuous. The inflorescence is a cyme of white flowers which is terminal and 5 cm long. The calyx is minute, pubescent and develops five lobes which are broadly ovate. The corolla tube is 0.5 cm long and develops five lobes which are broadly ovate, pubescent, 0.3–0.4 cm long and contortate. The fruit consists of a pair of follicles which are showy and 18–35 cm×1–1.5 cm. The seeds are 2 cm long, elliptic and ciliate at base and apex (Figure 1.9).

Medicinal Uses In China, the plant is used to treat lung infections, coughs and asthma.

Phytopharmacology Phytochemical study of the plant resulted in the identification of: monoterpenoid indole alkaloids including echitamine, echita-

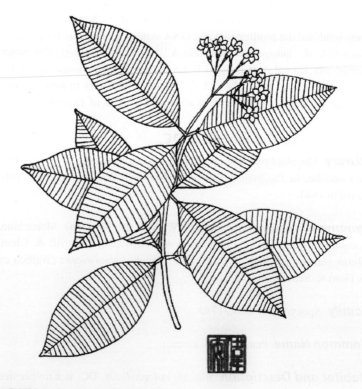

■ **FIGURE 1.9** *Winchia calophylla* A. DC.

midine,[112] alstogustine, *N*-(4)-methyl akuammicine, *N* -(4)-demethyl alstogustine, tubotawine, rhazimanine, 17-*O*-acetylechitamine, pseudoakuammigine, picrinine, nareline, stemmadenine,[113] *N*-(4)-demethyl-12-methoxyalstogustine, 17-carboxyl-*N*-(4)-methylechitamidine chloride, 17-carboxyl-12-methoxy-*N*-(4)-methylechitamidine chloride, 12-methoxyechitamidine, *N*-(4)-demethylechitamine, akuammicine, 12-methoxytubotaiwine, 1,2 dehydroaspidospermidine, 6,7-*seco*-angustilobine B, undulifoline, panarine, 19,20-*E*-vallesamine and 19,20-*Z*-vallesamine[114]; β-carboline alkaloids including rutaecarpine, evodiamine, dehydroevodiamine hydrochloride, 1,2,3,4- tetrahydro-1-oxo-carboline[113]; pyridine derivatives such as cantleyine, isocantleyine, venoterpine and 1-methyl-2-[(10Z)-10-pentadecanenyl]-4 (1 H)-quinolone[113]; simple phenols like paeonol, 4-hydroxy-3-methoxybenzoic acid, 3, 4-dihydroxybenzoic acid and 2,3-dihydroxybenzoic acid[113]; the iridoid loganin; the lignans sesamin and (–)-lyoniresinol[113]; the triterpenes lupenone, lupenyl acetate, betulinic acid, amyrin acetate, ursolic acid, ptiloepoxide, amyrin and cycloeucalenol[115]; and series of monoterpenoid glycosides.[116,117]

The anti-asthmatic property of the plant has been confirmed by Zu et al. (2005) who observed that loganin, cantleyine, and *N*-(4)-methyl akuammicine inhibited the contraction of guinea-pig preparations induced by histamine, with IC_{50} values of $0.6 \mu g/mL$, $0.1 \mu g/mL$ and $0.7 \mu g/mL$, respectively.[113]

Proposed Research Pharmacological study of 12-methoxyechitamidine and derivatives for the treatment of cancer.

Rationale Several lines of evidence suggest that Strychnos-type monoterpenoid alkaloids induce apoptosis in cancer cells and inhibit angiogenesis. The Strychnos alkaloid leucoridine A isolated from *Leuconotis griffithii* Hook. f. (family Apocynaceae Juss.) abrogated the survival of human nasopharyngeal carcinoma (KB) and human nasopharyngeal carcinoma vincristine-resistant (KB/VJ300) cells, with IC_{50} values equal to $0.5 \mu g/mL$ and $2.39 \mu g/mL$, respectively.[118] The Strychnos alkaloid brucine isolated from the seeds of *Strychnos nux-vomica* L. (family Loganiaceae R. Br. ex Mart.) inhibited the growth of HepG2 cells by 78.3% at a dose of $1 \mu M$ and induced DNA fragmentation, caspase 9 and 3 activation, Ca^{2+}-dependent mitochondrial membrane depolarization, and release of cytochrome c, decreased the level of anti-apoptotic Bcl-2 protein in HepG2 cells and increased the protein level of pro-apoptotic Bcl-2-associated X protein (Bax).[119] In another study Yin et al. (2007) showed that brucine (CS 1.23) inhibited the growth of human hepatoma (SMMC-7721) cells

■ **CS 1.23** Brucine

■ **CS 1.24** 12–Methoxyechitamidine

by 92.3% at a dose of 1 µM together with the induction of vacuole, chromatin, condensation, formation of subdiploid DNA, caspase 3 activation and cyclooxygenase 2 inhibition.[120] In addition, brucine inhibited the growth of human breast adenocarcinoma (MCF-7) cells with IC$_{50}$ values of 0.9 µM, with activation of caspase 3 and 9. Brucine also enhanced the survival of mice infested with Ehrlich's ascites carcinoma cells, and reduced the tumor growth and the secretion of vascular endothelial growth factor when given at a dose of 50 mg/kg/day.[121] The Strychnos-type monoterpenoid alkaloid 12-methoxyechitamidine (CS 1.24) isolated from *Winchia calophylla* A. DC. (family Apocynaceae Juss.) was cytotoxic against human lung adenocarcinoma epithelial (A549) cells, with an IC$_{50}$ value equal to 11.5 µM.[114]

REFERENCES

[48] Weiming C, Yaping Y, Xiaotian L. Alkaloids from roots of *Alstonia yunnanensis*. Planta Med 1983;49(1):62.

[49] Chen WM, Yan YP, Wang YJ, Liang XT. Isolation and identification of three new alkaloids from the roots of *Alstonia yunnanensis* Diels Indiana, USA. Acta Pharm Sin 1985;20(12):906–12.

[50] Chen WM, Yan YP, Ma XM. Isolation and identification of the alkaloids from the stems and leaves of *Alstonia yunnanensis*. Acta Pharm Sin 1986;21(3):187–90.

[51] Feng T, Li Y, Cai XH, Gong X, Liu YP, Zhang RT, et al. Monoterpenoid indole alkaloids from *Alstonia yunnanensis*. J Nat Prod 2009;72(10):1836–41.

[52] Wittmann T, Hyman A, Desai A. The spindle: a dynamic assembly of microtubules and motors. Nat Cell Biol 2001;3:E28–34.

[53] Jordan A, Hadfield JA, Lawrence NJ, McGown AT. Tubulin as a target for anticancer drugs: agents which interact with the mitotic spindle. Med Res Rev 1998;18:259–96.

[54] Wu M, Wu P, Xie H, Wu G, Wei X. Monoterpenoid indole alkaloids mediating DNA strand scission from *Turpinia arguta*. Planta Med 2011;77(3):284–6.

[55] Pham VC, Ma J, Thomas SJ, Xu Z, Hecht SM. Alkaloids from *Alangium javanicum* and *Alangium grisolleoides* that mediate Cu2+-dependent DNA strand scission. J Nat Prod 2005;68(8):1147–52.

[56] Lorence A, Nessler CL. Camptothecin, over four decades of surprising findings. Phytochem 2004;65(20):2735–49.

[57] Mansoor TA, Ramalho RM, Mulhovo S, Rodrigues CMP, Ferreira MJU. Induction of apoptosis in HuH-7 cancer cells by monoterpene and β-carboline indole alkaloids isolated from the leaves of *Tabernaemontana elegans*. Bioorganic and Med Chem Lett 2009;19(15):4255–8.

[58] Kolomeichuk SN, Bene A, Upreti M, Dennis RA, Lyle CS, Rajasekaran M, et al. Induction of apoptosis by vinblastine via c-Jun autoamplification and p53-independent down-regulation of p21. Mol Pharmacol 2008;73(1):128–36.

[59] Piazza GA, Alberts DS, Hixson LJ, Paranka NS, Li H, Finn T, et al. Sulindac sulfone inhibits azoxymethane-induced colon carcinogenesis in rats without reducing prostaglandin levels. Cancer Res 1997;57:2909–15.

[60] Fukuda R, Kelly B, Semenza GL. Vascular endothelial growth factor gene expression in colon cancer cells exposed to prostaglandin E2 is mediated by hypoxia inducible factor 1. Cancer Res 2003;63:2330–4.

[61] Yao M, Kargman S, Lam EC, Kelly CR, Zheng Y, Luk P, et al. Inhibition of cyclooxygenase-2 by rofecoxib attenuates the growth and metastatic potential of colorectal carcinoma in mice. Cancer Res 2003;63:586–92.

[62] Prescott SM, Fitzpatrick FA. Cyclooxygenase-2 and carcinogenesis. Biochim Biophys Acta 2000;1470:M69–78.

[63] Wangboonskul JD, Pummangura S, Chaichantipyuth C. Five coumarins and a carbazole alkaloid from the root bark of *Clausena harmandiana*. J Nat Prod 1984;47(6):1058–9.

[64] Chaichantipyuth C, Pummangura S, Naowsaran K, Thanyavuthi D, Anderson JE, Mclaughlin JL. Two new bioactive carbazole alkaloids from the root bark of *Clausena harmandiana*. J Nat Prod 1988;51(6):1285–8.

[65] Songsiang U, Thongthoom T, Boonyarat C, Yenjai C. Claurailas A–D, cytotoxic carbazole alkaloids from the roots of *Clausena harmandiana*. J Nat Prod 2011;74(2):208–12.

[66] Yenjai C, Sripontan S, Sriprajun P, Kittakoop P, Jintasirikul A, Tanticharoen M, et al. Coumarins and carbazoles with antiplasmodial activity from *Clausena harmandiana*. Planta Med 2000;66(3):277–9.

[67] Chu Y, Hsu MT. Ellipticine increases the superhelical density of intracellular SV40 DNA by intercalation. Nucleic Acids Res 1992;20:4033–8.

[68] Goodwin S, Smith AF, Horning EC. Alkaloids of *Ochrosia elliptica* Labill. J Am Chem Soc 1959;81(8):1903–8.

[69] Canals A, Purciolas M, Aymamí J, Coll M. The anticancer agent ellipticine unwinds DNA by intercalative binding in an orientation parallel to base pairs. Acta Crystallogr D Biol Crystallogr 2005;61(7):1009–12.

[70] Liu LF. DNA topoisomerase poisons as antitumor drugs. Annu Rev Biochem 1989;58:351–75.

[71] Fosse P, Rene B, Charra M, Paoletti C, Saucier JM. Stimulation of topoisomerase II-mediated DNA cleavage by ellipticine derivatives: structure–activity relationships. Mol Pharmacol 1992;42:590–5.

[72] Champoux JJ. DNA topoisomerases: structure, function, and mechanism. Annu Rev Biochem 2001;70:369–413.

[73] Ohashi M, Sugikawa E, Nakanishi N. Inhibition of p53 protein phosphorylation by 9-hydroxyellipticine: a possible anticancer mechanism. Jpn J Cancer Res 1995;86:819–29.

[74] Schwaller MA, Allard B, Lescot E, Moreau F. Protonophoric activity of ellipticine and isomers across the energy-transducing membrane of mitochondria. J Biol Chem 1995;270:22709–22713.

[75] Hagg M, Berndtsson M, Mandic A, Zhou R, Shoshan MC, Linder S. Induction of endoplasmic reticulum stress by ellipticine plant alkaloids. Mol Cancer Ther 2004;3:489–97.

[76] Ondetti MA, Deulofeu V. The structure of olivacine and u-alkaloid C (Guatambuine). Tetrahedron 1961;15(1–4):160–6.

[77] Awada A, Giacchetti S, Gerard B, Eftekhary P, Lucas C, De Valeriola D, et al. Clinical phase I and pharmacokinetic study of S 16020, a new olivacine derivative: report on three infusion schedules. Ann Oncol 2002;13(12):1925–34.

[78] Kamata J, Okada T, Kotake Y, Niijima J, Nakamura K, Uenaka T, et al. Synthesis and evaluation of novel pyrimido-acridone, -phenoxadine, and -carbazole as topoisomerase II inhibitors. Chem Pharm Bull 2004;52(9):1071–81.

[79] Thongthoom T, Promsuwan P, Yenjai C. Synthesis and cytotoxic activity of the heptaphylline and 7-methoxyheptaphylline series. Eur J Med Chem 2011;46(9):3755–61.

[80] Zhang YW, Yang R, Cheng Q, Ofuji K. Henrycinols A and B, two novel indole alkaloids isolated from *Melodinus henryi* CRAIB. Helvetica Chimica Acta 2003;86(2):415–9.

[81] Feng T, Cai XH, Li Y, Wang YY, Liu YP, Xie MJ, et al. Melohenines A and B, two unprecedented alkaloids from *Melodinus henryi*. Organic Letters 2009;11(21):4834–7.

[82] Feng T, Cai XH, Liu YP, Li Y, Wang YY, Luo XD. Melodinines A–G, monoterpenoid indole alkaloids from *Melodinus henryi*. J Nat Prod 2010;73(1):22–6.

[83] Zhou H, He HP, Wang YH, Hao XJ. A novel alkaloid from *Melodinus henryi*. Helvetica Chimica Acta 2010;93(10):2030–2.

[84] Mariee NK, Khalil AA, Nasser AA, Al-Hiti MM, Ali WM. Isolation of the antimicrobial alkaloid stemmadenine from Iraqi Rhazya stricta. J Nat Prod 1988;51(1):186–7.

[85] Shang JH, Cai XH, Feng T, Zhao YL, Wang JK, Zhang LY, et al. Pharmacological evaluation of *Alstonia scholaris*: anti-inflammatory and analgesic effects. J Ethnopharmacol 2010;129(2):174–81.

[86] Wehner F, Nören-Müller A, Müller O, Reis-Corrêa Jr. I, Giannis A, Waldmann H. Indoloquinolizidine derivatives as novel and potent apoptosis inducers and cell-cycle blockers. ChemBioChem 2008;9(3):401–5.

[87] Yuzurihara M, Ikarashi Y, Goto K, Sakakibara I, Hayakawa T, Sasaki H. Geissoschizine methyl ether, an indole alkaloid extracted from *Uncariae Ramulus* et Uncus, is a potent vasorelaxant of isolated rat aorta. Eur J Pharm 2002;444(3):183–9.

[88] Yano S, Horiuchi H, Horie S, Aimi N, Sakai S, Watanabe K. $Ca2^+$ channel blocking effects of hirsutine, an indole alkaloid from *Uncaria* genus, in the isolated rat aorta. Planta Med 1991;57:403–5.

[89] Watanabe K, Yano S, Horiuchi H. $Ca2^+$ channel-blocking effect of the yohimbine derivatives, 14β-benzoyloxyyohimbine and 14β-*p*-nitrobenzoyloxyyohimbine. J Pharm Pharmacol 1987;39(6):439–43.

[90] Godfraind T, Miller RC, Lima JS. Calcium entry blocking action of yohimbine and its isomers. Arch Int Pharmacodyn Ther 1982;260(2):280–1.

[91] Lamar JC, Poignet H, Beaughard M, Dureng G. Calcium antagonist activity of vinpocetine and vincamine in several models of cerebral ischaemia. Drug Dev Res 1988;14(3–4):297–304.

[92] Berridge MJ. Calcium signaling and cell proliferation. BioEssays 1995;17:491–500.

[93] Kahl CR, Means AR. Regulation of cell cycle progression by calcium/calmodulin-dependent pathways. Endocr Rev 2003;24:719–36.

[94] Conrad DM, Furlong SJ, Doucette CD, West KA, Hoskin DW. The Ca2+ channel blocker flunarizine induces caspase-10-dependent apoptosis in Jurkat T-leukemia cells. Apoptosis 2010;15(5):597–607.

[95] Hussain MM, Kotz H, Minasian L, Premkumar A, Sarosy G, Reed E, et al. Phase II trial of carboxyamidotriazole in patients with relapsed epithelial ovarian cancer. J Clin Oncol 2003;21:4356–63.

[96] Fujita E, Fujita T, Suzuki T. On the constituents of Nauclea orientalis L. I. Noreugenin and naucleoside, a new glycoside. (Terpenoids V). Chem Pharm Bull 1967;15(11):1682–6.

[97] Erdelmeier CAJ, Regenass U, Rali T, Sticher O. Indole alkaloids with in vitro antiproliferative activity from the ammoniacal extract of *Nauclea orientalis*. Planta Med 1992;58(1):43–8.

[98] Takayama H, Ohmori O, Sakai M, Funahashi M, Kitajima M, Santiarworn D, et al. Isolation, partial synthesis and stereochemical study of 10-hydroxystrictosamide, a constituent of *Nauclea orientalis* in Thailand. Heterocycles 1998;49(1):49–52.

[99] Zhang Z, ElSohly HN, Jacob MR, Pasco DS, Walker LA, Clark AM. New indole alkaloids from the bark of *Nauclea orientalis*. J Nat Prod 2001;64(8):1001–5.

[100] Sichaem J, Surapinit S, Siripong P, Khumkratok S, Jong-Aramruang J, Tip-Pyang S. Two new cytotoxic isomeric indole alkaloids from the roots of *Nauclea orientalis*. Fitoterapia 2010;81(7):830–3.

[101] He ZD, Ma CY, Zhang HJ, Tan GT, Tamez P, Sydara K, et al. Antimalarial constituents from *Nauclea orientalis* (L.) L. Chem Biodivers 2005;2(10):1378–86.

[102] Erdelmeier CAJ, Wright AD, Orjala J, Baumgartner B, Rali T, Sticher O. New indole alkaloid glycosides from *Nauclea orientalis*. Planta Med 1991;57(2):149–52.

[103] Takayama H. Chemistry and pharmacology of analgesic indole alkaloids from the rubiaceous plant, *Mitragyna speciosa*. Chem Pharm Bull 2004;52(8):916–28.

[104] Kocic I. Negative inotropic action of corynanthine in the guinea pig papillary muscle is attenuated by glibenclamide. Gen Pharmacol 1994;25(3):553–7.

[105] Roquebert J. Selectivity of raubasine stereoisomers for α1- and α2-adrenoceptors in the rat. Arch Int Pharmacodyn Ther 1986;282(2):252–61.

[106] Chang P, Koh YK, Geh SL, Soepadmo E, Goh SH, Wong AK. Cardiovascular effects in the rat of dihydrocorynantheine isolated from *Uncaria callophylla*. J Ethnopharmacol 1989;25(2):213–5.

[107] Bruzzone A, Piñero CP, Rojas P, Romanato M, Gass H, Lanari C, et al. α2-adrenoceptors enhance cell proliferation and mammary tumor growth acting through both the stroma and the tumor cells. Curr Cancer Drug Targets 2011;11(6):763–74.

[108] Partin JV, Anglin IE, Kyprianou N. Quinazoline-based alpha 1-adrenoceptor antagonists induce prostate cancer cell apoptosis via TGF-beta signaling and I kappa B alpha induction. Br J Cancer 2003;88:1615–21.

[109] Liou SF, Lin HH, Liang JC, Chen IJ, Yeh JL. Inhibition of human prostate cancer cells proliferation by a selective alpha1-adrenoceptor antagonist labedipinedilol-A involves cell cycle arrest and apoptosis. Toxicology 2009;256(1–2):13–24.

[110] Dincer HE, Gangopadhyay N, Wang R, Uhal BD. Norepinephrine induces alveolar epithelial apoptosis mediated by α-, β-, and angiotensin receptor activation. Am J Physiol Lung Cell Mol Physiol 2001;281(–3):L624–30.

[111] Ckless K, Schottfeldt L, Henriques JAP, Peres A, Nardi N, Wajner M. Reduction of mitogen-induced responsiveness of human and murine leukocytes in vitro by yohimbine and corynanthine. Int J Immunopathol Pharmacol 1996;9(2):59–65.

[112] Li CM, Zhang XM, Zhou YL, Huang LY, Tao GD. Studies on the indole alkaloids of *Winchia calophylla* A. DC. Acta Pharm Sin 1993;28(7):512–5.

[113] Zhu WM, He HP, Fan LM, Shen YM, Zhou J, Hao XJ. Components of stem barks of *Winchia calophylla* A. DC. and their bronchodilator activities. J Integr Plant Biol 2005;47(7):892–6.

[114] Gan LS, Yang SP, Wu Y, Ding J, Yue JM. Terpenoid indole alkaloids from *Winchia calophylla*. J Nat Prod 2006;69(1):18–22.

[115] Zhu WM, Shen YM, Hong X, Zuo GY, Yang XS, Hao XJ. Triterpenoids from the dai medicinal plant *Winchia calophylla*. Acta Botanica Sinica 2002;44(3):354–8.

[116] Zhu WM, Wang BG, Kang WY, Hong X, Zhou J, Hao XJ. Two new monoterpene diglycosides from *Winchia calophylla* A. DC. Chinese Chemical Letters 2003;14(10):1029–32.

[117] Zhu WM, Lu CH, Wang Y, Zhou J, Hao XJ. Monoterpenoids and their glycosides from *Winchia calophylla*. J Asian Nat Prod Res 2004;6(3):193–8.

[118] Gan CY, Etoh T, Hayashi M, Komiyama K, Kam TS. Leucoridines A–D, cytotoxic Strychnos-Strychnos bisindole alkaloids from Leuconotis. J Nat Prod 2010;73(6):1107–11.

[119] Deng X, Yin F, Lu X, Cai B, Yin W. The apoptotic effect of brucine from the seed of *Strychnos nux-vomica* on human hepatoma cells is mediated via Bcl-2 and Ca^{2+} involved mitochondrial pathway. Toxicol Sci 2006;91(1):59–69.

[120] Yin W, Deng XK, Yin FZ, Zhang XC, Cai BC. The cytotoxicity induced by brucine from the seed of *Strychnos nux-vomica* proceeds via apoptosis and is mediated by cyclooxygenase 2 and caspase 3 in SMMC 7221 cells. Food Chem Toxicol 2007;45(9):1700–8.

[121] Agrawal SS, Saraswati S, Mathur R, Pandey M. Cytotoxic and antitumor effects of brucine on Ehrlich ascites tumor and human cancer cell line. Life Sci 2011;89(5–6):147–58.

Isoquinoline Alkaloids and Derivatives

1.3.1 *Alangium Salviifolium* (L. f.) Wangerin

History The plant was first described by Walther Leonhard Wangerin, in *Das Pflanzenreich*, published in 1910.

Synonyms *Grewia salviifolia* L. f., *Alangium lamarckii* Thwaites

Family Cornaceae Bercht. & J. Presl, 1825

Common Names Sage-leaved alangium, ankol (Sanskrit), tu tan shu (Chinese)

Habitat and Description This is a tree that grows in the forests of China, India, Sri Lanka, Nepal, Cambodia, Vietnam, Burma, Laos, Thailand, Malaysia, Indonesia and the Philippines. The trunk is 20 m high. The stems are hairy and develop some thorns. The leaves are simple. The petiole is 1 cm long. The blade is oblong, papery, acute to slightly cordate at base, obtuse at apex, 8–20 cm × 2–7 cm and bears 3–5 pairs of secondary nerves. The inflorescence is an axillary cyme of whitish and fragrant flowers. The flower buds are cylindrical and linear. The calyx tube is 0.25 cm long and develops 5–10 lobes which are 0.3 cm long and toothed. The corolla develops four to six lobes which are linear, obtuse at apex, recurved, 1.5–3 cm long and tomentose beneath. The androecium is conspicuous and includes 10–30 stamens which are 0.5–1.5 cm long. The anthers are linear. The disc is discrete and develops some lobes. The style is glabrous, conspicuous, 0.8–3 cm long and develops a capitate stigma. The fruit is a red drupe which is globose, 1–2.5 cm × 0.6–1.5 cm, ribbed and marked at apex by a short tube (Figure 1.10).

Medicinal Uses In Thailand, the plant is used to treat asthma, coughs, hemorrhoids, diarrhea and to expel worms from the intestines. In India, the plant is used to expel worms from the intestines, to break fever, to induce emesis and defecation, and to treat leprosy.

Phytopharmacology The plant has been the subject of a number of phytochemical studies which have resulted in the isolation of a compelling array

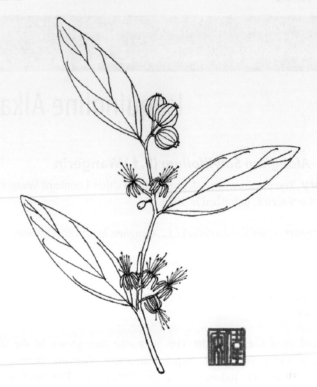

■ **FIGURE 1.10** *Alangium salviifolium* (L. f.) Wangerin

of isoquinoline alkaloids including: series of tetrahydroisoquinoline monoterpene alkaloids such as alangiside, 3-*O*-demethyl-2-*O*-methylalangiside,
isoalangiside, 3-*O*-demethyl-2-*O*-methylisoalangiside, methyliso-
alangiside, 6-*O*-methyl-*N*-deacetylisoipecosidic acid, 7-*O*-methyl-*N*-
deacetylisoipecosidic acid, 6,7-di-*O*-methyl-*N*- deacetylisoipecosidic acid,
6-*O*-methyl-*N*-deacetyl-6″-*O*-α-D-glucopyranosyl-isoipecosidic acid, 2′-*O*-
trans-feruloyldemethylalangisidine, 2′-*O*-trans-feruloylalangisidine, 2′-*O*-
trans-feruloyl-3-*O*-demethyl-2-*O*-methylalangiside, 2′-*O*-trans-sinapoylalangiside,
2′-*O*-trans-sinapoyl-3-*O*-demethyl-2-*O*-methylalangiside, 2′-*O*-trans-[4-(1,3-
dihydroxypropoxy)-3-methoxycinnamoyl] alangiside, 6′-*O*-β-D-gluco-
pyranosylalangiside, 3′-*O*-β-D-glucopyranosylalangiside, 6′-*O* α-D-
glucopyranosylalangiside, 6′-*O*-α-D-glucopyranosyl-3-*O*-demethyl-2-*O*-
methylalangiside, 6′-*O*-α-D-xylopyranosylalangiside, neoalangiside,
demethylneoalangiside, 6-*O*-methyl-*N*-deacetylisoipecosidic acid,
7-*O*-methyl-*N*-deacetylisoipecosidic acid, 6,7-di-*O*-methyl-*N*-deacetyl-
isoipecosidic acid and 6″-*O*-α-D-glucopyranosyl-6-*O*-methyl-*N*-
deacetylisoipecosidic acid[122–130]; benzoquinolizidine alkaloids such as

cephaeline, emetine, tubulosine, psychotrine, desmethylpsychotrine, deoxytubulosine, 1′,2′-dehydrotubulosine, alangine, isotubulosine, alangimarckine and isocephaeline[131–135]; the protoberberine alkaloid bharatamine[136] and the benzopyridoquinolizine alkaloids alamaridine, alangimaridine, alangimarine and isoalangimarine.[137–139] Other natural products identified from this plant include the sesquiterpene lacinilene C,[140] the triterpenes betulinic acid, lupeol, and isoalangidiol[141,142] and the sterol stigmasta-5,22,25-trien-3-β-ol.[143]

The parasiticidal and antimycobacterial properties of *Alangium salviifolium* (L. f.) Wangerin are most probably due to its alkaloidal content. Note for instance that deoxytubulosine inhibited the activity of two enzymes that are pivotal in the building of DNA and RNA: dihydrofolate reductase, at a dose of 30 μM, and thymidylate synthetase, at a dose of 80 μM.[133,144]

Proposed Research Pharmacological study of alangiumkaloid A and derivatives for the treatment of cancer.

Rationale Protoberberine alkaloids such as these have been shown to be potent inhibitors of topoisomerases I and II, which are two of the most important molecular targets currently used in clinical cancer treatment.[145,146] Coralyne inhibited the growth of human glioblastoma (SF268) and human lymphoblast (RPMI 8402) cells, with IC_{50} values equal to 0.1 μg/mL and 0.01 μg/mL, respectively; nitidine inhibited growth in the same cell lines, with IC_{50} values of 1.6 μg/mL and 0.2 μg/mL, respectively.[147] Apoptosis following DNA lesions induced by topoisomerase inhibitors such as protoberberine results from the binding of the protein kinases DNA-PK, ATM and ATR to DNA breaks, which results in the phosphorylation of kinases such as c-Abl.[148]

One such alkaloid is berberine (CS 1.25), which, at a dose of 100 μg/mL, inhibited the growth of human bladder cancer BIU-87 and T24 cells to 22.7% and 27.7% 72 h after treatment via inhibition of G protein Ras, G_0/G_1 phase arrest, caspase 9 and 3 activation and apoptosis.[149] In addition, berberine was cytotoxic to human gastric carcinoma (SNU-5) cells, with an IC_{50} equal to 48 μmol/L via G_2/M arrest and apoptosis.[150] The G_2/M arrest in human gastric carcinoma (SNU-5) cells (wild-type p53) involved induction of pro-apoptotic protein (p53) expression that led to the decrease of cyclin B and cyclin-dependent kinase 1 (CDK1) and to increased expression of cdc25c and Wee1. Apoptosis resulted from increased Ca^{2+} concentration, which decreased the level of anti-apoptotic Bcl-2 protein and increased the protein level of pro-apoptotic Bcl-2-associated X protein (Bax), cytochrome C release and caspase 3 activity.[150] Another

■ **CS 1.25** Berberine

■ **CS 1.26** Alangiumkaloid A

pro-apoptotic protoberberine is jatrorrhizine, which abrogated the survival of human erythromyeloblastoid leukemia (K562) via increase in reactive oxygen species and decreased mitochondrial membrane potential.[151] The protoberberine alangiumkaloid A (CS 1.26) isolated from *Alangium salviifolium* (L. f.) Wangerin (family Cornaceae Bercht. ex J. Presl) inhibited the growth of human malignant T-lymphoblastic (MOLT-3), human hepatocellular liver carcinoma (HepG2), human lung adenocarcinoma epithelial (A549) and human bile duct epithelial carcinoma (HuCCA-1) cells, with IC_{50} values equal to 13 μM, 27.7 μM, 76.5 μM and 58.6 μM.[152]

1.3.2 *Aristolochia Cucurbitifolia* Hayata

History The plant was first described by Bunzô Hayata, in *Icones Plantarum Formosanarum nec non et Contributiones ad Floram Formosanam*, published in 1915.

Family Aristolochiaceae Juss., 1789

Common Name Gua ye ma dou ling (Chinese)

Habitat and Description This herbaceous climber grows in the forests of Taiwan. The stems are terete, longitudinally striated and pubescent at apices. The leaves are simple. The petiole is 3 cm long and pubescent. The blade is cordate at base, variable in shape but mostly deeply palmate, with 5–7 lobes, 6–9 cm × 5–11 cm, the midlobe spathulate, papery and pubescent beneath. The inflorescence is solitary on a 3 cm pubescent pedicel. The calyx is 5–6 cm long, tomentose, pipe-shaped, yellow and develops three dark purple triangular lobes. Anthers are oblong, approximately 1.5 mm long. The gynostemium is three-lobed and supports a series of minute oblong anthers. The fruit is an ovoid capsule which is 6 × 1.5 cm and dehiscent. The seeds are triangular and 0.5 cm long (Figure 1.11).

Medicinal Uses In Taiwan, the plant is used to treat coughs and asthma, to break fever, to clean the lungs, to alleviate stomachache and to counteract snake poisoning.

Phytopharmacology Members of the family Aristolochiaceae are particular as they synthesize and accumulate substantial amounts of a unique series of isoquinoline aporphine derivatives, namely aristolochic acids, denitro aristolochic acid derivatives and aristolactams. Phytochemical study of the plant resulted in the isolation of: aristolochic acids I, IV, IVa and VIIa, the methyl esters of aristolochic acids I–IV and VII, sodium aristolochate I–III and IVa, sodium 7-hydroxylaristolochate and aristolochic acid C; denitro aristolochic acid derivatives including

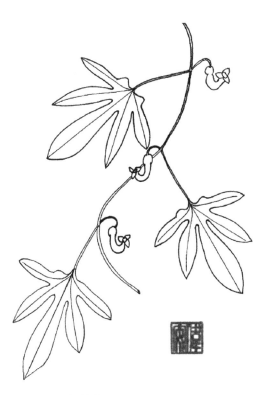

■ **FIGURE 1.11** *Aristolochia cucurbitifolia* Hayata

aristolic acid, aristolic acid methyl ester, aristofolin A, sodium aristofolin A, aristofolin B–D and sodium 7-hydroxyl-8-methoxyaristolate, and the aristolactam alkaloids aristolactams I and II, 9-methoxyaristolactam I, aristolactam C *N*-β-D-glucoside, aristolactam I, AII, BII, IIIa, 9-methoxyaristolactam-I, 9-methoxyaristolactam IV, piperolactam A and cepharanones A and C.[153,154] Other products found in the plant are series of phenylpropanoids such as *N-trans*-cinnamoyltyramine, 4-hydroxylcinnamic acid, 4-hydroxyl-3-methoxycinnamic acid and *cis*- and *trans*-4-hydroxyl-3-methoxycinnamic acid methyl ester.[154]

The medicinal properties of the plant have not been substantiated yet but one might think that the anti-inflammatory, analgesic and antidotal properties are due to compounds such as aristolochic acid, which is a well-established inhibitor of phospholipase A2.[155,156] Phenylpropanoids might account for the bechic property of the plant.

Proposed Research Pharmacological study of aristolactam derivatives for the treatment of cancer.

■ **CS 1.27** Aristolactam 21

■ **CS 1.28** Aristolactam 13b

■ **CS 1.29** Aristolactam AIIIa

Rationale Aristolochic acid I isolated from *Aristolochia cucurbitifolia* Hayata (family Aristolochiaceae Juss.) inhibited the growth of human epidermoid carcinoma (9-KB), mouse leukemia (P388), human lung adenocarcinoma epithelial (A549), human colon cancer (HT-29) and human promyelocytic leukemia (HL-60) cells, with IC_{50} values equal to 4.0 µg/mL, 0.7 µg/mL, 5.0 µg/mL, 8.3×10^{-4} µg/mL and 3.4 µg/mL, respectively.[154] Aristolochic acid I is cytotoxic but it is a carcinogenic substance which has been responsible for outbreaks of urothelial cancers in Europe.[157] Once aristolochic I is ingested, the nitro group is metabolized in an electrophilic cyclic *N*-acylnitrenium ion that reacts with purine bases to form covalent DNA adducts, which are associated with the activation of G protein Ras and therefore tumorigenesis.[158]

Hegde et al. (2010) synthesized series of derivatives from aristolochic acid and produced aristolactam 21 (CS 1.27), which inhibited the enzymatic activity of cyclin-dependent kinase 2 (CDK2), with an IC_{50} value of 35 nM.[159] The structure–activity relationship revealed that the lactam ring is essential for potent inhibition of cyclin kinase 2 (CDK2), hydroxyl groups at C6 or C8 positions enhance CDK2, as well as substitution of hydroxyl groups at the C3 and C4 positions, and protection of NH by a methyl group reduces activity.[159] Other strongly cytotoxic aristolactams synthesized are aristolactam 13b (CS 1.28), which was cytotoxic against mouse lymphocytic leukemia (L1210) cells, with an IC_{50} value of 1.6 µM,[160] and aristolactam 34, with IC_{50} values inferior to 0.5 µM against human lung adenocarcinoma epithelial (A549), human ovary adenocarcinoma (SK-OV-3), human epithelial carcinoma (A-431) and human colorectal adenocarcinoma (HCT-15) cells.[161]

Aristolactam AIIIa (CS 1.29) isolated from *Aristolochia argentina* Griseb. (family Aristolochiaceae Juss.) abrogated the survival of human epitheloid cervix carcinoma (Hela) and human lung adenocarcinoma epithelial (A549) cells, with IC_{50} values of 7.9 µmol/L and 15 µmol/L, with cycle arrest at the G_2/M phase, PARP (poly [ADP-ribose] polymerase) cleavage and inhibition of polo-like kinase 1 (PLK1), which has been determined as an attractive target for cancer therapy.[162–164] Balachandran et al. (2005) showed that several aristolactams were not toxic towards porcine kidney (LLC-PK) cells cultured in vitro and suggested that aristolactam might not be nephrotoxic. In fact, aristolochic acid derivatives owe their toxicity to the C10 nitro group and a C8 methoxy group.[165]

1.3.3 *Aristolochia Manshuriensis* Kom.

History The plant was first described by Vladimir Leontjevich Komarov, in *Trudy Imperatorskago S.Peterburgskago Botaničeskago Sada*, published in 1903.

Synonyms *Hocquartia manshuriensis* (Kom.) Nakai, *Isotrema manshuriensis* (Kom.) H. Huber

Family Aristolochiaceae Juss., 1789

Common Names Manchurian birthwort, Manchurian Dutchman's pipe, kanmokudsu (Japanese), kwangbanggi (Korean), guan mu tong (Chinese)

Habitat and Description This is a climber which grows in the forests of China and Korea. The stems are woody and terete. The petiole is fleshy and 6–8 cm long. The leaves are simple. The blade is broadly cordate, 15–30 cm × 13–28 cm and presents four to six pairs of secondary nerves. The inflorescence is simple and axillary with a pedicel which is 1.5–3 cm long. The calyx is light green, 4.5–5.5 cm long, pipe-shaped and develops three broadly deltoid yellowish and pubescent lobes. The gynostemium is three-lobed and supports a series of minute oblong anthers. The fruit is a cylindrical, tuberculate, 9–11 cm × 3–4 cm and dehiscent capsule. The seeds are triangular and 0.5 cm long (Figure 1.12).

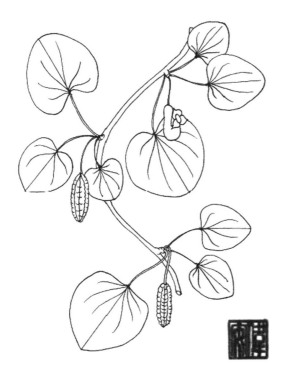

■ **FIGURE 1.12** *Aristolochia manshuriensis* Kom.

Medicinal Uses In Korea, the plant is used to break fever and to promote urination. In China, the plant is used to treat edema and diabetes, to stimulate the circulation of blood and the secretion of milk, and to induce menses.

Phytopharmacology The plant is known to contain aristolochic acids[166] such as aristolochic acids I, II, IIIa, IV, IVa and D, aristolochic acid I methyl ester, aristolochic acid IV methyl ester, aristoloside,[167] aristolochic acid II methyl ester, sodium aristolochate-I, aristolochic acid IVa methyl ester and aristolic acid II,[168] the denitro aristolochic acid derivative demethylaristofolin E, the aristolactams aristolactam I, aristolactam-II, 9-methoxyaristolactam I, aristolactam AII, manshurienines A and B,[168,169] series of phenylpropanoids such as coumaric acid and ferulic acid[168] and the sesquiterpenes isobicyclogermacrenal[170] and manshurolide.[171] The medicinal properties of the plant have not yet been substantiated. It would be interesting to look for antidiabetic agents from this plant.

■ **CS 1.30** SCH 546909

Proposed Research Pharmacological study of SCH 546909 and derivatives for the treatment of bone cancer.

Rationale A compelling feature of aristolactams and their synthetic derivatives is the broad range of cellular targets with which they interact to bring about cancer cell death. One of these targets is the serine/threonine kinase CDK2, which is a key enzyme in the regulation of the cell cycle and an attractive target for the development of anticancer drugs.[172] The aristolactam SCH 546909 (CS 1.30) isolated from *Aristolochia manshuriensis* Kom. (family Aristolochiaceae Juss.) inhibited the enzymatic activity of CDK2, with an IC_{50} value of 140 nM.[173] This compound is actually identical to aristolactam AIIIa isolated from *Aristolochia argentina* Griseb. (family Aristolochiaceae Juss.), which abrogated the survival of human epitheloid cervix carcinoma (Hela) and human lung adenocarcinoma epithelial (A549), with IC_{50} values of 7.98 μmol/L and 15 μmol/L, respectively, with cycle arrest at the G_2/M phase, PARP (poly [ADP-ribose] polymerase) cleavage and inhibition of polo-like kinase 1. As a consequence, it has been determined that the latter is an attractive target for cancer therapy.[162–164] Oh et al. (2011) synthesized the aristolactam KO-202125 (CS 1.31) derived from sauristolactam (CS 1.32) isolated from *Saururus chinensis* (Lour) Baill. (family Saururaceae Rich. ex T. Lestib.).[174] KO-202125 abrogated the survival of human lung adenocarcinoma epithelial (A549), human ovary adenocarcinoma (SK-OV-3), human colorectal adenocarcinoma (HCT-15), human epithelial carcinoma (A-431), human breast adenocarcinoma (MDA-MB-231) and human

■ **CS 1.31** KO-202125

■ **CS 1.32** Saurolactam

breast cancer (SK-BR-3) cells, with IC_{50} values of 1 μM, 1.3 μM, 1.4 μM, 1.15 μM, 1.17 μM and 1.79 μM, respectively, by interacting with epidermal growth factor receptor (EGFR), thus reducing its activity, and downregulation of the Akt pathway.[174]

Gefitinib is a commercial tyrosine kinase inhibitor that competes with ATP in the tyrosine-kinase domain of epidermal growth factor and which is used in therapy of non-small cell lung cancer.[175] Note that the aristolactam saurolactam from *Saururus chinensis* (Lour.) Baill. (family Saururaceae Rich. ex T. Lestib.) inhibited tartrate-resistant acid phosphatase activity induced by receptor activator of nuclear factor-κB ligand (RANKL).[176] Receptor activator of nuclear factor-κB ligand (RANKL) plays a key role in the establishment and propagation of bone metastases, and inhibitors of this receptor activator, like denosumab, are of value for the treatment of metastatic disease to the bone.[177,178]

1.3.4 *Cocculus Trilobus* (Thunb.) DC.

History The plant was first described by Augustin Pyramus de Candolle in *Regni Vegetabilis Systema Naturale*, published in 1818.

Synonyms *Cocculus orbiculatus* var. *orbiculatus*, *Menispermum trilobum* Thunb.

Family Menispermaceae Juss., 1789

Common Names Korean moonseed, mu fang ji (Chinese)

Habitat and Description *Cocculus trilobus* (Thunb.) DC. is a climber which grows on the forest margins of Laos, Malaysia, Indonesia, the Philippines, China, Taiwan, Korea and Japan. The stems are terete, woody and pubescent. The leaves are simple. The petiole is flexuous and 1–3 cm long. The blade is papery, broadly lanceolate, rounded at base, acute at apex and presents two pairs of secondary nerves emerging from the base. The inflorescence is an axillary cyme. The male flower presents six sepals which are ovate and 0.1–0.2 cm long, and six petals which are ovate, as long as the sepals and bifid at apex. The androecium consists of six stamens, which are half the length of the petals. The female flowers include six staminodes and six carpels. The fruit is a drupe which is purple, globose and 0.8 cm in diameter. The infructescence is grape-like (Figure 1.13).

Medicinal Uses In China, the plant is used to promote urination and to treat edema, infection of the genitals, tumors, paralysis, asthma and bronchitis.

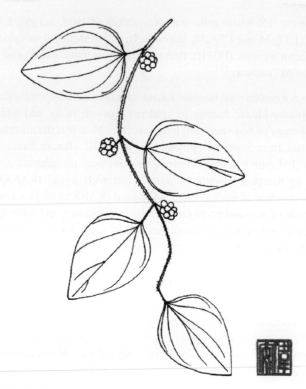

■ FIGURE 1.13 *Cocculus trilobus* (Thunb.) DC.

Phytopharmacology Phytochemical investigation of the plant resulted in the isolation of a broad array of isoquinoline alkaloids including: the erythrinan alkaloids coccutrine,[179] dihydroerysovine, cocculolidine and cocculine[180,181]; the bisbenzylisoquinoline alkaloids coclobine,[182] isotrilobine, trilobine, isotrilobine-*N*-2-oxide, nortrilobine,[183] (+)-coccuorbiculatine A, (+)-10-hydroxyisotrilobine, (+)-1,2-dehydroapateline, (+)-*O*-methylcocsoline[184]; the aporphines magnoflorine,[184] isoboldine,[186] (+)-laurelliptinhexadecan-1-one, (+)-laurelliptinoctadecan-1-one and (+)-norboldine[184]; the protoberberines (–)-4-methoxy-13,14-dihydrooxypalmatine, (–)-4-methoxypalmatine, oxy-palmatine,[184] and the oxoaporphine peruvianine.[184]

The diuretic, anti-inflammatory, antiseptic, muscular and anti-asthmatic properties have not been yet substantiated but these are obviously linked to the pharmacology of alkaloids. The morphinane alkaloids isosinococuline and sinococuline protected rodents against P-388 leukemia infestation at doses of 25 mg/kg/day and 40 mg/kg/day[187,188] and account for the antitumor property of the plant.

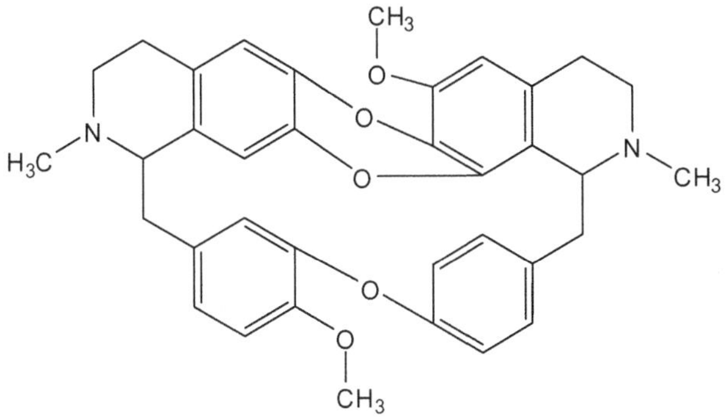

■ CS 1.33 Isotrilobine

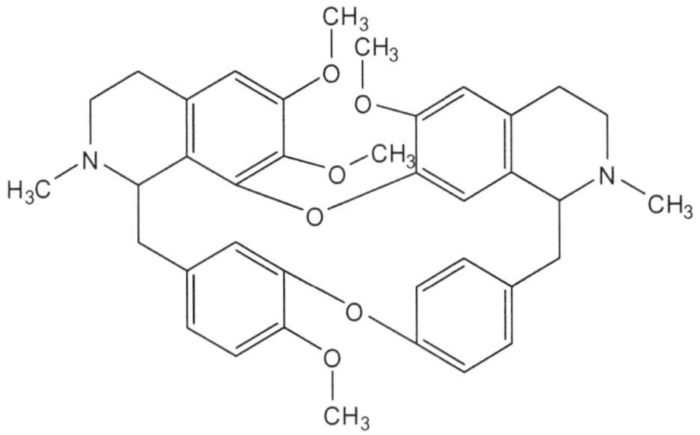

■ CS 1.34 Tetrandrine

Proposed Research Pharmacological study of isotrilobine and derivatives for the treatment of cancer.

Rationale The bisbenzylisoquinoline isotrilobine inhibited the proliferation of human hepatocellular liver carcinoma (HepG2), human hepatocellular carcinoma (Hep3B), human breast adenocarcinoma (MCF-7) and human breast adenocarcinoma (MDA-MB-231) cells, with IC_{50} values equal to 0.6 μg/mL, 0.75 μg/mL, 3.9 μg/mL and 1.6 μg/mL, respectively.[184] The mode of action of isotrilobine (CS 1.33) is unknown but current evidence suggests that bisbenzylisoquinoline alkaloids are potent inducers of apoptosis. One such alkaloid is the bisbenzylisoquinoline tetrandrine isolated from *Stephania tetrandra* S. Moore (family Menispermaceae Juss.), which, at a concentration of 10 μg/mL, reduced the cell viability of human leukemic (CEM-C7), human promyelocytic leukemia (HL-60), and human lymphoblastoid (BM-13674) cells by 30.7%, 24.8%, and 23.2%, respectively, via apoptosis with DNA fragmentation and blocking of calcium channels.[189] In addition, tetrandrine (CS 1.34) induced the apoptosis of HepG2 cells with an IC_{50} value equal to 9 μM, DNA fragmentation, activation of caspase 3, cleavage of PARP (poly [ADP-ribose] polymerase) and Ca^{2+}-channel blocking.[190]

Tetrandrine reduced the viability of rat glioma (RT2) cells with an IC_{50} of 11.3 μM and inhibited the expression of vascular endothelial growth factor (VEGF).[191] In addition, 150 mg/kg/day of tetrandrine prolonged the survival of rodents against intracerebral gliomas.[191] Another compelling feature of bis-benzylisoquinolines is their ability to reverse P-glycoprotein-mediated efflux of anticancer agents and therefore to abrogate drug resistance. Tetrandrine at a dose of 2.5 μM reversed the sensitivity of multidrug-resistant human epidermoid carcinoma (KBv200) cells to paclitaxel 10-fold[192] by a mechanism related to Ca^{2+} homeostasis. The exact apoptotic mode of action of tetrandrine is still undefined but it is known that tetrandrine induces pro-apoptotic protein (p53) expression in wild-type p53 cancer cells, increases the level of reactive oxygen species (ROS), releases mitochondrial cytochrome c, activates the enzymatic activity of caspases 8 and 9, inhibits the large K^+ (BK) channels,[193,194] activates mitogen-activated protein kinase (MAPK) p38[195] and Wnt/β-catenin signaling pathways,[196] and suppresses Akt activation.[197]

1.3.5 *Corydalis Ternata* (Nakai) Nakai

History The plant was first described by Takenoshin Nakai, in *Botanical Magazine (Tokyo)*, published in 1914.

Synonyms *Corydalis bulbosa* (Linnaeus) Candolle f. *ternata* Nakai, *Corydalis nakaii* Ishid., *Corydalis remota* Fischer ex Maximowicz var. *ternata* (Nakai) Makino, *Coridalys turtschaninovii* Besser var. *ternata* (Nakai) Ohwi.

Family Papaveraceae Juss., 1789

Common Names Three-leaf corydalis, deul hyun ho sag (Korean), san lie yan hu suo (Chinese)

Habitat and Description This is a little erect herb that grows in the watery fields and low mountain slopes of Korea and China. The tuber is oblong and 0.5–1.5 cm long. The leaves are trifoliolate. The folioles are elliptic, serrate and 0.8–2 cm × 0.3–1.5 cm. The inflorescence is a terminal raceme of 7–12 flowers. The sepals are minute. The petals are purplish, 1.5–2 cm long and form a zygomorphic tubular corolla. The androecium presents two stamens. The fruit is a capsule which is linear, 1.5 cm long and encloses 5–10 minute seeds (Figure 1.14).

Medicinal Uses In Korea, the plant is used to alleviate headaches and spasms, to heal gastric ulcers and to check uterine hemorrhages. In China,

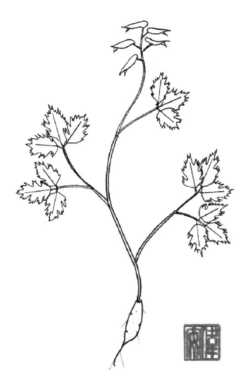

■ **FIGURE 1.14** *Corydalis ternata* (Nakai) Nakai

the plant is used to stimulate the circulation of blood, to assuage pain and to heal boils.

Phytopharmacology The plant is known to accumulate series of isoquinoline alkaloids, including: the protoberberine alkaloids berberine, coptisine,[198] protopine,[199] dehydrocorydaline,[200] demethylcorydalmine, tetrahydroberberine, corydalmine, thalifendine and cheilanthifoline,[201] and the benzylisoquinoline alkaloids epi-coryximine, coryternatines, A-C, (S)-reticuline and (R)-reticuline. Other chemical constituents include the triterpenoid saponins coryternic acid 3-*O*-D glucuronopyranoside and coryternic acid 3-*O*-D-glucuronopyranoside-6-*O*-methyl ester.[202] To date, the medicinal properties of the plant have not been substantiated but one could reasonably attribute the analgesic, antiseptic, antihemorrhagic and antispasmodic effects to protoberberine alkaloids.[203]

Proposed Research Pharmacological study of demethylcorydalmine for the treatment of brain tumors and/or leukemia.

■ **CS 1.35** LDD503

■ **CS 1.36** Demethylcorydalmine

Rationale A mounting body of evidence suggests that protoberberines inhibit the purinergic ATP-gated ion channel P2×7. Lee et al. (2011) synthesized the protoberberine derivative LDD503 (CS 1.35), which was able to inhibit by 95% the 2′,3′-O-(4-benzoyl-benzoyl) adenosine triphosphate (BzATP)-induced formation of pores in purinergic receptor P2×7-expressing human embryonic kidney (HEK293) cells at a dose of 10 μM.[204] P2×7 is an extracellular ATP-activated plasma membrane ion channel which has been recently identified as a key factor in the development and dissemination of metastases and as such as a potent target in cancer treatments.[205] In fact, P2×7 receptor antagonists have anticancer properties in vivo. Intravenous administration of the P2×7 receptor antagonist brilliant blue G inhibited the growth of brain tumors induced by mouse glioma (C6) cells in rodents by 52%[206] and P2×7 receptor antagonists prolonged the survival rate and attenuated the metastatic invasion in rodents infested with murine macrophage-like (P388D1) cells.[207] The precise molecular mode of action of P2×7 receptor antagonists on cancer remains undefined but one could draw an inference that the inhibition of P2×7 abrogates the ATP-induced increase in intracellular Ca^{2+} concentration and release of substance P, which are crucial for cancer cells' viability and growth.[208,209] Note that the secretion of substance P is Ca^{2+} dependent.[210] The protoberberine demethylcorydalmine (CS 1.36) exhibited cytotoxic properties against human lung adenocarcinoma epithelial (A549), human ovary adenocarcinoma (SK-OV-3), human skin melanoma (SK-MEL-2) and human colorectal adenocarcinoma (HCT-15) cells, with IC_{50} values equal to 8.3 μM, 5.1 μM, 7.8 μM and 2.8 μM, respectively.[80]

1.3.6 *Cyclea racemosa* Oliv.

History The plant was first described by Daniel Oliver, in *Hooker's Icones Plantarum*, published in 1890.

Synonym *Cyclea racemosa* fo. *emeiensis* H.S. Lo & S.Y. Zhao

Family Menispermaceae Juss., 1789

Common Name Lun huan teng (Chinese)

Habitat and Description *Cyclea racemosa* Oliv. is a woody climber which grows wild in China. The stems are slender and terete. The leaves are simple. The petiole is flexuous and up to 10 cm long. The blade is 4–9 cm × 3.5–8 cm, slightly peltate, pubescent beneath, round at base, acuminate at apex and presents 9–11 pairs of secondary nerves, the first pair originating from the base. The inflorescence is an elongated panicle which is 3–10 cm long. The male flowers present a calyx that develops four lobes which are ovate and 0.2 cm long. The corolla is cup-shaped and has two to six lobes. The androecium includes four anthers. The female flowers are tiny and include two sepals and two petals which are minute. The gynoecium is hairy and develops a three-lobed stigma. The fruit is a globose drupe which is 0.5 cm in diameter. The seeds are horse-shoe-shaped, 0.4 cm in diameter and tubercular (Figure 1.15).

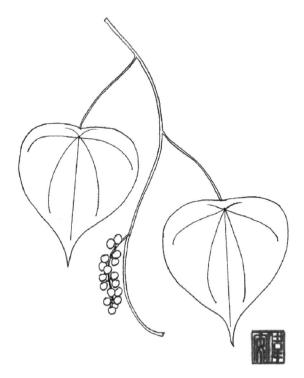

■ **FIGURE 1.15** *Cyclea racemosa* Oliv.

■ **CS 1.37** O-(4-Ethoxyl-butyl)-berbamine

Medicinal Uses In China, the plant is used to heal gastric ulcers and to assuage tooth pain.

Phytopharmacology Phytochemical evaluation has resulted so far in the isolation of series of bisbenzylisoquinoline alkaloids such as cycleaneonine,[211,212] racemosidines A–C and racemosinines A–C, as well as (–)-curine, 7-O-methylhayatidine and the protoberberine R-cyclanoline.[213] The properties mentioned above have not yet been substantiated but other bisbenzylisoquinoline alkaloids are known to be anti-inflammatory.[214]

Proposed Research Pharmacological study of racemosidine A and derivatives in the treatment of multidrug-resistant breast cancer.

Rationale The cytotoxic properties of bisbenzylisoquinoline alkaloids are compelling and have spurred efforts to develop series of derivatives which have appeared to be potent inhibitors of multidrug resistance. One such alkaloid is the calmodulin antagonist O-(4-ethoxyl-butyl)-berbamine (CS 1.37), which compromised the depolymerization of microtubules and impaired mitochondrial and endoplasmic reticulum function in human breast adenocarcinoma (MCF-7) cells, with an IC_{50} value of 3 μmol/L.[215] In addition, O-(4-ethoxyl-butyl)-berbamine administered with doxorubicin increased the median survival time of rodents infested with mouse ascites hepatoma (H22) cells via a mechanism implying an increase in the efficacy of doxorubicin.[216] In fact, the enhancement of doxorubicin antitumor effects by O-(4-ethoxyl-butyl)-berbamine activity was observed against multidrug-resistant breast carcinoma MCF-7/ADR cells, with apoptosis

■ CS 1.38 H1

induction following cytoplasmic accumulation of doxorubicin, down-regulation of P-glycoprotein expression (22.3% at 6 μM), downregulation of the cyclin B1-cdc2/p34 pathway and very strong G_2/M arrest.[217] O-(4-Nitrobenzyl)-berbamine, O-((6-chloropyridin-3-yl)methyl)-berbamine and O-(1H-indole-2-carbonyl)-berbamine inhibited the nuclear translocation of nuclear factor kappa-light-chain-enhancer of activated B cells (NF-κB) p65 in imatinib-resistant human erythromyeloblastoid leukemia (K562R) cells, with IC_{50} values of 0.3 μM, 0.4 μM, and 0.4 μM, respectively.[218] A brominated tetrandrine derivative named H1 (CS 1.38) abrogated the survival of human ileocecal adenocarcinoma (HCT-8), human ovarian cancer (A2780), human nasopharyngeal carcinoma (KB), human lung adeno-carcinoma epithelial (A549), MCF-7, human hepatocellular carcinoma (BEL-7402), human erythromyeloblastoid leukemia (K562) and human gastric cancer (BGC-823) cells, with a mean IC_{50} of 2.2 μM via a mechanism involving ROS production, collapse of the mitochondrial membrane potential, the release of mitochondrial cytochrome c, a decreased level of anti-apoptotic Bcl-2 protein, an increase in the protein level of pro-apoptotic Bcl-2-associated X protein (Bax) and caspase 9 activation.[219]

H1 also protected rodents against multidrug-resistant human epidermoid carcinoma (KBv200) infestation, with an inhibition of tumor mass of 74.9%.[219] The bisbenzylisoquinoline alkaloid racemosidine A (CS 1.39) isolated from *Cyclea racemosa* Oliv. abrogated the survival of human ileocecal adenocarcinoma (HCT-8), human liver cancer (BEL-7402) and human ovarian cancer (A2780) cells, with IC_{50} values equal to 2.8 μM, 6 μM and 6.7 μM,[213] respectively.

1.3.7 *Goniothalamus Amuyon* (Blco.) Merr.

History The plant was first described by Elmer Drew Merrill, in *Philippine Journal of Science*, published in 1915.

■ **CS 1.39** Racemosidine A

Synonym *Uvaria amuyon* Blanco, *Polyalthia sasakii* Yamamoto

Family Annonaceae Juss., 1789

Common Names Taiwan Goniothalamus, amuyon (the Philippines), tai wan ge na xiang (Chinese)

Habitat and Description This is a treelet that grows in the forests of Taiwan and the Philippines. The trunk is 5 m tall. The leaves are simple. The petiole is 0.6–0.8 cm long. The blade is oblong elliptic, 8–15 cm × 3–6 cm, papery, acute at base, acuminate or acute at apex and with 8–10 pairs of secondary nerves. The inflorescence is axillary. The calyx consists of three sepals which are ovate, pubescent beneath and 0.5 cm long. The corolla comprises six petals which are thick, pubescent beneath and up to 3.5 cm long. The androecium includes numerous stamens. The gynoecium comprises several carpels. The fruit consist of several ripe carpels which are 1–2 cm × 0.5–1 cm. Each carpel includes one or two seeds (Figure 1.16).

Medicinal Uses In Taiwan, the plant is used externally to treat scabies. In the Philippines, the plant is used to promote digestion and to treat rheumatism, edema and infections of the ears.

Phytopharmacology Phytochemical investigation of the plant resulted in the isolation of a range of isoquinoline alkaloids including: the

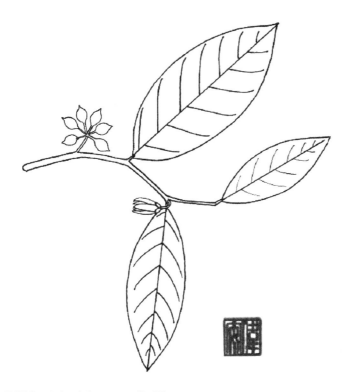

■ **FIGURE 1.16** *Goniothalamus amuyon* (Blco.) Merr.

protoberberines (−)-tetrahydropalmatine and palmatine; the aporphines (−)-anolobine, (−) anonaine, liriodenine[220] and lysicamine,[221] and the aristolactams piperlactam C, aristolactam FII,[221] velutinam and aristolactam BII.[222] Other principles found in this plant include the styryllactones goniodiol-7-monoacetate, goniodiol-8-monoacetate[223] 8-methoxygoniodiol, 8-chlorogoniodiol, goniothalamin epoxide, goniothalamin, (5*S*,6*R*,7*R*,8*R*)-goniotriol, (+)-9-deoxygoniopypyrone,[221] digoniodiol, deoxygoniopypyrone A, goniofupyrone A[224] and leiocarpin C[222]; acetogenins[225]; the lignan (+)-pinoresinol: the benzoic acid veratric acid; the phenylpropanoid cinnamic acid;[221] the tetrahydrofuran goniothalesacetate; the tetrahydropyran goniothaesdiol A[222] and the azaanthraquinones griffithazanone A and 4-methyl-2,9,10-(2*H*)-1-azaanthracencetrione.[222]

The parasiticidal, anti-inflammatory and antiseptic properties of the plant have yet to be confirmed experimentally but one could reasonably frame the hypothesis that these properties result from a synergistic action between isoquinoline alkaloids and styryllactones. It might be interesting to evaluate the activity of the plant and its constituents against *Sarcoptes scabiei*.

■ **CS 1.40** Nantenine

■ **CS 1.41** (−)-Anonaine

■ **CS 1.42** Actinodaphnine

Proposed Research Pharmacological study of liriodenine and derivatives for the treatment of cancer.

Rationale There are a number of reports on the cytotoxic activity of aporphine alkaloids.[226] The aporphine alkaloid nantenine (CS 1.40) from *Nandina domestica* Thunb. (family Berberidaceae Juss.) inhibited the growth of human colorectal carcinoma (HCT-116), human colon cancer (CaCo-2) and human colon fibroblast (CCD-18Co) cells, with IC_{50} values equal to HCT-11638.3 µM, 36.2 µM and 56.5 µM, respectively.[227] Analysis of the structure–activity relationship revealed that the presence of a hydroxy group in C1 and a bromine atom in C3 were detrimental to the cytotoxic activity whereas an intact aporphine framework and substitution of C1 with alkyloxy groups or a benzoate derivative improved cytotoxicity.[227] (−)-Anonaine (CS 1.41) isolated from *Michelia alba* DC. (family Magnoliaceae Juss.), which is alkoxylated at C1 and presents an intact aporphine scaffold, precipitated apoptosis in human epitheloid cervix carcinoma (Hela) cells through the generation of ROS, which in turn compromised the mitochondrial potential that triggered caspase 3 activation.[228] Quite similar apoptotic events were observed with the aporphine alkaloid actinodaphnine (CS 1.42), isolated from *Cinnamomum insularimontanum* Hayata (family Lauraceae Juss.), against the cells of the human hepatocellular carcinoma cell line Mahlavu.[229] Another aporphine alkaloid with similar structural requirements is d-dicentrine (CS 1.43), isolated from *Lindera megaphylla* Hemsl. (family Lauraceae Juss.), which protected severe combined immunodeficiency (SCID) rodents against human erythromyeloblastoid leukemia (K562) cell proliferation when administered parenterally at the dose of 100 µg twice a week for 4 weeks.[230]

Aporphine alkaloids such as dauriporphine and thaliblastine (CS 1.44) have the appealing ability to mitigate the multidrug resistance of cancer cells owing to P-glycoprotein pump efflux inhibition. Dauriporphine from *Sinomenium acutum* (Thunb.) Rehder & E.H. Wilson (family Menispermaceae Juss.) impeded P-glycoprotein-induced multidrug resistance with ED_{50} values of 0.03 µg/mL and 0.0001 µg/mL in multidrug-resistant human uterine sarcoma (MES-SA/Dx5) cells and human colorectal adenocarcinoma (HCT-15) cells, respectively.[231] The aporphine alkaloid thaliblastine isolated from members of the genus *Thalictrum* L. (family Ranunculaceae Juss.) at a dose of 2 µM elevated the amount of adriamycin in multidrug-resistant mouse leukemia (P388/R-84) cells through inhibition of P-glycoprotein and lowered the IC_{50} value of adriamycin towards these cells to 1.4 µM.[232,233] Lastly, planar aporphine

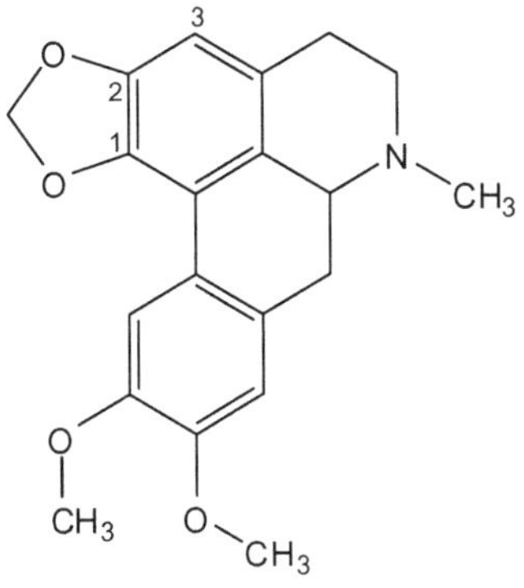

■ **CS 1.43** d-Dicentrine

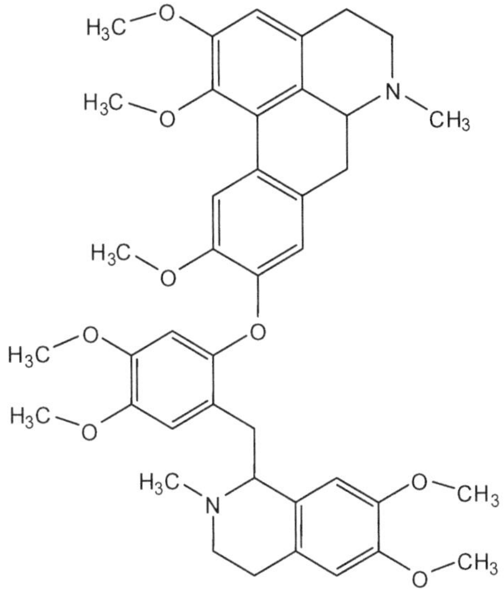

■ **CS 1.44** Thaliblastine

alkaloids like liriodenine (CS 1.45) have the tendency to strongly inter-calate into DNA and to inhibit the enzymatic activity of topoisomerase II.[234,235] In fact, liriodenine is of exceptional interest as it is not only a potent topoisomerase inhibitor but also a pro-apoptotic agent. Liriodenine, which occurs in *Goniothalamus amuyon* (Blco.) Merr. (family Annonaceae Juss.), induced the apoptosis of human hepatocellular liver carcinoma (HepG2) (wild-type p53) and human liver adenocarcinoma (SK-HEP-1) cells (wild-type p53) via a topoisomerase-independent mechanism, implying an increase in cytoplasmic nitric oxide (NO) that induced DNA damage-causing pro-apoptotic protein (p53) accumulation and apopto-sis.[236] Apoptosis was also provoked by liriodenine in human lung adeno-carcinoma epithelial (A549) cells via caspase activation, inhibition of the kinase activity of the cyclin B1/CDK1 complex and cleavage of PARP (poly [ADP-ribose] polymerase).[237]

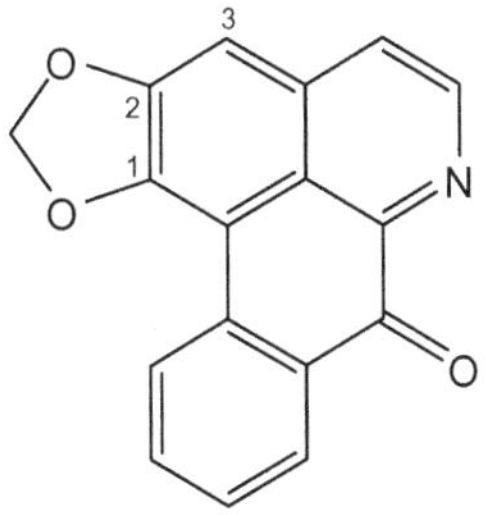

■ **CS 1.45** Liriodenine

1.3.8 *Macleaya Microcarpa* (Maxim.) Fedde

History The plant was first described by Friedrich Karl Georg Fedde, in *Botanische Jahrbücher für Systematik, Pflanzengeschichte und Pflanzengeographie*, published in 1905.

Synonym *Bocconia microcarpa* Maxim.

Family Papaveraceae Juss., 1789

Common Names Plume poppy, xiao guo bo luo hui (Chinese)

Habitat and Description *Bocconia microcarpa* (Maxim.) is a rhizomatous herb which grows to 2 m tall in the open lands of China. It is also a common garden ornamental. The stem is yellow, laticiferous, terete and glaucous. The leaves are simple. The petiole is 5–10 cm long. The blade is palmate, cordate at base, lobed at margin, glaucous beneath, 5–15 cm × 5–12 cm and presents five pairs of secondary nerves. The inflorescence is a conspicuous, terminal, feather-like and reddish panicle which is 15–30 cm long. The flowers are small. The perianth consists of four sepals which are 0.5 cm long and oblong. The androecium comprises 8–12 stamens, the anthers of which are oblong and 0.4 cm long. The gynoecium is 0.3 cm long and develops a bifid stigma. The fruit is a capsule which is narrowly obovoid, 0.5 cm in diameter and encloses a single seed (Figure 1.17).

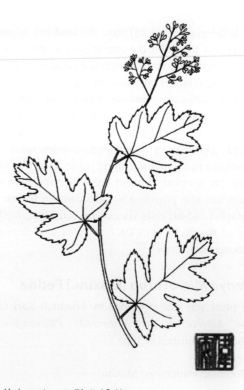

■ **FIGURE 1.17** *Macleaya microcarpa* (Maxim.) Fedde

Medicinal Uses In China, the plant is used to heal sores, boils and skin cancer, to treat ringworm infection, to kill maggot diseases and to sooth inflammation. The yellow latex exuded by the plant is extremely poisonous.

Phytopharmacology The plant is known to produce series of isoquinoline alkaloids, for example the protopine alkaloids allocryptopin, protopine and cryptopine,[238] the benzophenanthridine alkaloids sanguinarine, chelerythrine, chelirubine, chelilutine, macarpine,[238–241] dihydrosanguinarine, dihydrochelerythrine, 6-methoxydihydrosanguinarine, 6-methoxydihydrochelerythrine,[242] 6-(10-hydroxyethyl)dihydrochelerythrine, bis[6-(5,6-dihydrochelerythrinyl)]ether, pancorine, spallidamine and maclekarpine A-E[243] and traces of the protoberberine alkaloids coptisine and berberine.[241] The antibacterial, antifungal, insecticidal and anti-inflammatory properties of the plant have not been confirmed but one might logically suspect the involvement of isoquinoline alkaloids.

Proposed Research Pharmacological study of 6-methoxydihydrosanguinarine and derivatives for the treatment of cancer.

Rationale The benzophenanthridine alkaloid 6-methoxydihydrosanguinarine (CS 1.46) compromised the survival of human intestinal adenocarcinoma (HCT-8), human hepatocellular carcinoma (BEL-7402), human gastric cancer (BGC-823), human ovarian cancer (A2780) and human lung adenocarcinoma epithelial (A549) cells, with IC_{50} values equal to 0.5 μM, 0.5 μM, 0.6 μM, 0.5 μM and 0.6 μM, respectively.[243] 6-methoxydihydrosanguinarine isolated from a member of the genus *Hylomecon* Maxim. (family Papaveraceae Juss.) induced apoptosis in human hepatocellular liver carcinoma (HepG2), with an IC_{50} value of 3.8 μM.[244] It also increased the cytoplasmic levels of ROS in HepG2 (wild-type p53) and, hence, DNA fragmentation and subsequent expression of pro-apoptotic protein (p53). This in turn increased the protein level of pro-apoptotic Bcl-2-associated X

■ CS 1.46 6-Methoxydihydrosanguinarine

■ **CS 1.47** Sanguinarine

■ **CS 1.48** Chelerythrine

protein (Bax) and therefore the release of cytochrome c from mitochondria, which activated the cleaving of PARP (poly [ADP-ribose] polymerase) by caspase 3.[244] However, in human colon cancer (HT-29) cells (mutant p53) 5 μM of 6-methoxydihydrosanguinarine induced caspase activation and DNA fragmentation.[245] Similarly, the benzophenanthridine sanguinarine (CS 1.47), which is commonly found in *Sanguinaria canadensis* L. (family Papaveraceae Juss.), increased the cytoplasmic levels of ROS in human breast adenocarcinoma (MDA-MB-231) cells that resulted in a decrease in mitochondrial membrane potential followed by the release of cytochrome c, which activated caspases 9 and 3.[246] In fact, there is a growing body of evidence suggesting that benzophenanthridine alkaloids are capable of boosting the intracellular concentrations of ROS in cancer cells because of their iminium groups which react with the thiol groups of the main cellular antioxidant glutathione.[247] For instance, the increase of ROS in human epitheloid cervix carcinoma (Hela) cells exposed to the benzophenanthridine alkaloid chelerythrine (CS 1.48) isolated from *Chelidonium majus* L. (family Papaveraceae Juss.) had a detrimental effect on a phosphatase, which resulted in an elevation in mitogen-activated protein kinase (MAPK) cascade MEKK1 and MKK4 kinase activities and hence the activation of pro-apoptotic c-Jun N-terminal kinase (JNK) and the mitogen-activated protein kinase (MAPK) p38.[247] An analogous mechanism was observed in human promyelocytic leukemia (HL-60) cells where sanguinarine impeded the enzymatic activity of protein phosphatase 2, which caused an increase in the MAPK p38, triggering caspase 3/7 activation, and therefore apoptosis, with an IC_{50} value of 0.3 μM.[248] In human prostate cancer (PC-3) cells, sanguinarine at a dose of 5 μM evoked a dramatic depletion of glutathione which resulted in the build-up of ROS, which fragmented DNA and cleaved PARP by caspase 3.[249] In addition to this, Jan et al. (2009) observed that sanguinarine compromised the viability of human lung adenocarcinoma epithelial (A549) cells (60% at 5 μM) with DNA fragmentation, cleavage of PARP by caspase 3, activation of MAPK p38 and a fall in glutathione levels.[250] Human breast adenocarcinoma (MDA-MB-231) cells treated with sanguinarine (1–7.5 µM) had high levels of ROS that degraded anti-apoptotic Bcl-2 protein and induced DNA fragmentation and the subsequent release of cytochrome c from mitochondria, which provoked the cleaving of PARP by caspase 3.[251] Sanguinarine induced apoptosis of human prostate carcinoma (DU-145) cells, with an IC_{50} value of 1 µM, via the proteolysis of survivin by ubiquitin-proteasome[252] activated by high oxidative activity.[253]

REFERENCES

[122] Shoeb A, Raj K, Kapil RS, Popli SP. Alangiside, the monoterpenoid alkaloidal glycoside from *Alangium lamarckii* Thw. J Chem Soc Perkin 1975;1(13):1245–8.

[123] Itoh A, Tanahashi T, Nagakura N, Nayeshiro H. Tetrahydroisoquinoline-monoterpene glucosides from *Alangium lamarckii* and *Cephaelis ipecacuanha*. Phytochem 1994;36(2):383–7.

[124] Itoh A, Tanahashi T, Nagakura N. Three tetrahydroisoquinoline-monoterpene glucosides from *Alangium lamarckii*: the first occurrence of glucosides with the same absolute configurations as deacetylisoipecoside, a key intermediate in the biosynthesis of ipecac alkaloids. Chem Pharm Bull 1994;42(10):2208–10.

[125] Itoh A, Tanahashi T, Tabata M, Shikata M, Kakite M, Nagai M, et al. Tetrahydroisoquinoline-monoterpene and iridoid glycosides from *Alangium lamarckii*. Phytochem 2001;56(6):623–30.

[126] Itoh A, Tanahashi T, Nagakura N. Five tetrahydroisoquinoline-monoterpene glucosides and a tetrahydro-β-carboline-monoterpene glucoside from *Alangium lamarckii*. J Nat Prod 1995;58(8):1228–39.

[127] Itoh A, Tanahashi T, Nagakura N. Acylated tetrahydroisoquinoline-monoterpene glucosides from *Alangium lamarckii*. Phytochem 1996;41(2):651–6.

[128] Itoh A, Tanahashi T, Nagakura N. Five tetrahydroisoquinoline monoterpene glycosides with a disaccharide moiety from *Alangium lamarckii*. Phytochem 1997;46(7):1225–9.

[129] Ohmori O, Takayama H, Kitajima M, Aimi N. First synthesis of neoalangiside, a new tetrahydroisoquinoline-monoterpene glucoside with oxygen functions at unusual C1, C2 positions. Chem Pharm Bull 1999;47(10):1512–3.

[130] Itoh A, Tanahashi T, Tabata M, Shikata M, Kakite M, Nagai M, et al. Tetrahydroisoquinoline-monoterpene and iridoid glycosides from *Alangium lamarckii*. Phytochem 2001;56(6):623–30.

[131] Albright JD, Van Meter JC, Goldman L. Alkaloid studies. IV. Isolation of cephaeline and tubulosine from *Alangium lamarckii*. Lloydia 1965;28(3):212–7.

[132] Pakrashi SC, Achari B. Demethylcephaeline, a new alkaloid from Alangium lamarckii. Characterization of AL 60, the hypotensive principle from the stembark. Experientia 1970;26(9):933–4.

[133] Rao KN, Venkatachalam SR. Dihydrofolate reductase and cell growth activity inhibition by the β-carboline-benzoquinolizidine plant alkaloid deoxytubulosine from *Alangium lamarckii*: its potential as an antimicrobial and anticancer agent. Bioorg Med Chem 1999;7(6):1105–10.

[134] Itoh A, Ikuta Y, Tanahashi T, Nagakura N. Two alangium alkaloids from *Alangium lamarckii*. J Nat Prod 2000;63(5):723–5.

[135] Achari B, Ali E, Ghosh Dastidar PP. Further investigations on the alkaloids of *Alangium lamarckii*. Planta Med 1980;40:5–7.

[136] Pakrashi SC, Mukhopadhyay R, Ghosh Dastidar PP. Bharatamine – a unique protoberberine alkaloid from *Alangium lamarckii* THW., biogenetically derived from monoterpenoid precursor. Tetrahedron Lett 1983;24(3):291–4.

[137] Bhattacharjya A, Mukhopadhyay R, Pakrashi SC. Structure and synthesis of alamaridine, a novel benzopyridoquinolizine alkaloid from *alangium lamarckii*. Tetrahedron Lett 1986;27(10):1215–6.

[138] Itoh A, Tanahashi T, Nagakura N. Biogenetic conversion of tetrahydroisoquinoline-monoterpene glucosides into benzopyridoquinolizine alkaloids of *Alangium lamarckii*. J Nat Prod 1996;59(5):535–8.

[139] Jain S, Sinha A, Bhakuni DS. The biosynthesis of β-carboline and quinolizidine alkaloids of *Alangium lamarckii*. Phytochem 2002;60(8):853–9.

[140] Mukhopadhyay R, Dastidar PPG, Ali E, Pakrashi SC. Studies on Indian medicinal plants, 87. Lacinilene C – A rare sesquiterpene from *Alangium lamarckii*. J Nat Prod 1987;50(6):1185.

[141] Pakrashi SC, Bhattacharyya J, Mookerjee S, Samatan TB, Vorbrüggen H. Studies on Indian medicinal plants. XVIII. The non-alkaloidal constituents from the seeds of *Alangium lamarckii* Thw. Phytochem 1968;7(3):461–6.

[142] Bhattacharyya K, Chaudhuri S, Achari B, Mazumdar B, Mazumdar SK. Isoalangidiol monoacetate, a triterpene alcohol. Acta Crystallographica 1997;53(9):1299–301.

[143] Pakrashi SC, Achari B. Stigmasta-5,22,25-trien-3β-ol: a new sterol from *Alangium lamarckii* Thw. Tetrahedron Lett 1971;12(4):365–8.

[144] Narasimha Rao K, Bhattacharya RK, Venkatachalam SR. Thymidylate synthase activity and the cell growth are inhibited by the β-carboline-benzoquinolizidine alkaloid deoxytubulosine. J Biochem Mol Toxicol 1998;12(3):167–73.

[145] Gatto B, Sanders MM, Yu C, Wu HY, Makhey D, LaVoie EJ, et al. Identification of topoisomerase I as the cytotoxic target of the protoberberine alkaloid coralyne. Cancer Res 1996;56(12):2795–800.

[146] Kim SA, Kwon Y, Kim JH, Muller MT, Chung IK. Induction of topoisomerase II-mediated DNA cleavage by a protoberberine alkaloid, berberine. Biochem 1998;37(46):16316–16324.

[147] Sanders MM, Liu AA, Li TK, Wu HY, Desai SD, Mao Y, et al. Selective cytotoxicity of topoisomerase-directed protoberberines against glioblastoma cells. Biochem Pharmacol 1998;56(9):1157–66.

[148] Sordet O, Khan QA, Kohn KW, Pommier Y. Apoptosis induced by topoisomerase inhibitors. Curr Med Chem 2003;3(4):271–90.

[149] Yan K, Zhang C, Feng J, Hou L, Yan L, Zhou Z, et al. Induction of G1 cell cycle arrest and apoptosis by berberine in bladder cancer cells. Eur J of Pharmacol 2011;661(1–3):1–7.

[150] Lin JP, Yang JS, Lee JH, Hsieh WT, Chung JG. Berberine induces cell cycle arrest and apoptosis in human gastric carcinoma SNU-5 cell line. World J Gastroenterol 2006;12(1):21–8.

[151] Wang TX, Li K, Wang SG, Cong Y, Wang YY. Study on jatrorrhizine inducing K562 cells apoptosis. Chin Pharm J 2010;45(23):1822–6.

[152] Pailee P, Prachyawarakorn V, Mahidol C, Ruchirawat S, Kittakoop P. Protoberberine alkaloids and cancer chemopreventive properties of compounds from *Alangium salviifolium*. Eur J Org Chem 2011(20–21):3809–14.

[153] Wu TS, Leu YL, Chan YY. Constituents of the fresh leaves *Aristolochia cucurbitifolia*. Chem Pharm Bull 1999;47(4):571–3.

[154] Wu TS, Chan YY, Leu YL. The constituents of the root and stem of *Aristolochia cucurbitifolia* hayata and their biological activity. Chem Pharm Bull 2000;48(7):1006–9.

[155] Rosenthal MD, Vishwanath BS, Franson RC. Effects of aristolochic acid on phospholipase A2 activity and arachidonate metabolism of human neutrophils. Biochim Biophys Acta 1989;1001(1):1–8.

[156] Chandra V, Jasti J, Kaur P, Srinivasan A, Betzel Ch, Singh TP. Structural basis of phospholipase A2 inhibition for the synthesis of prostaglandins by the plant alkaloid aristolochic acid from a 1.7 A crystal structure. Biochem 2002;41(36):10914–10919.

[157] Nortier JL, Martinez MC, Schmeiser HH, Arlt VM, Bieler CA, Petein M, et al. Urothelial carcinoma associated with the use of a Chinese herb (*Aristolochia fangchi*). N Engl J Med 2000;342:1686–92.

[158] Schmeiser HH, Stiborovà M, Arlt VM. Chemical and molecular basis of the carcinogenicity of Aristolochia plants. Curr Opin Drug Discov Devel 2009;12(1):141–8.

[159] Hegde VR, Borges S, Pu H, Patel M, Gullo VP, Wu B, et al. Semi-synthetic aristolactams – inhibitors of CDK2 enzyme. Bioorg Med Chem Lett 2010;20(4):1384–7.

[160] Couture A, Deniau E, Grandclaudon P, Rybalko-Rosen H, Léonce S, Pfeiffer B, et al. Synthesis and biological evaluation of aristolactams. Bioorg Med Chem Lett 2002;12(24):3557–9.

[161] Choi YL, Kim JK, Choi SU, Min YK, Bae MA, Kim BT, et al. Synthesis of aristolactam analogues and evaluation of their antitumor activity. Bioorg Med Chem Lett 2009;19(11):3036–40.

[162] Priestap HA. Seven aristololactams from *Aristolochia argentina*. Phytochem 1985;24(4):849–52.

[163] Li L, Wang X, Chen J, Ding H, Zhang Y, Hu TC, et al. The natural product Aristolactam AIIIa as a new ligand targeting the polo-box domain of polo-like kinase 1 potently inhibits cancer cell proliferation. Acta Pharm Sin 2009;30(10):1443–53.

[164] Christoph DC, Schuler M. Polo-like kinase 1 inhibitors in mono- and combination therapies: a new strategy for treating malignancies. Expert Rev Anticancer Ther 2011;11(7):1115–30.

[165] Balachandran P, Wei F, Lin RC, Khan IA, Pasco DS. Structure activity relationships of aristolochic acid analogues: toxicity in cultured renal epithelial cells. Kidney Int 2005;67(5):1797–805.

[166] Ruecker G, Chung BS. Aristolochic acids from *Aristolochia manshuriensis*. Planta Med 1975;27(1):68–71.

[167] Nakanishi T, Iwasaki K, Nasu M, Miura I, Yoneda K. Aristoloside, an aristolochic acid derivative from stems of *Aristolochia manshuriensis*. Phytochem 1982;21(7):1759–62.

[168] Wu PL, Su GC, Wu TS. Constituents from the stems of *Aristolochia manshuriensis*. J Nat Prod 2003;66(7):996–8.

[169] Zhang YT, Jiang JQ. Alkaloids from *Aristolochia manshuriensis* (Aristolochiaceae). Helvetica Chimica Acta 2006;89(11):2665–70.

[170] Rücker G, Mayer R, Wiedenfeld H, Chung BS, Güllmann A. (+)-Isobicyclo-germacrenal from *aristolochia manshuriensis*. Phytochem 1987;26(5):1529–30.

[171] Rücker G, Ming CW, Mayer R, Will G, Güllmann A. Manshurolide, a sesquiterpene lactone from *Aristolochia manshuriensis*. Phytochem 1990;29(3):983–5.

[172] Meijer L. Cyclin-dependent kinases inhibitors as potential anticancer, antineurodegenerative, antiviral and antiparasitic agents. Drug Resist Updat 2000;3(2):83–8.

[173] Hegde VR, Borges S, Patel M, Das PR, Wu B, Gullo VP, et al. New potential antitumor compounds from the plant *Aristolochia manshuriensis* as inhibitors of the CDK2 enzyme. Bioorg Med Chem Lett 2010;20(4):1344–6.

[174] Oh M, Lee JY, Shin DH, Park JH, Oian T, Kim HJ, et al. The in vitro and in vivo anti-tumor effect of KO-202125, a sauristolactam derivative, as a novel epidermal growth factor receptor inhibitor in human breast cancer. Cancer Sci 2011;102(3):597–604.

[175] Jänne PA, Gurubhagavatula S, Yeap BY, Lucca J, Ostler P, Skarin AT, et al. Outcomes of patients with advanced non-small cell lung cancer treated with gefitinib (ZD1839, 'Iressa') on an expanded access study. Lung Cancer 2004;44:221–30.

[176] Kim MH, Ryu SY, Choi JS, Min YK, Kim SH. Saurolactam inhibits osteoclast differentiation and stimulates apoptosis of mature osteoclasts. J Cell Physiol 2009;221(3):618–28.

[177] Body JJ, Facon T, Coleman RE, Lipton A, Geurs F, Fan M, et al. A study of the biological receptor activator of nuclear factor-kappaB ligand inhibitor, denosumab, in patients with multiple myeloma or bone metastases from breast cancer. Clin Cancer Res 2006;12(4):1221–8.

[178] Castellano D, Sepulveda JM, García-Escobar I, Rodriguez-Antolín A, Sundlöv A, Cortes-Funes H. The role of RANK-Ligand inhibition in cancer: the story of Denosumab. Oncologist 2011;16(2):136–45.

[179] McPhail AT, Onan KD, Furukawa H, Ju-ichi M. Structure and stereochemistry of coccutrine, a new erythrina alkaloid from *Cocculus trilobus* D C. Tetrahedron Lett 1976;17(6):485–8.

[180] Wada K, Marumo S, Munakata K. An insecticidal alkaloid, cocculolidine from *cocculus trilobus* DC. Tetrahedron Lett 1966;7(42):5179–84.

[181] Ju-ichi M, Ando Y, Yoshida Y, Kunitomo J, Shingu T, Furukawa H. Alkaloids of *Cocculus trilobus* DC. Isolation and structure of erythrinan alkaloids. Yakugaku Zasshi 1978;98(7):886–90.

[182] Ito K, Furukawa H, Sato K, Takahashi J. Studies on the alkaloids of menispermaceous plants. CCL. Structure of coclobine, a new biscoclaurine alkaloid from *Cocculus trilobus* DC. Yakugaku Zasshi 1969;89(8):1163–6.

[183] Chen HS, Liang HQ, Liao SX. Studies on the chemical constituents of the root of *Cocculus Trilobus* DC. Acta Pharm Sin 1991;26(10):755–8.

[184] Chang FR, Wu YC. New bisbenzylisoquinolines, fatty acid amidic aporphines, and a protoberberine from formosan *Cocculus orbiculatus*. J Nat Prod 2005;68(7):1056–60.

[185] Nakano T. Studies on the alkaloids of menispermaceous plants. CXXXI. Isolation of magnoflorine from *Cocculus trilobus* DC. Pharm Bull 1956;4(1):69–70.

[186]   Wada K, Munakata K. Naturally occurring insect control chemicals: isoboldine, a feeding inhibitor, and cocculolidine, an insecticide in the leaves of *Cocculus trilobus* DC. J of Agr Food Chem 1968;16(3):471–4.

[187]   Itokawa H, Tsuruoka S, Takeya K. An antitumor morphinane alkaloid, sinococuline, from *Cocculus trilobus*. Chem Pharm Bull 1987;35(4):1660–2.

[188]   Itokawa H, Nishimura K, Hitotsuyanagi Y, Takeya K. Isosinococuline, a novel antitumor morphinane alkaloid from *Cocculus trilobus*. Bioorg Med Chem Lett 1995;5(8):821–2.

[189]   Teh BS, Chen P, Lavin MF, Seow WK, Thong YH. Demonstration of the induction of apoptosis (programmed cell death) by tetrandrine, a novel anti-inflammatory agent. Int J Immunopharmacol 1991;13(8):1117–26.

[190]   Yoo SM, Oh SH, Lee SJ, Lee BW, Ko WG, Moon CK, et al. Inhibition of proliferation and induction of apoptosis by tetrandrine in HepG2 cells. J Ethnopharmacol 2002;81(2):225–9.

[191]   Chen Y, Chen JC, Tseng SH. Tetrandrine suppresses tumor growth and angiogenesis of gliomas in rats. Int J Cancer 2009;124(10):2260–9.

[192]   Zhu X, Sui M, Fan W. In vitro and in vivo characterizations of tetrandrine on the reversal of P-glycoprotein-mediated drug resistance to paclitaxel. Anticancer Res 2005;25(3B):1953–62.

[193]   Wang G, Lemos JR, Iadecola C. Herbal alkaloid tetrandrine: from an ion channel blocker to inhibitor of tumor proliferation. Trends Pharmacol Sci 2004;25:120–3.

[194]   Jin Q, Kang C, Soh Y, Sohn NW, Lee J, Cho YH, et al. Tetrandrine cytotoxicity and its dual effect on oxidative stress induced apoptosis through modulating cellular redox states in Neuro 2a mouse neuroblastoma cells. Life Sci 2002;71:2053–66.

[195]   Wu JM, Chen Y, Chen JC, Lin TY, Tseng SH. Tetrandrine induces apoptosis and growth suppression of colon cancer cells in mice. Cancer Lett 2010;287(2):187–95.

[196]   He BC, Gao JL, Zhang BQ, Luo Q, Shi Q, Kim SH, et al. Tetrandrine inhibits Wnt/β-catenin signaling and suppresses tumor growth of human colorectal cancer. Mol Pharmacol 2011;79(2):211–9.

[197]   Liu C, Gong K, Mao X, Li W. Tetrandrine induces apoptosis by activating reactive oxygen species and repressing Akt activity in human hepatocellular carcinoma. Int J Cancer 2011;129(6):1519–31.

[198]   Lee HY, Kim CW. Isolation and quantitative determination of berberine and coptisine from tubers of *Corydalis ternata*. Korean J Pharmacog 1999;30(3):332–4.

[199]   Kim SR, Hwang SY, Jang YP, Park MJ, Markelonis GJ, Oh TH, et al. Protopine from *Corydalis ternata* has anticholinesterase and antiamnesic activities. Planta Med 1999;65(3):218–21.

[200]   Kim JA, Choi JY, Kim DC, Lee HS, Lee SH. Determination of dehydrocorydaline in the Corydalis tuber using HPLC-UVD. Korean J Pharmacog 2008;39(4):305–9.

[201]   Kim KH, Lee IK, Piao CJ, Choi SU, Lee JH, Kim YS, et al. Benzylisoquinoline alkaloids from the tubers of *Corydalis ternata* and their cytotoxicity. Bioorg Med Chem Lett 2010;20(15):4487–90.

[202]   Kim KH, Lee IK, Choi SU, Lee JH, Moon E, Kim SY, et al. New triterpenoids from the tubers of *Corydalis ternata*: structural elucidation and bioactivity evaluation. Planta Med 2011;77(13):1555–8.

[203]   Leitao da-Cunha EV, Fechine IM, Guedes DN, Barbosa-Filho JM, Sobral da Silva M. Protoberberine alkaloids. Alkaloids Chem Biol 2005;62:1–75.

[204] Lee GE, Lee WG, Lee SY, Lee CR, Park CS, Chang S, et al. Characterization of protoberberine analogs employed as novel human P2X7 receptor antagonists. Toxicol Appl Pharmacol 2011;252(2):192–200.

[205] Roger S, Pelegrin P. P2X7 receptor antagonism in the treatment of cancers. Expert Opin Investig Drugs 2011;20(7):875–80.

[206] Ryu JK, Jantaratnotai N, Serrano-Perez MC, McGeer PL, McLarnon JG. Block of purinergic P2X7R inhibits tumor growth in a C6 glioma brain tumor animal model. J Neuropathol Exp Neurol 2011;70(1):13–22.

[207] Ren S, Zhang Y, Wang Y, Lui Y, Wei W, Huang X, et al. Targeting P2X7 receptor inhibits the metastasis of murine P388D1 lymphoid neoplasm cells to lymph nodes. Cell Biol Int 2010;34(12):1205–11.

[208] Sun SH. Roles of P2X7 receptor in glial and neuroblastoma cells: the therapeutic potential of P2X7 receptor antagonists. Mol Neurobiol 2010;41(2–3):351–5.

[209] Raffaghello L, Chiozzi P, Falzoni S, Di Virgilio F, Pistoia V. The P2X7 receptor sustains the growth of human neuroblastoma cells through a substance P-dependent mechanism. Cancer Res 2006;66(2):907–14.

[210] Kopp UC, Cicha MZ. PGE2 increases substance P release from renal pelvic sensory nerves via activation of N-type calcium channels. Am J Physiol 1999;276:R1241–R1248.

[211] Lai S, Zhao TF, Wang XK. Cycleaneonine, a new bisbenzylisoquinoline alkaloid from *Cyclea racemosa* Oliv. Acta Pharm Sin 1988;23(5):356–60.

[212] Xian-Kai W, Tong-Fang Z, Sheng L, Shizuri Y, Yamamura S. Three cissampareine-type bisbenzylisoquinoline alkaloids from *Cyclea* species. Phytochem 1993;33(5):1249–52.

[213] Wang JZ, Chen QH, Wang FP. Cytotoxic bisbenzylisoquinoline alkaloids from the roots of *Cyclea racemosa*. J Nat Prod 2010;73(7):1288–93.

[214] Kondo Y, Takano F, Hojo H. Inhibitory effect of bisbenzylisoquinoline alkaloids on nitric oxide production in activated macrophages. Biochem Pharmacol 1993;46(11):1887–92.

[215] Zou LL, Zhou Y, Yang M, Dai XH, Qi J, Xiong DS, et al. The growth inhibiting effect of calmodulin antagonist O-(4-ethoxyil-butyl)-berbamine on leukemia cells K562 and K562/A02. Chin Pharmacol Bull 2009;25(10):1313–7.

[216] Fang BJ, Yu ML, Yang SG, Liao LM, Liu JW, Zhao RCH. Effect of O-4-ethoxyl-butyl-berbamine in combination with pegylated liposomal doxorubicin on advanced hepatoma in mice. World J Gastroenterol 2004;10(7):950–3.

[217] Liu R, Zhang Y, Chen Y, Qi J, Ren S, Xushi MY, et al. A novel calmodulin antagonist O-(4-ethoxyl-butyl)-berbamine overcomes multidrug resistance in drug-resistant MCF-7/ADR breast carcinoma cells. J Pharml Sci 2010;99(7):3266–75.

[218] Xie J, Ma T, Gu Y, Zhang X, Qiu X, Zhang L, et al. Berbamine derivatives: a novel class of compounds for anti-leukemia activity. Eur J Med Chem 2009;44(8):3293–8.

[219] Wei N, Liu GT, Chen XG, Liu Q, Wang FP, Sun H. H1, a derivative of Tetrandrine, exerts anti-MDR activity by initiating intrinsic apoptosis pathway and inhibiting the activation of Erk1/2 and Akt1/2. Biochem Pharmacol 2011;82(11):1593–603.

[220] Lu ST, Wu YC, Leou SP. Alkaloids of formosan Fissistigma and Goniothalamus species. Phytochem 1985;24(8):1829–34.

[221] Lan YH, Chang FR, Yu JH, Yang YL, Chang YL, Lee SJ, et al. Cytotoxic styryl-pyrones from *Goniothalamus amuyon*. J Nat Prod 2003;66(4):487–90.

[222] Lan YH, Chang FR, Yang YL, Wu YC. New constituents from stems of *Goniothalamus amuyon*. Chem Pharm Bull 2006;54(7):1040–3.

[223] Wu YC, Chang FR, Duh CY, Wang SK, Wu TS. Cytotoxic styrylpyrones of *Goniothalamus amuyon*. Phytochem 1992;31(8):2851–3.

[224] Lan YH, Chang FR, Liaw CC, Wu CC, Chiang MY, Wu YC. Digoniodiol, deoxygoniopypyrone A, and goniofupyrone A: three new styryllactones from *Goniothalamus amuyon*. Planta Med 2005;71(2):153–9.

[225] Li X, Chang CJ. Antitumor cytotoxicity and stereochemistry of polyketides from *Goniothalamus amuyon*. Nat Prod Lett 1996;8(3):207–15.

[226] Stévigny C, Bailly C, Quetin-Leclercq J. Cytotoxic and antitumor poten-tialities of aporphinoid alkaloids. Curr Med Chem Anticancer Agents 2005;5(2):173–82.

[227] Ponnala S, Chaudhary S, González-Sarrias A, Seeram NP, Harding WW. Cytotoxicity of aporphines in human colon cancer cell lines HCT-116 and Caco-2: an SAR study. Bioorg Med Chem Lett 2011;21(15):4462–4.

[228] Tseng WC, Lu FJ, Hung RP, Chen CH, Chen CH. (–)-Anonaine induces apop-tosis through Bax- and caspase-dependent pathways in human cervical cancer (HeLa) cells. Food Chem Toxicol 2008;46(8):2694–702.

[229] Hsieh TJ, Liu TZ, Lu FJ, Hsieh PY, Chen CH. Actinodaphnine induces apopto-sis through increased nitric oxide, reactive oxygen species and down-regulation of NF-κB signaling in human hepatoma Mahlavu cells. Food Chem Toxicol 2006;44(3):344–54.

[230] Huang RL, Chen CC, Huang YL, Ou JC, Hu CP, Chen CF, et al. Anti-tumor effects of d-dicentrine from the root of *Lindera megaphylla*. Planta Med 1998;64(3):212–5.

[231] Min YD, Choi SU, Lee KR. Aporphine alkaloids and their reversal activity of multidrug resistance (MDR) from the stems and rhizomes of *Sinomenium acu-tum*. Arch Pharm Res 2006;29(8):627–32.

[232] Chen G, Ramachandran C, Krishan A. Thaliblastine, a plant alkaloid, circum-vents multidrug resistance by direct binding to P-glycoprotein. Cancer Res 1993;53(11):2544–7.

[233] Chen G, Teicher BA, Frei E. Differential interactions of Pgp inhibitor thali-blastine with adriamycin, etoposide, taxol and anthrapyrazole CI941 in sensi-tive and multidrug-resistant human MCF-7 breast cancer cells. Anticancer Res 1996;16(6B):3499–505.

[234] Woo SH, Sun NJ, Cassady JM, Snapka RM. Topoisomerase II inhibition by aporphine alkaloids. Biochem Pharmacol 1999;57(10):1141–5.

[235] Woo SH, Reynolds MC, Sun NJ, Cassady JM, Snapka RM. Inhibition of topoi-somerase II by liriodenine. Biochem Pharmacol 1997;54(4):467–73.

[236] Hsieh TJ, Liu TZ, Chern CL, Tsao DA, Lu FJ, Syu YH, et al. Liriodenine inhib-its the proliferation of human hepatoma cell lines by blocking cell cycle progres-sion and nitric oxide-mediated activation of p53 expression. Food Chem Toxicol 2005;43(7):1117–26.

[237] Chang HC, Chang FR, Wu YC, Lai YH. Anti-cancer effect of liriodenine on human lung cancer cells. Kaohsiung J Med Sci 2004;20(8):365–71.

[238] Slavík J, Slavíková L. Papaveraceae alkaloids VI. On alkaloids of *Macleaya microcarpa* (Maxim) Fedde. Chem. Listy 1954;48:106–10.

[239] Abizov EA, Tolkachev ON, Kopylova IE, Luferov AN. Distribution of the sum of sanguinarine and chelerythrine in the above-ground part of *Macleaya microcarpa*. Pharm Chem J 2003;37(8):413–4.

[240] Suchomelová J, Bochořáková H, Paulová H, Musil P, Táborská E. HPLC quantification of seven quaternary benzo[c]phenanthridine alkaloids in six species of the family Papaveraceae. J Pharm Biomed Anal 2007;44(1):283–7.

[241] Pěnčíková K, Urbanová J, Musil P, Táborská E, Gregorová J. Seasonal variation of bioactive alkaloid contents in *Macleaya microcarpa* (Maxim.) fedde. Molecules 2011;16(4):3391–401.

[242] Qin H, Wang P, Li Z, Liu X, He W. The establishment of the control substance and 1 H nuclear magnetic resonance fingerprint of *Macleaya microcarpa* (Maxim.) Fedde. Fenxi Huaxue 2004;32(9):1165–70.

[243] Deng AJ, Qin HL. Cytotoxic dihydrobenzophenanthridine alkaloids from the roots of *Macleaya microcarpa*. Phytochem 2010;71(7):816–22.

[244] Yin HQ, Kim YH, Moon CK, Lee BH. Reactive oxygen species-mediated induction of apoptosis by a plant alkaloid 6-methoxydihydrosanguinarine in HepG2 cells. Biochem Pharmacol 2005;70(2):242–8.

[245] Lee YJ, Yin HQ, Kim YH, Li GY, Lee BH. Apoptosis inducing effects of 6-methoxydihydrosanguinarine in HT29 colon carcinoma cells. Arch Pharm Res 2004;27(12):1253–7.

[246] Choi WY, Kim GY, Lee WH, Choi YH. Sanguinarine, a benzophenanthridine alkaloid, induces apoptosis in MDA-MB-231 human breast carcinoma cells through a reactive oxygen species-mediated mitochondrial pathway. Chemother 2008;54(4):279–87.

[247] Yu R, Mandlekar S, Tan TH, Kong AN. Activation of p38 and c-Jun N-terminal kinase pathways and induction of apoptosis by chelerythrine do not require inhibition of protein kinase C. J Biol Chem 2000;275:9612–9.

[248] Kimura KI, Aburai N, Yoshida M, Ohnishi M. Sanguinarine as a potent and specific inhibitor of protein phosphatase 2C in vitro and induces apoptosis via phosphorylation of p38 in HL60 cells. Biosci, Biotech Biochem 2010;74(3):548–52.

[249] Debiton E, Madelmont JC, Legault J, Barthomeuf C. Sanguinarine induced apoptosis is associated with an early and severe cellular glutathione depletion. Cancer Chemother Pharmacol 2003;51:474–82.

[250] Jang BC, Park JG, Song DK, Baek WK, Yoo SK, Jung KH, et al. Sanguinarine induces apoptosis in A549 human lung cancer cells primarily via cellular glutathione depletion. Toxicol in vitro 2010;23(2):281–7.

[251] Kim S, Lee TJ, Leem J, Keong SC, Park JW, Taegu KK. Sanguinarine-induced apoptosis: generation of ROS, down-regulation of Bcl-2, c-FLIP, and synergy with TRAIL. J Cell Biochem 2008;104(3):895–907.

[252] Sun M, Lou W, Chun JY, Cho DS, Nadiminty N, Evans CP, et al. Sanguinarine suppresses prostate tumor growth and inhibits survivin expression. Genes Cancer 2010;1(3):283–92.

[253] Koch A, Steffen J, Krüger E. TCF11 at the crossroads of oxidative stress and ubiquitin proteasome system. Cell Cycle 2011;10(8):1200–7.

Terpenoid Alkaloids

1.4.1 *Daphniphyllum Glaucescens* **Blume**

History The plant was first described by Carl Ludwig von Blume, in *Bijdragen tot de flora van Nederlandsch Indië*, published in 1825.

Synonyms *Daphniphyllum lancifolium* Hook. f., *Daphniphyllum scortechinii* Hook. f

Family Daphniphyllaceae Müll. Arg., 1869

Habitat and Description This is a dioecious tree which grows in the rainforests of Sri Lanka, Southeast Asia, Hong Kong, Taiwan, Korea and Japan. It is grown as an ornamental. The trunk is 20 m tall. The leaves are simple and clustered. The petiole is 1.2–4 cm long and straight. The blade is elliptic, 2.5–6 cm × 5–15 cm, pointed at base, blunt at apex, glaucous beneath and leathery. The inflorescence is an axillary raceme. The male flowers are minute and present a five-lobed calyx. The androecium includes 10 stamens. In female flowers the ovary is ovoid with thick recurved stigmas. The fruit is an ellipsoid drupe which is 1 cm long, with a few knobbles, seated on the persistent calyx and topped by prominent twin recurved styles (Figure 1.18).

Medicinal Uses In Taiwan, the plant is used externally to heal ulcers.

Phytopharmacology The plant is known to contain calycine and glaucescine,[254] the quaternary alkaloids daphniglaucins A, B,[255] E–H and K[256] and the tertiary alkaloids daphniglaucins C,[257] D and J.[256] These triterpene alkaloids are known as Daphniphyllum alkaloids. The medicinal property of the plant has not yet been substantiated.

Proposed Research Pharmacological study of daphniglaucine C and derivatives for the treatment of cancer.

Rationale Several lines of evidence suggest that Daphniphyllum alkaloids are compelling candidates for the development of anticancer agents.

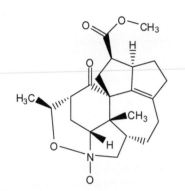

■ **CS 1.49** Daphniyunnine D

■ **CS 1.50** Calyciphylline A

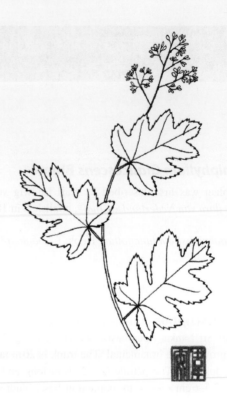

■ **FIGURE 1.18** *Daphniphyllum glaucescens* Blume

■ **CS 1.51** Daphnicyclidin D

Daphniyunnine D (CS 1.49) isolated from *Daphniphyllum yunnanense* C.C. Huang ex T.L. Ming (Daphniphyllaceae Müll. Arg.) was cytotoxic against mouse leukemia (P388) and human lung adenocarcinoma epithelial (A549) cells, with IC$_{50}$ values of 3 μM and 0.6 μM, respectively.[258] Calyciphyllines A (CS 1.50) and B from *Daphniphyllum calycinum* Benth. (Daphniphyllaceae Müll. Arg.) negated the survival of mouse lymphocytic leukemia (L1210) cells, with IC$_{50}$ values equal to 2.1 μg/mL and 4.2 μg/mL, respectively.[259] From the same plant, daphcalycine and daphnicyclidin D (CS 1.51) mitigated the viability of human nasopharyngeal carcinoma (KB) cells, with IC$_{50}$ values of 13 μg/mL and 7 μg/mL.[260] Daphnezomines P, Q, R, and S (CS 1.52) from *Daphniphyllum humile* Maxim. ex Franch. & Sav. (Daphniphyllaceae Müll. Arg.) were cytocidal to mouse lymphocytic leukemia (L1210) cells, with IC$_{50}$ values of 8.5 μg/mL, 9.2 μg/mL, 4.8 μg/mL and 2.7 μg/mL, respectively.[261]

Daphmanidins A (CS 1.53) and B from *Daphniphyllum teijsmanii* Zoll. ex Teijsm. & Binn. (Daphniphyllaceae Müll. Arg.) induced the death of L1210 cells, with IC$_{50}$ values of 8 μg/mL and 7.6 μg/mL, respectively.[262]

■ **CS 1.52** Daphnezomine S

■ **CS 1.53** Daphmanidin A

■ **CS 1.54** Epothilone B

■ **CS 1.55** Daphniglaucine C

Daphniglaucins A and B from *Daphniphyllum glaucescens* Blume (Daphniphyllaceae Müll. Arg.) were lethal for L1210 cells, with IC_{50} values of 2.7 µg/mL and 3.9 µg/mL, respectively, and also for KB cells, with IC_{50} values of 2 µg/mL and 10 µg/mL /mL, respectively.[255]

The molecular events that precipitate cancer cell death following Daphniphyllum alkaloids exposure is to date unexplored but one may have noted that these alkaloids share a key chemical feature with the most common agents affecting tubulin: the coexistence of a bulky and complex alkylated moiety with amides and/or heterocyclic pyrrole groups as seen with taxol, epothilone B (CS 1.54), vinblastine, rhizoxin, IDN-5109 and BMS-247550. In fact, daphniglaucine C (CS 1.55) isolated from *Daphniphyllum glaucescens* Blume (Daphniphyllaceae Müll. Arg.) was cytotoxic against L1210, with an IC_{50} value of 0.1 µg/mL, and inhibited the polymerization of tubulin, with an IC_{50} value of 25 µM.[257]

1.4.2 *Nuphar Pumila* (Timm) DC.

History The plant was first described by Augustin Pyramus de Candolle, in *Regni Vegetabilis Systema Naturale*, published in 1821.

Synonyms *Nuphar lutea* subsp. *pumila* (Timm) E.O. Beal, *Nuphar shimadae* Hayata, *Nuphar subpumila* Miki, *Nuphar tenella* Rchb., *Nymphaea lutea* subsp. *pumila* (Timm) Bonnier & Layens, Nymphaea lutea var. pumila Timm, Nymphaea pumila (Timm) Hoffm.

Family Nymphaeaceae Salisb., 1805

Common Name Dwarf water lily, ping peng cao (Chinese)

Habitat and Description This is an aquatic herb which grows in the lakes and ponds of China, Taiwan, Japan, Korea, Mongolia, Russia and Europe. The plant grows from a rhizome. The petiole is terete, flexuous and 20–50 cm long. The blade is orbicular to ovate, strongly cordate at base, and 5–15 cm×6–10 cm. The inflorescence is solitary and aerial. The calyx comprises five sepals which are sulfur yellow, elliptic, 1–2.5 cm long and recurved inward. The petals are numerous, sulfur yellow, stamen-like, linear, 5–7 cm long and emarginate at apex. The anthers are numerous, yellow, 0.1–0.5 cm long and with strap-like filaments. The gynoecium is pyriform and topped with a star-shaped, stigmatic disc which is 0.4–0.8 cm across. The fruit is pyriform, aerial, dull green, 1–2 cm in diameter, somewhat lobed, topped with a disc and seated in the vestigial calyx. The seeds are glossy, dark brown and 0.5 cm long (Figure 1.19).

Medicinal Uses In China, the plant is used to invigorate, to check bleeding, to promote urination and to allay pain in the joints of the elderly.

Phytopharmacology Phytochemical study of the plant has resulted in the isolation of: the quinolizidine sesquiterpene alkaloids nupharopumiline,[263] deoxynupharidine, 7-*epi*-deoxynupharidine, nupharidine, 7-*epi*-nupharidine, nupharopumiline and nupharolutine[264,265]; the dimeric quinolizidine sesquiterpene thioalkaloids 6-hydroxythiobinupharidine, 6,6′-dihydroxythiobinupharidine, 6-hydroxythionuphlutine B, 6′-hydroxy-thionuphlutine B, neothiobinupharidine,[265] thionuphlutine B, 6,6′-hydroxy-thionuphlutine B,[266] thiobinupharidine, syn-thiobinupharidine sulfoxide, thionuphlutine B β-sulfoxide and neothiobinupharidine β-sulfoxide.[267]

The anti-inflammatory property is confirmed as being provided by deoxynupharidine, which also inhibited the secretion of interleukin 1 (IL-1) and tumor necrosis factor (TNF) by murine peritoneal macrophages in vitro.[264]

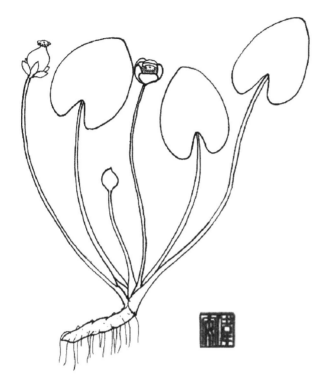

■ **FIGURE 1.19** *Nuphar pumila* (Timm) DC.

In addition, 6-hydroxythiobinupharidine, 6,6′-dihydroxythiobinupharidine, 6-hydroxythionuphlutine B and 6′-hydroxythionuphlutine B inhibited in vitro antibody formation in mice splenocytes.[265,266]

Proposed Research Pharmacological study of 6-hydroxythionuphlutine B and derivatives for the treatment of cancer.

Rationale The dimeric quinolizidine sesquiterpene thioalkaloids 6-hydroxy-thiobinupharidine, 6,6′-hydroxythiobinupharidine and 6-hydroxythionuphlutine B isolated from *Nuphar pumila* (Timm) DC. (Nymphaeaceae Salisb.) slowed the proliferation of mouse melanoma (B16) cells on collagen-coated filters by 84.8%, 83.2% and 61.4%, respectively, at a dose of 10μM.[267] In addition, 6-hydroxythiobinupharidine (CS 1.56) at a dose of 5mg/kg/day protected rodents against metastasis to the lungs of mouse melanoma (B16) cells.[267] The structure–activity relationship suggested that the thiohemiaminal structure with a hydroxyl group in C6 was required for cytotoxicity [267] and that the number of hydroxyl groups increased the cytotoxicity.[266] It was also shown that 10μM of 6-hydroxythiobinupharidine,

■ **CS 1.56** 6-Hydroxythiobinupharidine

■ **CS 1.57** 6,6'-Hydroxythiobinupharidine

■ **CS 1.58** 6-Hydroxythionuphlutine B

6,6'-hydroxythiobinupharidine (CS 1.57) and 6-hydroxythionuphlutine B mitigated the growth of human histiocytic lymphoma (U937) cells by 94.3%, 81.6% and 94.1%, respectively, with DNA fragmentation and caspase 8 and 3 activation.[268] Furthermore, a mixture of 6-hydroxythiobinupharidine and 6-hydroxythiobinuphlutine B (CS 1.58) isolated from *Nuphar lutea* (L.) Sm. (Nymphaeaceae Salisb.) induced apoptosis via a depletion of nuclear factor kappa-light-chain-enhancer of activated B cells (NF-κB), transcription factor (Rel B) and p52 subunits, which was followed by caspase 9 activation and PARP (Poly [ADP-ribose] polymerase) cleavage in Hodgkin's disease (L-428) cells.[269] The effect of dimeric quinolizidine sesquiterpene thioalkaloids on NF-κB inhibition is still unclear but a number of lines of evidence point to the presence of intact cysteine thiol groups in the DNA-binding domain of NF-κB subunits as being a structural requirement for DNA-binding. In normal cellular conditions, NF-κB thiols are protected from oxidation by thioredoxin.[270]

If NF-κB thiols are oxidized then NF-κB is unable either to bind to DNA or to command the synthesis of anti-apoptotic proteins. The fact that most thioredoxin inhibitors such as aurothiomalate, sulforaphane (CS 1.59) and aurothioglucose are organosulfurs raises the intriguing possibility that dimeric quinolizidine sesquiterpene thioalkaloids may induce apoptosis through thioreduxin inhibition and subsequent NF-κB inactivation. In fact, NF-κB activation plays a critical role in the progression of lymphoid malignancies and thyroid, ovarian, breast and hepatocellular carcinomas,[271,272] and the development of NF-κB inhibitors is of compelling interest for the development of anticancer drugs.

■ **CS 1.59** Sulfonarane

1.4.3 *Pachysandra Terminalis* **Siebold & Zucc.**

History The plant was first described by Philipp Franz Balthasar von Siebold, in *Abhandlungen der Mathematisch-Physikalischen Classe der Königlich Bayerischen Akademie der Wissenschaften*, published in 1845.

Family Buxaceae Durmort., 1822

Common Names Japanese pachysandra, Japanese-spurge, fukki-so (Japanese), ding hua ban deng guo (Chinese)

Habitat and Description This is a dwarf shrub which grows in China, Korea and Japan. The plant is a garden ornamental. The stem is terete and woody. The leaves are simple. The petiole is 1–3 cm long. The blade is rhombic, 2.5–5 cm × 1.5–3 cm, coriaceous, attenuate and cuneate at base and acute and serrate at apex and displays four to six pairs of secondary nerves. The inflorescence is a terminal spike which is 2–5 cm long. The male flowers comprise four oblong tepals, which are light brown and 0.3 cm long, and four conspicuous stamens with 0.7-cm-long, fleshy, pure white filaments. The female flowers include two exerted styles enclosed in four tepals and bracts. The fruit is a drupe which is globose, 0.5 cm across and topped by two vestigial styles which are reflexed (Figure 1.20).

Medicinal Uses In Japan, the plant is used to alleviate stomachache.

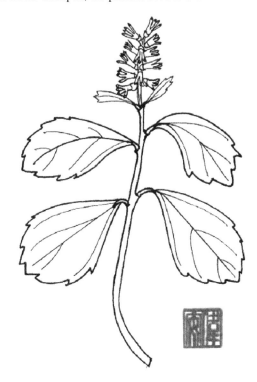

■ **FIGURE 1.20** *Pachysandra terminalis* Siebold & Zucc.

Phytopharmacology Phytochemical study of the plant resulted in the identification of pregnane steroidal alkaloids including pachysandrines A and B,[273,274] pachysamines A and B,[275] epipachysamines B, C,[276] D–F,[24] pachystermines A and B,[278] pachysantermine A,[279] terminaline[276] and spiropachysine,[280] and a series of triterpenes: friedelin, epifriedelanol, cycloartenol,[281] pachysandiol-B, pachysonol, triterpene triol,[282–284] pachysantriol,[285] 3-*O*-acetyl-16-*O*-*p*-bromobenzoylpachysandiol B[286] and taraxerol.[287]

The medicinal properties of the plant are confirmed and are attributed to the CNS action of pregnane steroidal alkaloids, such as epipachysamine A, which were shown to protect rodents against gastric acid secretion induced by 2-deoxy-D-glucose (200 mg/kg, i.v.).[288,289]

Proposed Research Pharmacological study of epipachysamine E and derivatives for the treatment of multidrug-resistant lymphoid malignancies.

Rationale Steroidal alkaloids are interesting because they interact with a broad array of cellular targets to bring about cancer cell death. Such alkaloids are epipachysamine E (CS 1.60) and pachysamine E (CS 1.61) from

■ **CS 1.60** Epipachysamine E

■ **CS 1.61** Pachysamine E

Pachysandra terminalis Siebold & Zucc. (Buxaceae Durmort.), which inhibited the growth of mouse leukemia (P388) and adriamycin-resistant mouse leukemia (P388/ADR) cells, with IC_{50} values equal to 0.5 μg/mL, 0.6 μg/mL and 0.4 μg/mL and 0.4 μg/mL, respectively.[290] Epipachysamine E from the same species was lethal to mouse melanoma (B16) and Shionogi carcinoma (SC 115) cells, with IC_{50} values of 2.5 μg/mL and 3.4 μg/mL, respectively, and reversed the resistance of adriamycin-resistant mouse leukemia (P388/ADR) cells to etoposide, mitoxanthrone and adriamycin, with IC_{50} values of 0.18 μg/mL, 0.24 μg/mL, 0.017 μg/mL and 0.24 μg/mL at a dose of 0.5 μg/mL.[290] Other examples of cytotoxic steroidal alkaloids are sarsaligenines A and B isolated from *Sarcococca saligna* (D. Don) Müll. Arg. (family Buxaceae Dumort.), which selectively repressed the multiplication of human promyelocytic leukemia (HL-60) cells, with IC_{50} values of 2.8 μM and 3.6 μM, respectively.[291]

A mounting body of evidence suggests that steroidal alkaloids are potent inhibitors of acetylcholinesterase. One such alkaloid is hookerianamide H (CS 1.62) isolated from *Sarcococca hookeriana* Baill. (family Buxaceae Dumort.), which abated the enzymatic activity of acetylcholinesterase and butyrylcholinesterase, with IC_{50} values equal to 2.9 μM and 1.9 μM, respectively.[292] The canonical function of acetylcholinesterase is to terminate the cholinergic neurotransmission but it is also involved in cell adhesion,[293] in the increase of cells' sensitivity to apoptotic induction[294] and in prompting apoptosome formation through interaction with caveolin 1[295] and the consequent induction of apoptosis.

Lavy et al. (2001) provided strong evidence that steroidal alkaloids such as the teratogenic cyclopamine (CS 1.63) from the genus *Veratrum californicum* Durand (family Melanthiaceae Batsch ex Borkh) and tomatidine

■ **CS 1.62** Hookerianamide H

■ **CS 1.63** Cyclopamine

(CS 1.64) from *Solanum lycopersicum* L. (family Solanaceae Juss.) reverse P-glycoprotein-mediated efflux of anticancer agents in (OVCAR-8) cells.[296] The molecular process involved there remains unexplained but it is conceivable that lipophilic steroidal alkaloids, which consist of both a planar polycyclic framework and nitrogen atoms, diffuse freely in the cell and bind to P-glycoprotein at the cytoplasmic level.[296] Furthermore a steroidal skeleton appears to be a cardinal chemical feature since progesterone inhibits P-glycoprotein-dependent ATPase activity and, thereby, drug efflux.[296] However, the steroidal alkaloid wrightiamine A (CS 1.65) mitigated the viability of vincristine-resistant murine leukemia (P388/VCR) cells, with an IC_{50} value of 2μg/ml, but did not reverse the resistance of the cells to vincristine.[297] Moreover, steroidal alkaloids interact with DNA. Verazine (CS 1.66) isolated from *Solanum surinamense* Steud. (family Solanaceae Juss.) was deleterious to Madison lung tumor (M-109) cells, with an IC_{50} value of 12.5μg/mL, via suppression of DNA synthesis.[298] In addition, a 4-methoxyphenylpyrazoline derivative of solanidine (CS 1.67) abrogated the survival of human promyelocytic leukemia (HL-60) cells (with an IC_{50} value of 1.7μM) by impairing the enzymatic activity of ribonucleotide reductase and thus curtailing DNA assemblage in the S phase of the cell

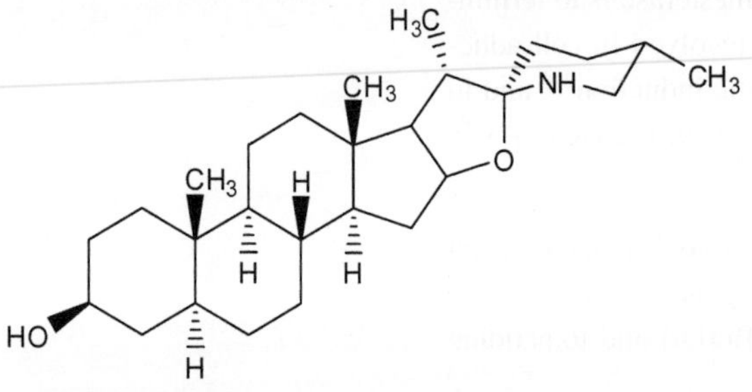

■ **CS 1.64** Tomatidine

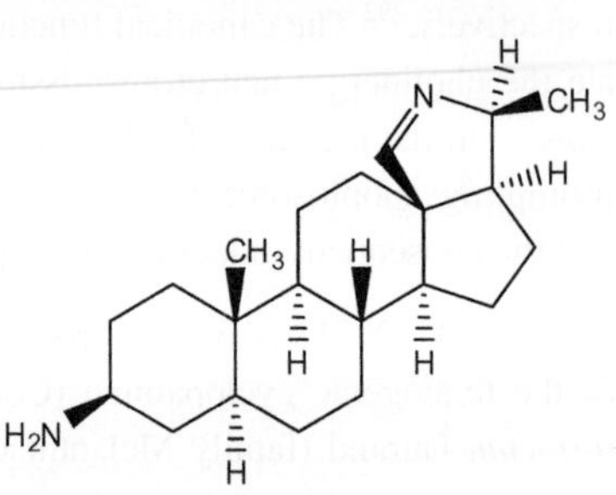

■ **CS 1.65** Wrightiamine A

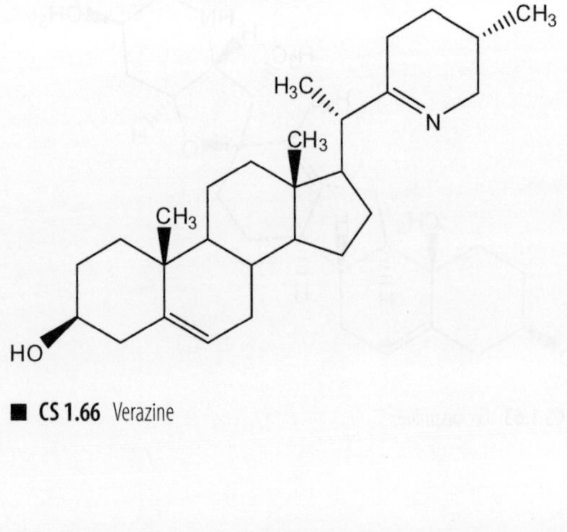

■ **CS 1.66** Verazine

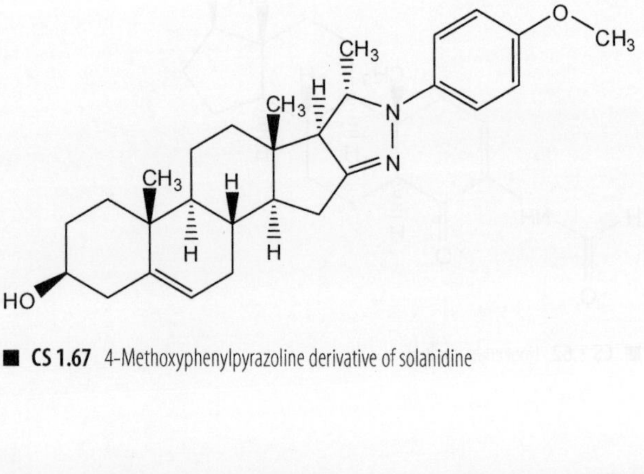

■ **CS 1.67** 4-Methoxyphenylpyrazoline derivative of solanidine

cycle.[299] Likewise, the steroidal alkaloid solasodine (CS 1.68) isolated from *Solanum umbelliferum* Eschsch. (family Solanaceae Juss.) exhibited significant activity toward DNA repair-deficient yeast mutants, probably as a result of DNA alkylation by the formation of an electrophilic iminium group.[300]

Steroidal alkaloids have an appealing capacity to disrupt the hedgehog signaling pathway and are of considerable interest for the treatment of brain, lung, mammary gland, pancreas, lung and skin cancers.[301,302] A derivative of cyclopamine (CS 1.69) from *Veratrum californicum* Durand (family Melanthiaceae Batsch ex Borkh) at 10 μM abated hedgehog-driven proliferation of mouse cerebellar granule neuron precursors by repressing the transmembrane protein Smoothened (SMO).[303] Lastly, the steroidal alkaloid (+)-(20S)-3-(benzoylamino)-20-dimethylamino)-5α-pregn-2-en-4β-ol (CS 1.70) isolated from *Pachysandra procumbens* Michx. (family Buxaceae Dumort.) curbed the enzymatic activity of estrone sulfatase, with an IC$_{50}$ value of 0.1 μM, and as a result could be of interest for the treatment of hormone-dependent carcinomas.[304]

■ **CS 1.68** Solasodine

■ **CS 1.69** Derivative of cyclopamine

■ **CS 1.70** (+)-(20S)-3-(Benzoylamino)-20-dimethylamino)-5α-pregn-2-en-4β-ol

REFERENCES

[254] Arthur HR, Chan RPK, Loo SN. Alkaloids of *Daphniphyllum calycinum* and *D. glaucescens* of Hong Kong. Phytochem 1965;4(4):627–9.

[255] Kobayashi J, Takatsu H, Shen YC, Morita H. Daphniglaucins A and B, novel polycyclic quaternary alkaloids from *Daphniphyllum glaucescens*. Org Lett 2003;5(10):1733–6.

[256] Takatsu H, Morita H, Shen YC, Kobayashi J. Daphniglaucins D–H, J, and K, new alkaloids from *Daphniphyllum glaucescens*. Tetrahedron 2004;60(30):6279–84.

[257] Morita H, Takatsu H, Shen YC, Kobayashi J. Daphniglaucin C, a novel tetracyclic alkaloid from *Daphniphyllum glaucescens*. Tetrahedron Lett 2004;45(5):901–4.

[258] Zhang H, Yang SP, Fan CQ, Ding J, Yue JM. Daphniyunnines A–E, alkaloids from *Daphniphyllum yunnanense*. J Nat Prod 2006;69(4):553–7.

[259] Morita H, Kobayashi J. Calyciphyllines A and B, two novel hexacyclic alkaloids from *Daphniphyllum calycinum*. Org Lett 2003;5(16):2895–8.

[260] Jossang A, El Bitar H, Cuong Pham V, Sévenet T. Daphcalycine, a novel hepta-cycle fused ring system alkaloid from *Daphniphyllum calycinum*. J Org Chem 2003;68(2):300–4.

[261] Morita H, Takatsu H, Kobayashi J. Daphnezomines P, Q, R and S, new alkaloids from *Daphniphyllum humile*. Tetrahedron 2003;59(20):3575–9.

[262] Kobayashi J, Ueno S, Morita H. Daphmanidin A, a novel hexacyclic alkaloid from *Daphniphyllum teijsmanii*. J Org Chem 2002;67(18):6546–9.

[263] Peura P, Lounasmaa M. Nupharopumiline, a new quinolizine alkaloid from *Nuphar pumila*. Phytochemistry 1977;16(7):1122–3.

[264] Zhang L, Huang Y, Qian Y, Xiao P. Effect of deoxynupharidine on immune func-tion in vitro. Acad Med Sin 1995;17(5):343–8.

[265] Yamahara J, Shimoda H, Matsuda H, Yoshikawa M. Potent immunosuppressive principles, dimeric sesquiterpene thioalkaloids, isolated from *Nupharis rhizoma*, the rhizome of *Nuphar pumilum* (nymphaeaceae): structure-requirement of nuphar-alkaloid for immunosuppressive activity. Biol Pharm Bull 1996;19(9):1241–3.

[266] Matsuda H, Shimoda H, Yoshikawa M. Dimeric sesquiterpene thioalkaloids with potent immunosuppressive activity from the rhizome of *Nuphar pumilum*: struc-tural requirements of nuphar alkaloids for immunosuppressive activity. Bioorg Med Chem 2001;9(4):1031–5.

[267] Matsuda H, Morikawa T, Oda M, Asao Y, Yoshikawa M. Potent anti-metastatic activity of dimeric sesquiterpene thioalkaloids from the rhizome of *Nuphar pum-ilum*. Bioorg Med Chem Lett 2003;13(24):4445–9.

[268] Matsuda H, Yoshida K, Miyagawa K, Nemoto Y, Asao Y, Yoshikawa M. Nuphar alkaloids with immediately apoptosis-inducing activity from *Nuphar pumi-lum* and their structural requirements for the activity. Bioorg Med Chem Lett 2006;16(6):1567–73.

[269] Ozer J, Eisner N, Ostrozhenkova E, Bacher A, Eisenreich W, Benharroch D, et al. Nuphar lutea thioalkaloids inhibit the nuclear factor κB pathway, potenti-ate apoptosis and are synergistic with cisplatin and etoposide. Cancer Biol Ther 2009;8(19):1860–8.

[270] Matthews JR, Wakasugi N, Virelizier JL, Yodoi J, Hay RT. Thioredoxin regu-lates the DNA binding activity of NF-kappa B by reduction of a disulphide bond involving cysteine 62. Nucleic Acids Res 1992;20:3821–30.

[271]   Pacifico F, Leonardi A. NF-kappaB in solid tumors. Biochem Pharmacol 2006;72:1142–52.

[272]   Arsura M, Cavin LG. Nuclear factor-kappaB and liver carcinogenesis. Cancer Lett 2005;229:157–69.

[273]   Tomita M, Uyeo Jr. S, Kikuchi T. Studies on the alkaloids of *Pachysandra terminalis* sieb. et zucc.: structure of Pachysandrine A and B. Tetrahedron Lett 1964;5(18):1053–61.

[274]   Tomita M, Uyeo Jr. S, Kikuchi T, Kapadi AH, Dev S. Studies on the alkaloids of *Pachysandra terminalis* Sieb. et Zucc.: structure of Pachysandrine A and B. Tetrahedron Lett 1964;5(28):1902.

[275]   Tomita M, Uyeo Jr. S, Kikuchi T. Studies on the alkaloids of *Pachysandra terminalis* sieb. et zucc. (2).: structure of pachysamine - A and -B. Tetrahedron Lett 1964;5(25):1641–4.

[276]   Kikuchi T, Uyeo S, Nishinaga T. Studies on the alkaloids of *Pachysandra terminalis* sieb. et zucc. (4).: structure of epipachysamine-B, -C and terminaline. Tetrahedron Lett 1965;6(24):1993–9.

[277]   Kikuchi T, Uyeo S, Nishinaga T. Studies on the alkaloids of *Pachysandra terminalis* sieb. et zucc. (5).: structure of epipachysamine-d, -e, and -f. Tetrahedron Lett 1965;6(36):3169–74.

[278]   Kikuchi T, Uyeo S. Studies on the alkaloids of *Pachysandra terminalis* sieb. et zucc. (6).: structure of pachystermine - a and -b, novel type alkaloids having β-lactam ring system. Tetrahedron Lett 1965;6(39):3473–85.

[279]   Kikuchi T, Uyeo S. Studies on the alkaloids of *Pachysandra terminalis* sieb. et zucc. (7).: structure of pachysantermine-A, a novel intramolecular ester alkaloid. Tetrahedron Lett 1965;6(39):3487–90.

[280]   Kikuchi T, Nishinaga T, Inagaki M, Koyawa M. Structure of spiropachysine a novel alkaloid from *Pachysandra terminalis* sieb. et zucc. Tetrahedron Lett 1968;9(17):2077–81.

[281]   Kikuchi T, Toyoda T, Arimoto M, Takayama M, Yamano M. Studies on the neutral constituents of *Pachysandra terminalis* Sieb. et Zucc. I. isolation and characterization of sterols and triterpenes. Yakugaku Zasshi 1969;89(10):1358–66.

[282]   Kikuchi T, Takayama M, Toyoda T, Arimoto M, Niwa M. Structures of pachysandiol-B and pachysonol, new triterpenes from *Pachysandra terminalis* sieb. et zucc. (I). Tetrahedron Lett 1971;12(19):1535–8.

[283]   Kikuchi T, Niwa M. Isolation and structure determination of a new triterpene-triol from pachysandra terminalis sieb. et zucc. Tetrahedron Lett 1971;12(41):3807–8.

[284]   Kikuchi T, Takayama M, Toyoda T. Studies on the neutral constituents of *Pachysandra terminalis* Sieb. et Zucc. V. Structures of pachysandiol B and pachysonol, new friedelin type triterpenes. Chem Pharm Bull 1973;21(10):2243–51.

[285]   Kikuchi T, Niwa M. Studies on the neutral constituents of Pachysandra terminalis Sieb. et Zucc. VI. Isolation and determination of structure of pachysantriol, a new friedelin type triterpene. Yakugaku Zasshi 1973;93(10):1378–82.

[286]   Masaki N, Niwa M, Kikuchi T. Studies on the neutral constituents of *Pachysandra terminalis* Sieb. et Zucc. Part IV. X-ray structure of 3-*O*-acetyl-16-*O*-*p*-bromobenzoylpachysandiol B: new conformation of a friedelin-type triterpene. J Chem Soc 1975;2(6):610–4.

[287] Kikuchi T, Takayama M. Studies on the neutral constituents of *Pachysandra terminalis* Sieb. et Zucc. II. identification of taraxerol and isolation of compound-3A and 3B. Yakugaku Zasshi 1970;90(8):1051–3.

[288] Maeda-Hagiwara M, Watanabe K, Watanabe H. Effects of gastric acid secretion of a steroidal alkaloid, epipachysamine-A, extracted from *Pachysandra terminalis* Sieb. et Zucc. J Pharmacobiodyn 1984;7(4):263–7.

[289] Watanabe H, Watanabe K, Shimadzu M. Anti-ulcer effect of steroidal alkaloids extracted from *Pachysandra terminalis*. Planta Med 1986;1:56–8.

[290] Funayama S, Noshita T, Shinoda K, Haga N, Nozoe S, Hayashi M, et al. Cytotoxic alkaloids of *Pachysandra terminalis*. Biol Pharm Bull 2000;23(2):262–4.

[291] Yan YX, Sun Y, Chen JC, Wang YY, Li Y, Qiu MH. Cytotoxic steroids from *Sarcococca saligna*. Planta Med 2011;77(15):1725–9.

[292] Devkota KP, Lenta BN, Choudhary MI, Naz Q, Fekam FB, Rosenthal PJ, et al. Cholinesterase inhibiting and antiplasmodial steroidal alkaloids from *Sarcococca hookeriana*. Chem Pharm Bull 2007;55(9):1397–401.

[293] Darboux I, Barthalay Y, Piovant M, Hipeau-Jacquotte R. The structure-function relationships in *Drosophila neurotactin* show that cholinesterasic domains may have adhesive properties. EMBO J 1996;15:4835–43.

[294] Jin QH, He HY, Shi YF, Lu H, Zhang XJ. Overexpression of acetylcholinesterase inhibited cell proliferation and promoted apoptosis in NRK cells. Acta Pharmacol Sin 2004;25:1013–21.

[295] Park SE, Kim ND, Yoo YH. Acetylcholinesterase plays a pivotal role in apoptosome formation. Cancer Res 2004;64:2652–5.

[296] Lavie Y, Harel-Orbital T, Gaffield W, Liscovitch M. Inhibitory effect of steroidal alkaloids on drug transport and multidrug resistance in human cancer cells. Anticancer Res 2001;21(2A):1189–94.

[297] Kawamoto S, Koyano T, Kowithayakorn T, Fujimoto H, Okuyama E, Hayashi M, et al. Wrightiamines A and B, two new cytotoxic pregnane alkaloids from *Wrightia javanica*. Chem Pharm Bull 2003;51(6):737–9.

[298] Abdel-Kader MS, Bahler BD, Malone S, Werkhoven MCM, Van Troon F, David DNA-damaging steroidal alkaloids from *Eclipta alba* from the Suriname rainforest. J Nat Prod 1998;61(10):1202–8.

[299] Minorics R, Szekeres T, Krupitza G, Saiko P, Giessrigl B, Wölfling J, et al. Antiproliferative effects of some novel synthetic solanidine analogs on HL-60 human leukemia cells in vitro. Steroids 2011;76(1–2):156–62.

[300] Kim YC, Che QM, Gunatilaka AAL, Kingston DGI. Bioactive steroidal alkaloids from *Solanum umbelliferum*. J Nat Prod 1996;59(3):283–5.

[301] Katano M. Hedgehog signaling pathway as a therapeutic target in breast cancer. Cancer Lett 2005;227(2):99–104.

[302] Evangelista M, Tian H, De Sauvage FJ. The Hedgehog signaling pathway in cancer. Clin Cancer Res 2006;12(20):5924–8.

[303] Winkler JD, Isaacs A, Holderbaum L, Tatard V, Dahmane N. Design and synthesis of inhibitors of hedgehog signaling based on the alkaloid cyclopamine. Organic Lett 2009;11(13):2824–7.

[304] Chang LC, Bhat KPL, Fong HHS, Pezzuto JM, Kinghorn AD. Novel bioactive steroidal alkaloids from *Pachysandra procumbens*. Tetrahedron 2000;56(20):3133–8.

Other Alkaloids

1.5.1 *Dysoxylum Acutangulum* Miq.

History The plant was first described by Miquel Friedrich Anton Wilhelm, in *Flora van Nederlandsch Indie, Eerste Bijvoegsel*, published in 1860.

Family Meliaceae Juss., 1789

Habitat and Description *Dysoxylum acutangulum* Miq. is a massive timber tree which grows in the rainforests of Thailand, Malaysia, Indonesia, the Philippines, Papua New Guinea and Northern Australia. The trunk is buttressed and 20 m tall. The bark is smooth, yellowish and lenticelled. The leaves are paripinnate. The rachis is 30 cm long and bears four pairs of folioles which are elliptical, acute at base, acuminate at apex and without apparent secondary nerves. The inflorescence is a cauliflorous panicle which is 10 cm long. The flowers are small and comprise four minute sepals and four petals which are yellowish. The androecium includes eight anthers within the tube throat. The disc is cupuliform. The fruit is a capsule which is four-valved, orange and contains a few black seeds (Figure 1.21).

Medicinal Uses In Indonesia, the seeds are used to poison fish.

Phytopharmacology The poisonous property of the seeds is due to the sesquiterpene (+)-8-hydroxycalamenene.[305] Other chemical constituents isolated from the plant include the triterpenes acutaxylines A and B[306] and a series of unusual piperidine chromone alkaloids including chrotacumines A–D and rohitukine as well as the chromone glycoside schumanniofioside A and the chromone noreugenin.[307]

Proposed Research Pharmacological study of 2-(2-bromophenyl)-5,6,7-trihydroxy-8-((2-hydroxy-4-methylpiperazin-1-yl)methyl)-4H-chromen-4-one and derivatives for the treatment of multidrug-resistant cancers.

Rationale The piperidine chromone alkaloid rohitukine from *Dysoxylum acutangulum* Miq. was cytotoxic against human promyelocytic leukemia

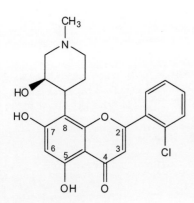

■ **CS 1.71** Rohitukine

■ **FIGURE 1.21** *Dysoxylum acutangulum* Miq.

(HL-60) (CS 1.71) and human colorectal carcinoma (HCT-116) cells, with IC$_{50}$ values equal to 7.5 μM and 8.8 μM, respectively.[307] Rohitukine from *Dysoxylum binectariferum* (Roxb.) Hook. f. ex Bedd. (Meliaceae Juss.) was converted into a chlorophenyl derivative named flavopiridol (CS 1.72), a compound that attracted a tremendous amount of interest because of its ability to inhibit the enzymatic activity of cyclin-dependent kinases (CDK) which control cell-cycle progression.[308] Flavopiridol induced apoptosis in human chronic lymphocytic leukemia (WSU-CLL) and human chronic B-cell leukemia (I83CLL) cells by downregulation of the anti-apoptotic protein Bcl-2 at a concentration of 400 nmol/L.[309] Other investigations showed that 0.1 μmol/L of flavopiridol induced dramatic reductions in the levels of the anti-apoptotic myeloid leukemia cell differentiation protein (Mcl-1) and X-linked inhibitor of apoptosis protein (XIAP) in B-CLL cells.[310] In fact the main mode of action of flavopiridol is the inactivation of the positive transcription elongation factor b (PTEF-b) CDK9/ cyclin T complex. In normal conditions, this complex phosphorylates the C-terminal repeat domain (CTD) of RNA polymerase, which undertakes

■ **CS 1.72** Flavopiridol

the transcription of various anti-apoptotic proteins[311–313] such as phosphatidylinositol 3-kinase (PI3K)[314] and survivin.[315] In brief, flavopiridol suppresses gene transcription by interfering with CDK9/cyclin T function[316] and consequently induces apoptosis[317] with precipitated loss of mitochondrial membrane potential followed by caspase 3 activation and cleavage of PARP (Poly [ADP-ribose] polymerase).[318,319] Other observations on the apoptotic mechanism of flavopiridol include, notably, the disruption of nuclear factor kappa-light-chain-enhancer of activated B cells (NF-κB) signaling, by inhibiting the IκB kinase,[320] and the accumulation of pro-apoptotic protein (p53) in the wild-type p53 cancer cells.[321]

Flavonoid alkaloids in flowering plants are not so common and not always cytotoxic. The piperidine flavonoid *O*-demethylbuchenavianine (CS 1.73) from *Buchenavia capitata* (Vahl) Eichl. (family Combretaceae R. Br.) elicited mild levels of cytotoxicity against melanoma cells, with IC$_{50}$ values ranging from 10^{-5} to 10^{-6} M.[322] This lack of natural resource has prompted the development of semisynthetic analogs. The amine flavonoid PD98059 promoted apoptotic cell death through the downregulation of extracellular-signal-regulated kinase (ERK) activity and anti-apoptotic Bcl-2 protein levels in human epitheloid cervix carcinoma (Hela) cells.[323] The structure–activity relationship suggested that flavopiridol docks into the CDK adenosine triphosphate (ATP)-binding sites via the chromone scaffold and the C8 piperidine nitrogen-Asp 145.[324] In addition, Zhang et al. (2008) synthesized and tested an interesting series of nitrogen-containing flavonoid analogues (3d - see CS 1.74 - and 3f - see CS 1.75) and made the striking finding that the incorporation of the C8 nitrogen atom in a morpholine [and especially a (4-methylpiperazin-1-yl)methyl] heterocycle boosted cyclin-dependent kinase 1 (CDK1) activity.[325] In addition, the acidic C5, C6, C7 hydroxyl, and C-4 carbonyl groups are vital for inhibitory activity since they occupy the ATP binding pocket.[325] In fact, the C5 hydroxyl and the C4 carbonyl are involved in critical hydrogen bonds with E-81 and L-83 in the adenine binding pocket.[325] The lipophilic substitutes on position C2 are important because

■ **CS 1.73** *O*-Demethylbuchenavianine

■ **CS 1.74** 3d

■ **CS 1.75** 3f

■ **CS 1.76** 2b

■ **CS 1.77** 2-(2-Bromophenyl)-5,6,7-trihydroxy-8-((2-hydroxy-4-methylpiperazin-1-yl)methyl)-4H-chromen-4-one

they occupy the lipophilic binding pocket of CDK1/Cyclin B whereas ATP does not have these lipophilic substitutes.[325] CDK1 inhibition by compounds 3d and 3f elicited IC_{50} values at 0.28 μM and 0.27 μM, respectively, which were superior to that of flavopiridol.[325] Bandgar et al. (2010) prepared a series of pyrrolidine chromone derivatives, one of which, organobromine 2b (CS 1.76), repressed the growth of human breast adenocarcinoma (MCF-7), human prostate cancer (PC-3), human promyelocytic leukemia (HL-60), human non-small cell lung cancer (H460) and human colorectal carcinoma (HCT-116) cells more potently than flavopiridol, with percentages of 55% 51% 62% 53% and 61%, respectively,[326] although lacking a chromone framework. Flavopiridol has been the subject of several clinical trials which showed poor responses although some levels of positive activity were found against chronic lymphocytic leukemia. Therefore, one might be tempted to think that rohitukine could be further modified structurally into a more potent anticancer agent: 2-(2-bromophenyl)-5,6,7-trihydroxy-8-((2-hydroxy-4-methylpiperazin-1-yl)methyl)-4H-chromen-4-one (CS 1.77), via hydroxylation in C6, substitution of C8 by 1-ethyl-2-hydroxy-4-methyl-piperazine and substitution of C2 by 2-(2-bromophenyl).

1.5.2 *Ficus Septica* Burm.f.

History The plant was first described by Nicolaas Laurens Burman in *Flora Indica: cui Accedit Series Zoophytorum Indicorum, nec non Prodromus Florae Capensis*, published in 1768.

Synonyms *Ficus haulii* Blanco, *Ficus kaukauensis* Hayata, *Ficus leucantatoma* Poir., *Ficus oldhamii* Hance

Family Moraceae Gaudich., 1835

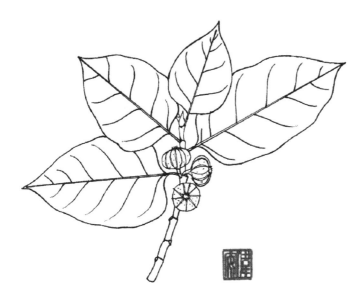

■ **FIGURE 1.22** *Ficus septica* Burm.f.

Common Names Hauili fig tree, leng guo rong (Chinese)

Habitat and Description This is a tree that grows on the forest edges in Indonesia, Taiwan, Japan, Papua New Guinea, Northern Australia and the Pacific Islands. The bark is light brown. The plant produces a white latex that causes blisters. The stems are terete, stout and articulate. The stipules are red, lanceolate, 2–3 cm long and membranaceous. The leaves are simple. The petiole is stout, somewhat flattened above and 2–5 cm long. The blade is broadly elliptic, 15–25 cm × 10–15 cm, thin, wedge-shaped at base, acuminate at apex and presents 6–12 pairs of secondary nerves. The inflorescence is axillary. The male flowers are minute and comprise a calyx which develops three lobes and one stamen. The female flowers are minute, on long pedicles and present a globose gynaecium. The fruit is a longitudinally ridged fig which is 1–2.5 cm across, green, lenticelled, depressed globose and contains numerous tiny achenes (Figure 1.22).

Medicinal Uses In Indonesia, the plant is used to treat inflammation, to heal putrefied wounds, to break fever and to counteract insect, snake and fish poisoning. In the Philippines, the plant is used to treat rheumatism and to alleviate headaches. In Papua New Guinea, the plant is used to treat colds, fungal and bacterial infections and to break fevers.

Phytopharmacology The plant elaborates series of phenanthroindolizidine alkaloids, notably tylophorine, tylocrebrine, septitine,[327] antofine,[328] isotylocrebrine, dehydrotylophorine, 10S,13aR-tylophorine N-oxide, 13aR-antofine N-oxide, ficuseptines B–D, 10R,13aR-tylophorine N-oxide, 10S, 13aR-tylocrebrine N-oxide, 10S,13aR-isotylocrebrine N-oxide, 10S, 13aS-isotylocrebrine N-oxide,[329] 14R-hydroxyisocrebrine N-oxide, ficuseptine A, 14-hydroxy-3,4,6,7-tetramethoxyphenanthroindolizidine, 14-hydroxy-2,3,4,6,7-entamethoxyphenanthroindolizidine, septicine and secoantofine,[330] the indolizidine alkaloid ficuseptine,[331] the aminocaprophenone alkaloids ficuseptamines A and B and the pyrrolidine alkaloids ficuseptamine C, phyllosterone and norruspoline.[330] Non-alkaloid constituents of the plant include the cyclic-monoterpene isoflavones, ficusins A and B,[332] the acetophenones 4-hydroxy-3-methoxyacetophenone, 3,4,5-trimethoxyacetophenone and pungenin, the triterpenoids: α-amyrin acetate and β-amyrin acetate, the steroids β-sitosterol and stigmasterol.[330,333]

The medicinal properties of the plant are most probably explainable by the presence of phenanthroindolizidines, which are known elicit antifungal, antibacterial, and anti-inflammatory activities.[330,334,335] Note that antofine and ficuseptine displayed antibacterial activity against *Escherichia coli* and *Bacillus subtilis*.[331]

Proposed Research Pharmacological study of phenanthroindolizidine derivatives for the treatment of cancer.

Rationale Phenanthroindolizidine alkaloids are remarkably cytotoxic. *Ficus septica* Burm.f (family Moraceae Gaudich.) produces a series of cytotoxic phenanthroindolizidines,[336,337] such as 10S,13aR-tylocrebrine N-oxide (CS 1.78), 10S,13aR-isotylocrebrine N-oxide (CS 1.79) and

■ **CS 1.78** 10S,13aR-Tylocrebrine N-oxide

■ **CS 1.79** 10S,13aR-Isotylocrebrine N-oxide

tylophorine (CS 1.80) which, at 10 μM, abrogated the viability of human epithelial tumor (HONE)-1 and human stomach cancer (NUGC) cells by 92%, 87%, and 80%, respectively, and 94%, 93%, and 85%, respectively.[329] The phenanthroindolizidine alkaloids hypoestestatins 1 and 2 isolated from *Hypoestes verticillaris* (L. f.) Sol. ex Roem. & Schult. (family Acanthaceae Juss.) inhibited mouse leukemia (P388) cells, with ED_{50} values of 10^{-5} μg/mL.[338] Cryptopleurine from *Boehmeria cylindrica* (L.) Sw. (family Urticaceae Juss.) exhibited potent cytotoxic effects against nasopharynx carcinoma cells.[339] (–)-10α-Antofine *N*-oxide (CS 1.81) and (–)-10α,13aR-α-hydroxyantofine *N*-oxide isolated from *Cynanchum vincetoxicum* (L.) Pers. (family Apocynaceae Juss.) inhibited the growth of human cervix carcinoma (KB-3-1) and multidrug-resistant human cervix carcinoma (KB-V1) cells, with IC_{50} values of 0.1 μM.[340] Antofine isolated from *Cynanchum paniculatum* (Bunge) Kitag. (family Apocynaceae Juss.) inhibited the growth of the following cell lines (with IC_{50} values in parentheses): human colorectal carcinoma HCT-116 (6.3 nM), 5-FU resistant human colorectal carcinoma HCT-116 (8.3 nM), human colon cancer HT-29 (10.8 nM), human lung adenocarcinoma epithelial A549 (9.6 nM), human lung adenocarcinoma epithelial A549-PA (8.9 nM), human caucasian bronchioalveolar carcinoma NCI-H358 (7 nM), human breast adenocarcinoma MDA-MB-231 (12.2 nM), human breast carcinoma T-47D (19.7 nM), human gastric cancer SNU-638 (8.9 nM), human fibrosarcoma (HT1080) (7.1 nM) and, human promyelocytic leukemia HL-60 (10.7 nM).[341] Trans-(+)-3,14-α-dihydroxy-4,6,7-trimethoxyphenanthroindolizidine *N*-oxide and (–)-7-hydroxy-2,3,6-trimethoxyphenanthroindolizidine from *Tylophora tanakae* Maxim.

■ **CS 1.80** (–)-Tylophorine

■ **CS 1.81** (–)-10α-Antofine *N*-oxide

■ **CS 1.82** Tylocrebrine

■ **CS 1.83** (−)-Cryptopleurine

ex Franch. & Sav. (family Apocynaceae Juss.) inhibited the growth of human gastric carcinoma (TMK-1) cells, with IC_{50} values equal to 6 ng/mL and 3 ng/mL, respectively.[342] Tylocrebrine abrogated the survival of mouse lymphocytic leukemia (L1210) cells[343] and underwent a phase I clinical trial, which was dropped because of the development of heavy central nervous system sides effects (e.g. ataxia) in patients.[344]

The precise molecular events triggered by phenanthroindolizidine alkaloids to induce cell death are to date unclear but some evidence indicates that DNA synthesis, cyclin D1, cyclin A2, downregulation and nuclear factor kappa-light-chain-enhancer of activated B cells (NF-κB) inhibition are to some extent involved. Tylocrebrine has been shown to inhibit protein and DNA synthesis irreversibly in human epitheloid cervix carcinoma (Hela).[345] Tylophorine and tylocrebrine (CS 1.82) and the phenanthroquinolizidine alkaloid cryptopleurine (CS 1.83) inhibited incorporation of leucine into protein in Ehrlich ascites-tumor cells by 50% at 1×10^{-6} M, 2×10^{-7} M, and 2×10^{-8} M, respectively.[346,347] These nucleic effects may result from enzyme inhibition and/or DNA intercalation. Pergularinine (CS 1.84) and tylophorinidine (CS 1.85) from *Pergularia pallida* (Roxb.) Wight & Arn. (family Apocynaceae Juss.) inhibited the enzymatic activity of thymidylate synthase, with IC_{50} values equal to 40 μM and 45 μM,[348] and of dihydrofolate reductase, with IC_{50} values equal to 40 μM and 32 μM, respectively.[349] Antofine (CS 1.86) (tylophorine B) formed noncovalent intercalative association with DNA at the A–T level in human cervix carcinoma (KB-3-1) and multidrug-resistant human cervix carcinoma (KB-V1) cells and compelled cell death, with IC_{50} values of 16 nM and 14 nM, respectively.[350]

■ **CS 1.84** Pergularinine

■ **CS 1.85** (+)-Tylophorinidine

■ **CS 1.86** (−)-Antofine

■ **CS 1.87** 24a

Benzophenanthridine alkaloids have the capacity to interfere with the transcription of DNA into RNA and therefore also to affect protein synthesis. Tylophorine from *Tylophora indica* (Burm. f.) Merr. (family Apocynaceae Juss.) inhibited the transcription of cyclin A2 and induced G_1 arrest in carcinoma cells;[351] antofine lowered cyclin D1 levels in human colorectal carcinoma (HCT-116) cells by hampering β-catenin/Tcf transcriptional dynamics.[341] Note that tylophorinidine from *Tylophora atrofolliculata* F.P. Metcalf (family Apocynaceae Juss.) inhibited the transcription induced by NF-κB and activator protein-1 (AP-1) in human hepatocellular liver carcinoma (HepG2) cells, with IC_{50} values equal to 10 nM and 6.5 nM, respectively.[352] Wei et al. (2007) synthesized *N*-(2,3-methylenedioxy-6-methoxy-phenanthr-9-ylmethyl)-L-4-piperidinemethanol (24a) (CS 1.87), which reduced the viability of cells from the following cell lines (IC_{50}

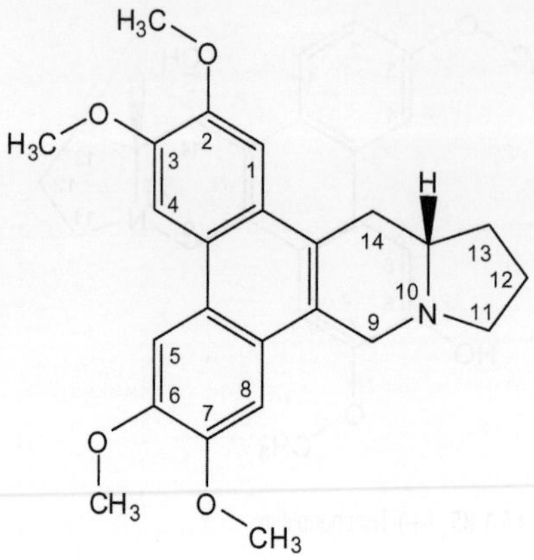

■ **CS 1.88** DCB-3500

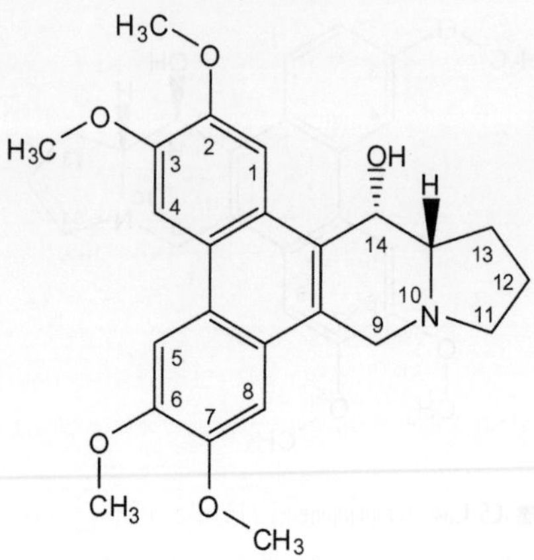

■ **CS 1.89** DCB-3503

values in parentheses): human lung adenocarcinoma epithelial A549 (0.08 μM), human nasopharyngeal carcinoma KB (0.24 μM), vincristine-resistant human nasopharyngeal carcinoma KB-VIN (0.09 μM), human prostate carcinoma DU-145 (0.03 μM) and human breast cancer ZR-751 (0.03 μM).[353] Gao et al. (2004) synthesized series of tylophorine analogs, one of which, DCB-3500 ((+)-S-tylophorine) (CS 1.88), curbed the proliferation of KB and HepG2 cells, with IC_{50} values equal to 12 nM and 11 nM, respectively,[354] via inhibition of NF-κB, with an IC_{50} value of 30 nM. In addition, DCB-3503 (CS 1.89) given to rodents at a dose of 6 mg/kg reduced the growth of tumor xenografts.[354] Phenanthroindolizidine alkaloids pose an interesting but yet unresolved challenge to medicinal chemists, who aim to enhance anticancer activity and mitigate neurotoxicity. In fact, the chemical evidence on the structure–activity relationship of phenanthroindolizidine alkaloid derivatives is somehow incongruous. Staerk et al. (2002) reported that a rigid phenanthrene structure with hydroxyl groups at C6 and C7 is indispensable for cytotoxicity against KB cells,[355] since antofine, 6-O-desmethylantofine (CS 1.90) from *Cynanchum vincetoxicum* (L.) Pers. (family Apocynaceae Juss.) and 7-desmethyltylophorine from *Tylophora tanakae* Maxim. ex Franch. & Sav. (family Apocynaceae Juss.) inhibited the growth of human cervix carcinoma (KB-3-1) (CS 1.91) cells, with IC_{50} values of 16 nM, 7 nM and 15 nM, respectively.[355] Fu et al. (2007) confirmed Staerk and colleagues' hypothesis by showing that the replacement of the 6-methoxy group in antofine by a hydroxyl group leading to 6-O-desmethylantofine boosted the cytocidal effect.[356] However, Wei et al. (2006) noted that a methoxy in C6 and a methylenedioxy ring at the C2 and C3 position, as well as an N-hydrophilic group at C9 and a

■ **CS 1.90** *6-O-Desmethylantofine*

■ **CS 1.91** *7-Desmethyltylophorine*

■ **CS 1.92** *N-(2,3-Methylenedioxy-6-methoxy-phenanthr-9-ylmethyl)-L-2-piperidinemethanol*

■ **CS 1.93** *N-(2,3-Methylenedioxy-6-methoxy-phenanthr-9-ylmethyl)-5-aminopentanol*

rigid and planar phenanthrene, were cardinal features for cytotoxic activity.[357] They produced series of 2,-methylenedioxy phenanthrene derivatives consisting of *N*-(2,3-methylenedioxy-6-methoxy-phenanthr-9-ylmethyl)-L-2-piperidinemethanol and *N*-(2,3-methylenedioxy-6-methoxy-phenanthr-9-ylmethyl)-5-aminopentanol, which compromised the growth of A549 (CS 1.92 and CS 1.93) cells, with IC_{50} values of 0.1 μM and 0.2 μM, respectively.[357] Later, Wei et al. (2007) suggested that a C-9 cyclic piperidine ring with a para terminal hydroxyl was favorable for cytotoxicity.[353] Gao et al. (2007) obtained evidence that the presence of a methoxy or absence of a methoxy at C6 was without significant effect whereas the presence of a hydroxyl group at C2, a methoxy at C3 and a C14-(S) hydroxyl fostered cytotoxicity.[352] The same team found that a hydroxyl group at C14 improved the pharmacokinetiCS of benzophenanthridine alkaloid derivatives but the presence of *N*-oxide led to loss of cytotoxicity.[352] Later, Gao et al. (2008) disproved the hypothesis of Wei et al. (2006) by obtaining measurements showing that replacement of the methoxy groups at C2 and C3 by a cyclic methylenedioxy moiety and opening of the indolizidine ring abated the cytocidal properties of phenanthroindolizidine alkaloids whilst the absence of methoxy at C7 and the presence of a quinolizidine framework increased the cellular toxicity.[358] They synthesized (−)-cryptopleurine, which inhibited the NF-κB pathway in HepG2 cells, with an IC_{50} value equal to 1.4 nM.[358] Su et al. (2008) pointed out that the presence of a methoxy group at C2 increased the cytotoxicity of benzophenanthridine alkaloid derivatives, which was abrogated by planarization of the indolizidine moiety, and they synthesized antofine, which inhibited the growth of human breast adenocarcinoma (MCF-7), A549, KB and −KB-VIN cells, with IC_{50} values of 0.00016 μg/mL, 0.0045 μg/mL, 0.00043 μg/mL, and 0.00083 μg/mL, respectively.[359] Ikeda et al. (2011) made the interesting finding that the side effects could be diminished by removing the hydroxyl group at C14.[360]

REFERENCES

[305]   Nishizawa M, Inoue A, Sastrapradja S, Hayashi Y. (+)-8-Hydroxycalamenene: a fish-poison principle of *Dysoxylum acutangulum* and *D. alliaceum*. Phytochem 1983;22(9):2083–5.

[306]   Ismail IS, Nagakura Y, Hirasawa Y, Hosoya T, Lazim MIM, Lajis NH, et al. Acutaxylines A and B, two novel triterpenes from *Dysoxylum acutangulum*. Tetrahedron Lett 2009;50(34):4830–2.

[307]   Ismail IS, Nagakura Y, Hirasawa Y, Hosoya T, Lazim MIM, Lajis NH, et al. Chrotacumines A-D, chromone alkaloids from *Dysoxylum acutangulum*. J Nat Prod 2009;72(10):1879–83.

[308]   Carlson BA, Dubay. MM, Sausville EA, Brizuela L, Worland PJ. Flavopiridol induces G1 arrest with inhibition of cyclin-dependent kinase (CDK) 2 and CDK4 in human breast carcinoma cells. Cancer Res 1996;56:2973–8.

[309]   König A, Schwartz GK, Mohammad RM, Al-Katib A, Gabrilove JL. The novel cyclin-dependent kinase inhibitor flavopiridol downregulates anti-apoptotic Bcl-2 protein and induces growth arrest and apoptosis in chronic B-cell leukemia lines. Blood 1997;90(11):4307–12.

[310]   Kitada S, Zapata JM, Andreeff M, Reed JC. Protein kinase inhibitors flavopiridol and 7-hydroxy-staurosporine down- regulate antiapoptosis proteins in B-cell chronic lymphocytic leukemia. Blood 2000;96(2):393–7.

[311]   Senderowicz AM. The cell cycle as a target for cancer therapy: basic and clinical findings with the small molecule inhibitors flavopiridol and UCN-01. Oncologist 2002;7(3):12–19.

[312] Chao SH, Fujinaga K, Marion JE, Taube R, Sausville EA, Senderwicz AM, et al. Flavopiridol inhibits P-TEFb and blocks HIV-1 replication. J Biol Chem 2000;275:28345–28348.

[313] Wittmann S, Bali P, Donapaty S, Nimmanapalli R, Guo F, Yamaguchi H, et al. Flavopiridol down-regulates antiapoptotic proteins and sensitizes human breast cancer cells to epothilone B-induced apoptosis. Cancer Res 2003;63(1):93–9.

[314] Yu C, Rahmani M, Dai Y, Conrad D, Krystal G, Dent P, et al. The lethal effects of pharmacological cyclin-dependent kinase inhibitors in human leukemia cells proceed through a phosphatidylinositol 3-kinase/Akt-dependent process. Cancer Res 2003;63(8):1822–33.

[315] Wall NR, O'Connor DS, Plescia J, Pommier Y, Altieri DC. Suppression of survivin phosphorylation on Thr34 by flavopiridol enhances tumor cell apoptosis. Cancer Res 2003;63(1):230–5.

[316] Sausville EA. Complexities in the development of cyclin-dependent kinase inhibitor drugs. Trends Mol Med 2002;8:S32–7.

[317] Andera L, Wasylyk B. Transcription abnormalities potentiate apoptosis of normal human fibroblasts. Mol Med 1997;3(12):852–63.

[318] Pepper C, Thomas A, Fegan C, Hoy T, Bentley P. Flavopiridol induces apoptosis in B-cell chronic lymphocytic leukaemia cells through a p38 and ERK MAP kinase-dependent mechanism. Leukemia Lymphoma 2003;44(2):337–42.

[319] Litz J, Carlson P, Warshamana-Greene GS, Grant S, Krystal GW. Flavopiridol potently induces small cell lung cancer apoptosis during S phase in a manner that involves early mitochondrial dysfunction. Clin Cancer Res 2003;9(12):4586–94.

[320] Takada Y, Aggarwal BB. Flavopiridol inhibits NF-κB activation induced by various carcinogens and inflammatory agents through inhibition of IκBα kinase and p65 phosphorylation: abrogation of cyclin D1, cyclooxygenase-2 and matrix metalloprotease-9. J Biol Chem 2004;279:4750–9.

[321] Blagosklonny MV. Flavopiridol, an inhibitor of transcription: implications, problems and solutions. Cell Cycle 2004;3(12):1537–42.

[322] Beutler JA, Cardellina II JH, McMahon JB, Boyd MR, Cragg GM. Anti-HIV and cytotoxic alkaloids from *Buchenavia capitata*. J Nat Prod 1992;55(2):207–13.

[323] Lee MW, Bach JH, Lee HJ, Lee DY, Joo WS, Kim YS, et al. The activation of ERK1/2 via a tyrosine kinase pathway attenuates trail-induced apoptosis in HeLa cells. Cancer Invest 2005;23(7):586–92.

[324] Murthi KK, Dubay M, McClure C, Brizuela L, Boisclair MD, Worland PJ, et al. Structure–activity relationship studies of flavopiridol analogues. Bioorg Med Chem Lett 2000;10(10):1037–41.

[325] Zhang S, Ma J, Bao Y, Yang P, Zou L, Li K, et al. Nitrogen-containing flavonoid analogues as CDK1/cyclin B inhibitors: synthesis, SAR analysis, and biological activity. Bioorg Med Chem 2008;16(15):7128–33.

[326] Bandgar BP, Totre JV, Gawande SS, Khobragade CN, Warangkar SC, Kadam PD. Synthesis of novel 3,5-diaryl pyrazole derivatives using combinatorial chemistry as inhibitors of tyrosinase as well as potent anticancer, anti-inflammatory agents. Bioorg Med Chem 2010;18(16):6149–55.

[327] Russel JH. Alkaloids of Ficus species. Occurrence of indolizidine alkaloids in *Ficus septica*. Die Naturwissenschaften 1963;50(12):443–4.

[328] Herbert RB, Moody CJ. Alkaloids of *Ficus septica*. Phytochem 1972;11(3):1184.

[329] Damu AG, Kuo PC, Shi LS, Li CY, Kuoh CS, Wu PL, et al. Phenanthroindolizidine alkaloids from the stems of *Ficus septica*. J Nat Prod 2005;68(7):1071–5.

[330] Ueda JY, Takagi M, Shinya K. Aminocaprophenone- and pyrrolidine-type alkaloids from the leaves of *Ficus septica*. J Nat Prod 2009;72(12):2181–3.

[331] Baumgartner B, Erdelmeier CAJ, Wright AD, Rali T, Sticher O. An antimicrobial alkaloid from *Ficus septica*. Phytochem 1990;29(10):3327–30.

[332] Aida M, Hano Y, Nomura T. Ficusins A and B, two new cyclic-monoterpene-substituted isoflavones from *Ficus septica* Burm. f. Heterocycles 1995;41(12):2761–8.

[333] Tsai I-L, Chen J-H, Duh C-Y, Chen I-S. Chemical constituents from the leaves of Formosan *Ficus septica*. Chin Pharm J 2000;52(4):195–201.

[334] Rao KN, Venkatachalam SR. Inhibition of dihydrofolate reductase and cell growth activity by the phenanthroindolizidine alkaloids pergularinine and tylophorinidine: the in vitro cytotoxicity of these plant alkaloids and their potential as antimicrobial and anticancer agents. Toxicol In Vitro 2000;14(1):53–9.

[335] Yang CW, Chen WL, Wu PL, Tseng HY, Lee SJ. Anti-inflammatory mechanisms of phenanthroindolizidine alkaloids. Mol Pharmacol 2006;69(3):749–58.

[336] Wu P-L, Rao KV, Su C-H, Kuoh C-S, Wu T-S. Phenanthroindolizidine alkaloids and their cytotoxicity from the leaves of *Ficus septica*. Heterocycles 2002;57(12):2401–8.

[337] Damu AF, Kuo PC, Shi LS, Li CY, Su CR, Wu TS. Cytotoxic phenanthroindolizidine alkaloids from the roots of *Ficus septica*. Planta Med 2009;75(10):1152–6.

[338] Pettit GR, Goswami A, Cragg GM, Schmidt JM, Zou JC. Antineoplastic agents, 103. The isolation and structure of hypoestestatins 1 and 2 from the East African *Hypoöstes verticillaris*. J Nat Prod 1984;47(6):913–9.

[339] Farnsworth NR, Hart NK, Johns SR, Lamberton JA, Messmer W. Alkaloids of *Boehmeria cylindrica* (family Urticaceae): identification of a cytotoxic agent, highly active against Eagle's 9 KB carcinoma of the nasopharynx in cell culture, as cryptopleurine. Aust J Chem 1969;22:1805–7.

[340] Stærk D, Christensen J, Lemmich E, Duus JØ, Olsen CE, Jaroszewski JW. Cytotoxic activity of some phenanthroindolizidine *N*-oxide alkaloids from *Cynanchum vincetoxicum*. J Nat Prod 2000;63(11):1584–6.

[341] Min HY, Chung HJ, Kim EH, Kim S, Park EJ, Lee SK. Inhibition of cell growth and potentiation of tumor necrosis factor-α(TNF-α)-induced apoptosis by a phenanthroindolizidine alkaloid antofine in human colon cancer cells. Biochem Pharmacol 2010;80(9):1356–64.

[342] Ohyama M, Komatsu H, Watanabe M, Enya T, Koyama K, Sugimura T, et al. Cytotoxic phenanthroindolizidine alkaloids in *Tylophora tanakae* against human gastric carcinoma cells. Proc Jpn Acad Ser B 2000;76(10):161–5.

[343] Gellert E, Rudzats R. Antileukemia activity of tylocrebrine. J Med Chem 1964;7(1964):361–2.

[344] Suffness M, Douros J.. In: Anticancer agents based on natural product models. London: Academic Press; 1980. 465–487

[345] Huang MT, Grollman AP. Mode of action of tylocrebrine: effects on protein and nucleic acid synthesis. Mol Pharmacol 1972;8(5):538–50.

[346] Donaldson GR, Atkinson MR, Murray AW. Inhibition of protein synthesis in Ehrlich ascites-tumor cells by the phenanthrene alkaloids tylophorine, tylocrebrine and cryptopleurine. Biochem Biophys Res Comm 1968;31(1):104–9.

[347] Grant P, Sanchez L, Jimtnez A. Cryptopleurine resistance: genetic locus for a 40s ribosomal component in *Saccharomyces cerevisiae*. J Bacteriol 1974;120:1308–14.

[348] Rao KN, Bhattacharya RK, Venkatachalam SR. Inhibition of thymidylate synthase and cell growth by the phenanthroindolizidine alkaloids pergularinine and tylophorinidine. Chem Biol Interact 1997;106(3):201–12.

[349] Rao KN, Venkatachalam SR. Inhibition of dihydrofolate reductase and cell growth activity by the phenanthroindolizidine alkaloids pergularinine and tylophorinidine: the in vitro cytotoxicity of these plant alkaloids and their potential as antimicrobial and anticancer agents. Toxicol in vitro 2000;14(1):53–9.

[350] Xi Z, Zhang R, Yu Z, Ouyang D, Huang R. Selective interaction between tylophorine B and bulged DNA. Bioorg Med Chem Lett 2005;15(10):2673–7.

[351] Wu CM, Yang CW, Lee YZ, Chuang TH, Wu PL, Chao YS, et al. Tylophorine arrests carcinoma cells at G1 phase by downregulating cyclin A2 expression. Biochem Biophys Res Comm 2009;386(1):140–5.

[352] Gao W, Bussom S, Grill SP, Gullen EA, Hu YC, Huang X, et al. Structure-activity studies of phenanthroindolizidine alkaloids as potential antitumor agents. Bioorg Med Chem Lett 2007;17(15):4338–42.

[353] Wei L, Shi Q, Bastow KF, Brossi A, Morris-Natschke SL, Nakagawa-Goto K, et al. Antitumor agents 253. Design, synthesis, and antitumor evaluation of novel 9-substituted phenanthrene-based tylophorine derivatives as potential anticancer agents. J Med Chem 2007;50(15):3674–80.

[354] Gao W, Lam W, Zhong S, Kaczmarek C, Baker DC, Cheng YC. Novel mode of action of tylophorine analogs as antitumor compounds. Cancer Res 2004;64(2):678–88.

[355] Stærk D, Lykkeberg AK, Christensen J, Budnik BA, Abe F, Jaroszewski JW. In vitro cytotoxic activity of phenanthroindolizidine alkaloids from *Cynanchum vincetoxicum* and *Tylophora tanakae* against drug-sensitive and multidrug-resistant cancer cells. J Nat Prod 2002;65(9):1299–302.

[356] Fu Y, Lee SK, Min HY, Lee T, Lee J, Cheng M, et al. Synthesis and structure-activity studies of antofine analogues as potential anticancer agents. Bioorg Med Chem Lett 2007;17(1):97–100.

[357] Wei L, Brossi A, Kendall R, Bastow KF, Morris-Natschke SL, Shi Q, et al. Antitumor agents 251. Synthesis, cytotoxic evaluation, and structure-activity relationship studies of phenanthrene-based tylophorine derivatives (PBTs) as a new class of anticancer agents. Bioorg Med Chem 2006;14:6560–9.

[358] Gao W, Chen AP, Leung CH, Gullen EA, Fürstner A, Shi Q, et al. Structural analogs of tylophora alkaloids may not be functional analogs. Bioorg Med Chem Lett 2008;18(2):704–9.

[359] Su CR, Damu AG, Chiang PC, Bastow KF, Morris-Natschke SL, Lee KH, et al. Total synthesis of phenanthroindolizidine alkaloids (±)-antofine, (±)-deoxypergularinine, and their dehydro congeners and evaluation of their cytotoxic activity. Bioorg Med Chem 2008;16(11):6233–41.

[360] Ikeda T, Yaegashi T, Matsuzaki T, Yamazaki R, Hashimoto S, Sawada S. Synthesis of phenanthroindolizidine alkaloids and evaluation of their antitumor activities and toxicities. Bioorg Med Chem Lett 2011;21(19):5978–81.

Chapter 2

Terpenes

■ INTRODUCTION

A crucial event in programmed cancer cell death is the presence of cytochrome c in the cytoplasm following mitochondrial insult by Bcl-2-associated X protein (Bax) or truncated BH3 interacting domain death agonist (tBid). In 'wild-type' p53 cancer cells terpenes induce the activation of Bax by stimulating p53, as observed with tanshinone IIA from *Salvia miltiorrhiza* Bunge (2.2.1), oridonin from *Isodon rubescens* (Hemsl.) H. Hara (2.2.4) and ganoderic acid T from *Ganoderma lucidum* (Curtis) P. Karst. (2.3.2). However, since most malignancies are p53 mutant, the ability of terpenes to induce Bax mitochondrial insults in these cells is therefore of tremendous oncological interest. Such p53-independent mechanisms are exemplified with poricotriol A from *Poria cocos* F.A. Wolf (2.3.2), alisol B acetate from *Alisma orientale* (Sam.) Juz. (2.3.1), cucurbitacin E from *Cucurbita pepo* L. (2.3.3), limonin from *Citrus reticulata* Blanco. (2.3.5), obacunone from *Citrus aurantium* L. (2.3.5), and maslinic acid from *Olea europaea* L. (2.3.7).

In cancer cells with mutated p53, Bax activation is often the result of activation of caspase 8, which cleaves pro-apoptotic BH3 interacting domain death agonist (Bid) into tBid, which translocates into the mitochondria where it interacts with Bax, as observed with yomogin from *Artemisia princeps* Pamp. (2.1.1) and myriadenolide from *Alomia myriadenia* Sch. Bip. ex Baker. (2.2.3). In 'wild-type' p53 malignancies, the pro-apoptotic protein (p53) is inhibited by protein kinase B (Akt), which has an immense influence on cancer cell survival. Interestingly, terpenes such as atractylenolide II from *Atractylodes macrocephala* Koidz. (2.1.1), cacalol from *Parasenecio delphiniphyllus* (H. Lév.) Y.L. Chen (2.1.5), polyporenic acid C from *Poria cocos* F.A. Wolf (2.3.2) and alisol B acetate from *Alisma orientale* (Sam.) Juz. (2.3.1) lowered the phosphorylation and therefore the activation of Akt. The inhibition of Akt results in myriad pro-apoptotic events, such as the inhibition of Raf kinases, mitogen-activated protein kinase (MEK) and extracellular-signal-regulated kinases (ERK), as provoked by the terpenes β-eudesmol (2.1.1), atractylenolide I and II from

Christophe Wiart: Lead Compounds from Medicinal Plants for the Treatment of Cancer
DOI: http: //dx.doi.org/10.1016/B978-0-12-398371-8.00002-7

Atractylodes macrocephala Koidz. (2.1.1), oridonin from *Isodon rubescens* (Hemsl.) H. Hara (2.2.4) and cucurbitacin B from *Marah oregana* (Torr. & A. Gray) Howell (2.3.3). Both Akt and Raf are stimulated by the G protein Ras upon binding of growth factor to tyrosine kinase receptor as part of the 'extrinsic' pathway. In fact, oridonin (2.2.4) reduced epidermal growth factor receptor (EGFR) tyrosine phosphorylation and therefore inhibition of Ras, Raf and ERK. Dehydrotrametenolic acid from *Poria cocos* F.A. Wolf (family Polyporaceae Corda) (2.3.2) inhibited G protein Ras and blocked both the Ras/Raf/MEK/ERK and Ras/phosphatidylinositol 3-kinase (PI3K)/ phosphatase and tensin homolog (PTEN)/Akt/mammalian target of rapamycin (mTOR) pathways. Lupeol from *Euphorbia fischeriana* Steud. (2.2.2) inhibits Ras and therefore the Ras/ERK and the Ras/PI3K/Akt pathways.

Further compelling oncological evidence about terpenes is that they impose apoptosis in mutant p53 and p53 null malignancies by inhibiting Janus kinase 2 (JAK2) and/or signal transducer and activator of transcription 3 (STAT3) and therefore hypotranscription of anti-apoptotic Bcl-2, Bcl-xL, myeloid leukemia cell differentiation protein (Mcl-1) and proproliferative cyclin D1, as observed with acetyl-11-keto-β-boswellic acid from *Boswellia serrata* Roxb. ex Colebr. (2.3.7), jolkinolide B from *Euphorbia fischeriana* Steud. (2.2.2), cucurbitacin B from *Marah oregana* (Torr. & A. Gray) Howell (2.3.3), cucurbitacin E from *Cucurbita pepo* L. (2.3.3) and cucurbitacin I from *Bryonia alba* L. (2.3.3).

Another cellular mechanism by which terpenes induce apoptosis is the blockade of target nuclear factor kappa-light-chain-enhancer of activated B cells (NF-κB) and therefore the transcription of anti-apoptotic Bcl-2, Bcl-xL, survivin, anti-apoptotic induced myeloid leukemia cell differentiation protein (Mcl-1), c-Myc and cyclin D1, observed with coronarin D (2.2.3), oridonin (2.2.4), nimbolide from *Azadirachta indica* A. Juss (2.3.5), betulinic acid from *Pulsatilla chinensis* (Bunge) Regel (2.3.6), celastrol isolated from members of the genus *Kokoona* Thwaites (2.3.7) and pristimerin members of the genus *Maytenus* Molina (2.3.7). Terpenes compromise the cell cycle in several types of malignancies by targeting the mitotic spindle and cyclins. For instance, pristimerin, 22β-hydroxytingenone and tingenone from the *Maytenus* Molina (2.3.7) and celastrol from the genus *Kokoona* Thwaites (2.3.7) inhibit the polymerization of tubulin into microtubules and therefore abrogate the division of cancer cells. Carnosol from *Rosmarinus officinalis* L (2.2.1) and ferruginol from *Persea nubigena* L.O. Williams (2.2.1) inhibit cyclin-dependent kinase 2 (CDK2), 23,24-dihydrocucurbitacin B (2.3.3) inhibits cyclin B and betulinic acid from *Pulsatilla chinensis* (Bunge) Regel (2.3.6) induces the degradation of cyclin D1.

DNA itself is a pro-apoptotic target for terpenes which inhibit the enzymatic activity of topoisomerase II, such as ganoderic acids X and T (2.3.2), lupeol from *Euphorbia fischeriana* Steud. (2.2.2) and betulin from *Pulsatilla chinensis* (Bunge) Regel (2.3.6), and boost the generation of highly reactive and deleterious oxygen species (ROS) in the cytoplasm of cancer cells, as commanded by δ-elemene from *Curcuma wenyujin* Y.H. Chen & C. Ling (2.1.2), atractylenolide I from (2.1.1), parthenolide *Tanacetum parthenium* (L.) Sch. Bip. (2.1.3), oridonin from *Isodon rubescens* (Hemsl.) H. Hara (2.2.4), cucurbitacin B from *Marah oregana* (Torr. & A. Gray) Howell (2.3.3), toosendanin from *Melia toosendan* Siebold & Zucc. (2.3.5) and betulinic acid *Pulsatilla chinensis* (Bge) Regel (2.3.6). Additionally, limonin from both *Citrus reticulata* Blanco. (2.3.5) and *Melia toosendan* Siebold & Zucc. (2.3.5) increases the cytoplasmic levels of Ca^{2+}, as do thapsigargin from *Thapsia garganica* L. (2.1.4) and alisol B (2.3.1). This effect results from the inhibition of sarco-endoplasmic reticulum Ca^{2+}-ATPases (SERCAs). Increase of cytoplasmic levels of Ca^{2+} triggers calmodulin-dependent cytocidal mechanisms. In this chapter atractylenolide II,β-elemene, inulacappolide, chinensiolide B, cacalol, tanshinone IIA,17-acetoxyjolkinolide B, rotundifuran, oridonin, alisol B, schisandronic acid, cucurbitacin E, brucein D, limonin, lup-28-al-20(29)-en-3-one and tingenone, and their derivatives, are identified as lead terpenes for the treatment cancer.

Topic **2.1**

Sesquiterpenes

2.1.1 *Atractylodes Macrocephala* **Koidz.**

History This plant was first described by Gen'ichi Koidzumi in *Florae Symbolae Orientali-Asiaticae*, published in 1930.

Synonyms *Atractylis lancea* var. *chinensis* (Bunge) Kitam., *Atractylis macrocephala* (Koidz.) Hand.-Mazz., *Atractylis macrocephala* (Koidz.) Nemoto, *Atractylis macrocephala* var. *hunanensis* Y. Ling

Family Asteraceae Bercht. & J. Presl, 1820

Common Names Largehead actractylodes, so jutsu (Japanese), bai shu (Chinese)

Habitat and Description Actractylodes is a rhizomatous herb that grows to 60 cm tall in the grasslands of China. The rhizome is anthropomorphic. The leaves are simple. The petiole is 3–5 cm long. The leaf blade is oblanceolate, spinulose at margin, 4.5–7.5 cm × 1.5–2 cm. The inflorescence is a massive capitulum. The corolla is purple and 1.5 cm long. The achenes are obconic, ovoid and 0.7.5 cm long. The pappus is whitish and 1.5 cm long and develops a series of plumose bristles which are arranged in one row and basally connate into a ring (Figure 2.1).

Medicinal Uses In China the plant is used to promote digestion, spleen function and diuresis and to treat diarrhea, dropsy, anxiety, eczema and cancer.

Phytopharmacology The plant is known to contain: the sesquiterpenes biatractylolide,[1] atractylenolide I,[2] atractylenolides II and III,[2,3]

■ **FIGURE 2.1** *Atractylodes macrocephala* Koidz.

atractylenolides V, VI and VII;[4] the alkaloid 14-acetoxy-12-senecioy-loxytetradeca-2E,8E,10E-trien-4,6-diyn-1-ol;[2,3] polyacetylenes including esters of tetradeca-2,8,10-triene-4,6-diyne-1,12,14-triol,[5] 14-acetoxy-12-α-methylbutyl-2E,8E,10E-trien-4,6-diyn-1-oland14-acetoxy-12-β-methylbutyl-2E,8E,10E-trien-4,6-diyn-1-ol,[6] and flavonoids such as apigenin and luteolin.[7]

To date the medicinal uses of the plants are not substantiated. Note that several anti-inflammatory principles have been identified, such as atrac-tylenolides I and III.[2,3,6]

Proposed Research Pharmacological study of atractylenolide II and derivatives for the treatment cancer.

Rationale The eudesmane sesquiterpene γ-eudesmol (CS 2.1) from *Cananga odorata* (Lam.) Hook. f. & Thomson (family Annonaceae, Juss.) inhibited the growth of human hepatocellular liver carcinoma (HepG2) cells with an IC$_{50}$ value of 1.5 μg/mL.[8]

β-Eudesmol suppressed the proliferation of porcine brain microvascu-lar endothelial cells (with an IC$_{50}$ value of 53.3 μM), which limited the incorporation of bromodeoxyuridine into DNA and reduced extracel-lular signal-regulated kinase (ERK),[9] thereby effecting the inhibition of the E twenty-six (ETS)-like transcription factor 1 (ELK1), STAT3 and the nuclear transcription factor cAMP response element binding protein (CREB). This was further confirmed by the fact that β-eudesmol (CS 2.2) blocked the activation of CREB in human umbilical vein endothelial (HUVEC) cells at a dose of 100 μM.[10] Such a mechanism may account for the cytotoxic effect of pumilaside A (CS 2.3), a eudesmane sesquiter-pene glycoside isolated from the seeds of *Litchi chinensis* Sonn. (family Sapindaceae, Juss.), which inhibited the survival of human lung adenocar-cinoma epithelial (A549), human hepatocellular liver carcinoma (HepG2) and human epitheloid cervix carcinoma (Hela) cells, with IC$_{50}$ values equal to 6.2 μM, 0.01 μM and 0.01 μM, respectively.[11]

The furano-eudesmane diterpenes 1β,8β-dihydroxyeudesm-3,7(11)-dien-8α,12-olide, 1β,8β-dihydroxyeudesm-4(15), 7(11)-dien-8α,12-olide, 1β-acetoxyeudesman-4(15), 7(11)-dien-8α,12-olide and 1β-acetoxy, 8β-hydroxyeudesman-4(15) (CS 2.4–2.7) and 7(11)-dien-8α,12-olide from *Smyrnium olusatrum* L. (family Apiaceae, Lindl.) negated the growth of mouse leukemia (P388) cells, with IC$_{50}$ values equal to 60 μg/mL, 65 μg/mL, 42 μg/mL and 58 μg/mL,[12] respectively, by a mechanism which may imply the inactivation of protein kinase B (Akt). Wang et al. (2002)

■ **CS 2.1** γ-Eudesmol

■ **CS 2.2** β-Eudesmol

■ **CS 2.3** Pumilaside A

■ **CS 2.4** 1β,8β-Dihydroxyeudesm-3,7(11)-dien-8α,12-olide

■ **CS 2.5** 1β,8β-Dihydroxyeudesm-4(15),7(11)-dien-8α,12-olide

■ **CS 2.6** 1β-Acetoxyeudesman-4(15),7(11)-dien-8αα,12-olide

■ **CS 2.7** 1β-Acetoxy,8β-hydroxyeudesman-4(15),7(11)-dien-8α,12-olide

■ **CS 2.8** Atractylenolide I

■ **CS 2.9** Atractylenolide III

■ **CS 2.10** Atractylenolide II

observed that the furanone-eudesmane sesquiterpene atractylenolide I at 30 μg/mL induced apoptosis in human promyelocytic leukemia (HL-60) and mouse leukemia (P388) cells.[13] Atractylenolide I (CS 2.8) abrogated the survival of HL-60 cells, with an IC_{50} of 10.6 μg/mL, together with an increase of reactive oxygen species (ROS), whereas atractylenolide III (CS 2.9) did not, suggesting that the double bond between C8 and C9 is crucial for activity.[14] However, atractylenolide III reduced the viability of A549 cells to 65% at a dose of 100 μM, with caspases 9 and 3 activation and cleavage of PARP (Poly [ADP-ribose] polymerase).[15] Atractylenolides I and II (CS 2.10) inhibited the growth of mouse melanoma (B16) cells with IC_{50} values equal to 76.4 μM and 84 μM via decrease in ERK and inactivation of Akt.[16] In fact, inactivation of Akt compels the upregulation of p21 and p27, which inhibit CDK2 and thereby cell division.[17,18] In addition, the inhibition of Akt by atractylenolide II was followed by increased levels of pro-apoptotic protein (p53) via probable inhibition of murine double minute (MDM2) protein and therefore elevation of anti-apoptotic Bcl-2 protein.[19] Apoptosis in B16 cells provoked by atractylenolide II involved caspases 3, 8 and 9 activation and increase of the pro-apoptotic mitogen-activated protein kinase (MAPK) p38[19] on probable account of Akt downregulation.

The dihydrofurano-eudesmane sesquiterpene encelin isolated from *Saussurea parviflora* (Poir.) DC. (family Asteraceae, Bercht. & J. Presl) abated the survival of human hepatic (L02), human hepatoma (SMMC-7721) and human epithelial ovarian cancer (HO-8910) cells, with IC_{50} values equal to 1.4 μg/mL, 0.5 μg/mL and 0.8 μg/mL, respectively.[20] Eudesmane sesquiterpenes with a reduced furanone ring do not appear to act with Akt but induce apoptosis via activation of caspase 8 and mitochondrial insult.

■ **CS 2.11** Yomogin

■ **CS 2.12** Gerin

■ **CS 2.13** 7*a*-Hydroxygerin

One such sesquiterpene is yomogin (CS 2.11) from *Artemisia princeps* Pamp. (family Asteraceae, Bercht. & J. Presl), which compelled the apoptosis of mouse leukemia (P388), mouse lymphocytic leukemia (L1210), human histiocytic lymphoma (U937) and HL-60 cells, with IC_{50} values equal to 18.5 µM, 16.3 µM, 8.3 µM and 4.2 µM, respectively, by a mechanism involving caspase 8-dependent, pro-apoptotic Bid and pro-apoptotic Bax translocation to the mitochondria followed by the release of cytochrome c. The latter binds to apoptotic peptidase activating factor 1 (Apaf-1) and pro-caspase 9 in the presence of ATP to form the apoptosome that activates caspase 9 and, therefore, caspase 3.[21] Note that the opening of the dihydrofuranone ring reduces slightly the apoptotic property of eudesmane sesquiterpenes: the eudesmane sesquiterpenes gerin (CS 2.12), 7*a*-hydroxygerin (CS 2.13), 8-*O*-deacetylgerin (CS 2.14) and encelin (CS 2.15) from *Saussurea cauloptera* Hand.-Mazz. (family Asteraceae, Bercht. & J. Presl) exhibited some cytocidal activities against human gastric cancer (SGC-7901) cells, with IC_{50} values equal to 35.2 µM, 42.3 µM, 22.5 µM and 14.1 µM, respectively.[22]

■ **CS 2.14** 8-*O*-Deacetylgerin

■ **CS 2.15** Encelin

2.1.2 *Curcuma Wenyujin* Y.H. Chen & C. Ling

History The plant was first described by Yao Hua Chen in *Acta Pharmaceutica Sinica*, published in 1981.

Family Zingiberaceae Martinov, 1820

Common Name Wen yu jin (Chinese)

Habitat and Description *Curcuma wenyujin* is a ginger which grows to 1.5 m tall in China. The rhizome is fleshy, off-white and ovoid. The leaves are simple and basal. The petiole is 30 cm long. The blade is oblong 35–70 cm × 15–20 cm, cuneate at base and acute at apex. The inflorescence is a terminal spike that emerges from separate shoots developing from the

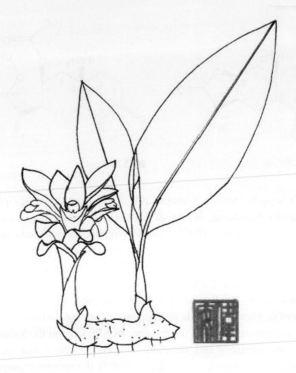

■ **FIGURE 2.2** *Curcuma wenyujin* Y.H. Chen & C. Ling

rhizome. The spike is 25 cm long and produces several ovate bracts which are 3–10 cm long and reddish green. The calyx is tubular, splitting on one side, white and 1 cm long. The corolla is infundibuliform, 2.5 cm long and hairy inside. The lateral staminodes are yellow, oblong and 1.5 cm long. The labellum is reflexed, yellowish, ovate, emarginate and 2 cm long. The filament is short, thick and bears a spurred and versatile anther. The gynaecium is trilocular and hairy. The fruit is a three-valved dehiscent capsule (Figure 2.2).

Medicinal Uses In China the plant is used to treat cancer of the uterus, jaundice and epilepsy, to regulate menses, to check blood in the urine and to alleviate stomachache.

Phytopharmacology The plant hosts myriad sesquiterpenes including germacrane sesquiterpenes: curdione, neocurdione, (1S,10S),(4S,5S)-germacrone-1(10),4-diepoxide, (1R,10R)-(−)-1,10-dihydrocurdione,[23] (4S, 5S)-(+)-germacrone-4.5-epoxide,[24] (+)-(4S,5S)-germacrone-4,5-epoxide,

(+)-(1S,4S,5S,10S)-germacrone-1(10)-4-diepoxide, (1E,4E,8R)-8-hydroxy-germacra-1(10),4,7(11)-trieno-12,8-lactone, (1E,4Z)-8-hydroxy-6-oxoger-macra-1(10),4,7(11)-trieno-12,8-lactone, curdionolides A–C,[25] curzere-none, curzerene and germacrone;[26] the elemane sesquiterpenes β-elemene and β-elemenone;[26] the carabrane sesquiterpene curcumenone and (4S)-dihydrocurcumenone;[27] the guaiane sesquiterpenes curcumol,[28] cur-cumenol, 4-*epi*-curcumenol;[29] the eudesmane sesquiterpenes curcolide, curcodione and (Z)-1β,4α-dihydroxy-5α,8β (H)-eudesm-7(11)-en-12,8-olide,[27] the xanthane sesquiterpene curcumadionol,[27] the pimarane cur-cumrinol A and B and the labdane curcumrinol C.[30]

The anticancer property of the plant is mainly owed to the synergistic effects of furanodiene[31] and β-elemene.[32]

Proposed Research Pharmacological study of β-elemene and deriva-tives for the treatment of brain cancer.

Rationale *Curcuma wenyujin* Y.H. Chen & C. Ling contains the elemane sesquiterpene β-elemene which, although structurally unsophisti-cated, provokes powerful apoptosis in a bewildering array of human can-cer cells (IC$_{50}$ values in parentheses): human brain glioblastoma (A-172; 65 µg/mL), human glioblastoma-astrocytoma, epithelial-like (U-MG87; 88 µg/mL), human brain astrocytoma (CCF-STTG1; 82 µg/mL), human colorectal adenocarcinoma (CCL-222; 47 µg/mL), human colon carcin-oma (CCL-25; 67 µg/mL), human pulmonary adenosquamous carcinoma (NCI-H596; 95 µg/mL), human small-cell lung carcinoma (NCI-H69; 52 µg/mL), human epitheloid cervix carcinoma (Hela; 63 µg/mL), human cervical cancer (ME-180; 68 µg/mL), human cervical carcinoma (HTB-33; 68 µg/mL), human breast adenocarcinoma (MCF-7; 93 µg/mL) and human ductal breast epithelial tumour (T47D; 63 µg/mL) cells.[32] The pre-cise apoptotic molecular events triggered by β-elemene are not yet fully understood but several lines of evidence suggest the involvement of DNA damage by ROS. β-Elemene mitigated the survival of human large-cell lung cancer (NCI-H460), human lung adenocarcinoma epithelial (A549), human normal fibroblast-like (CCD-19 Lu) and human bronchial epithelial (NL20) cells, with IC$_{50}$ values equal to 42 µg/mL, 48 µg/mL, 98 µg/mL and 126 µg/mL, respectively, via stimulation of the protein kinase CHK2 (by DNA damage). This results in inhibition of M-phase inducer phosphatase 3 (CDC25C) and inactivation of cyclin-dependent kinase 1 (CDK1), increase in cyclin-dependent kinase inhibitor p27 and hence cyclin-dependent kinase 2 (CDK2)/cyclin E inhibition.[33] Additionally, β-elemene induced the release of cytochrome c, activated caspases 9 and 3 and decreased anti-apoptotic Bcl-2 protein (via DNA damage-induced ATM phosphory-lation of c-Abl).[33] β-Elemene compromised the survival of human pros-tate carcinoma (DU-145) and human prostate cancer (PC-3) cells, with IC_{50} values equal to 75 μg/mL and 105 μg/mL, respectively, owing to the release of cytochrome c, the activation of caspases 9 and 3, PARP (Poly [ADP-ribose] polymerase) cleavage and decrease of anti-apoptotic Bcl-2 protein.[32] N-(β-Elemene-13-yl)tryptophan methyl ester at a concentration of 30 μmol/L significantly decreased the viability in both human promyelo-cytic leukemia (HL-60) and human promyelocytic leukemia (NB4) cells, with an increase of H_2O_2, a fall in mitochondrial membrane potential and the activation of caspase 3,[34] probably via the release of cytochrome c, which binds to Apaf-1, and pro-caspase 9 in the presence of ATP to form the apoptosome which activates caspase 9 and therefore caspase 3. Another molecular sequence involved in β-elemene-induced apoptosis is the inhibi-tion of mammalian target of rapamycin (mTOR), which may result from DNA damage.[35] β-Elemene (CS 2.16) inhibited the viability of human gas-tric carcinoma (MGC803) cells, with an IC_{50} value of 45 μg/mL, accompa-nied by inhibition of mTOR and therefore a fall in downstream p70S6K1 and decreased levels of survivin.[36]

δ-Elemene (CS 2.17) from *Curcuma wenyujin* Y.H. Chen & C. Ling reduced the viability of human colon carcinoma (DLD-1) cells by 50% at a concentration of 222.4 μM via generation of ROS, a fall in mitochon-drial membrane potential and translocation of pro-apoptotic Bax in the mitochondria, which resulted in the release of cytochrome c and apopto-sis inducing factor (AIF), activation of caspase 3 and cleavage of PARP.[37] The IC_{50} value for δ-elemene against human epitheloid cervix carcinoma (Hela) cells was equal to 157.9 μM, with DNA fragmentation, an increase in ROS, a fall in mitochondrial membrane potential, caspase 3 activation and cleavage of PARP.[38] δ-Elemene induced the death of 80% of human mucoepidermoid lung carcinoma (NCI-H292) cells at a dose of 400 μM, with DNA fragmentation, a fall in mitochondrial membrane potential, increased levels of expression of p38 MAPK (and hence activation of inducible nitric oxide synthetase (iNOS)) and a decrease in NF-κB and therefore a decreased level of anti-apoptotic Bcl-2 protein and anti-apop-totic Bcl-xL transcription.[39] Increased levels p38 MAPK may result from the upregulation of MKK3 and MKK6.[40]

Xu et al. (2006) synthesized a series of β-elemene derivatives, namely N-(β-elemene-13-yl)tryptophan methyl ester (CS 2.18), 13,14-bis(cis-3,5-dimethyl-1-piperazinyl)-β-elemene (IIi) (CS 2.19), 13, 14-bis[2-(2-thiophenyl)ethylamino]-β-elemene (IIm) (CS 2.20) and

■ **CS 2.16** β-Elemene

■ **CS 2.17** δ-Elemene

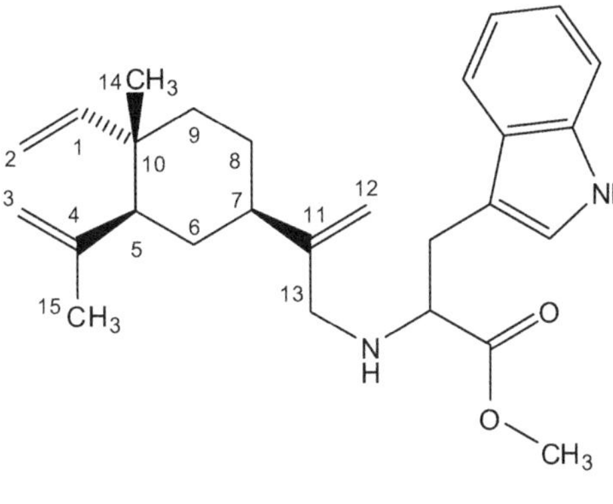

■ **CS 2.18** *N*-(β-elemene-13-yl)tryptophan methyl ester

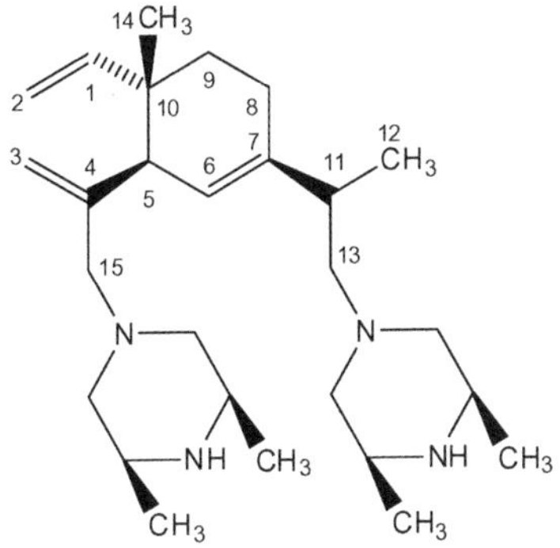

■ **CS 2.19** IIi

13,14-bis(cyclohexamino)-β-elemene (IIn), which inhibited the growth of Hela cells, with IC$_{50}$ values equal to 3.2 μM, 5.9 μM and 8.8 μM compared with 230 μM for β-elemene (by a mechanism implying the inhibition of mTOR).[41]

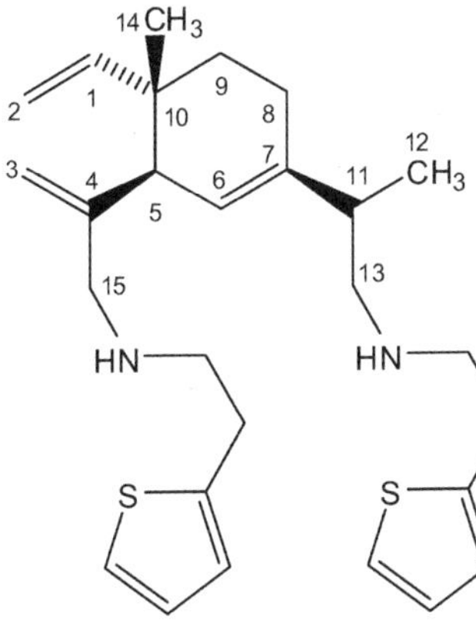

2.1.3 *Duhaldea Cappa* (Buch.Ham. ex D. Don) Pruski & Anderb.

History This plant was first described by John Francis Pruski in *Compositae Newsletter*, published in 2003.

Synonyms *Conyza cappa* Buch.Ham. ex D. Don, *Duhaldea cappa* (DC.) Anderb., *Duhaldea cappa* Anderb., *Duhaldea chinensis* var. *cappa* (Buch. Ham. ex D. Don) Steetz, *Inula cappa* (Buch.Ham. ex D. Don) DC., *Inula cappa* DC., *Inula eriophora* DC., *Inula oblonga* DC., *Inula pseudocappa* DC

Family Asteraceae Bercht. & J. Presl, 1820

Common Names Sheep's ear, yang er ju (Chinese)

Habitat and Description *Duhaldea cappa* is a shrub that grows to 1.5 m tall in the grasslands of China, Vietnam, Thailand, Malaysia, Pakistan, Bhutan, Nepal and India. The root is serpentine. The stems are hairy. The leaves are simple. The petiole is 0.5 cm long. The blade is dull green above, elliptic, 8–20 cm × 2.5–6 cm, coriaceous, rounded at base, serrate, acute at apex and hairy and light green beneath. The inflorescence is a terminal corymb of golden yellow capitula. The ray florets are yellow, 0.5 cm long and tubular. The disk floret corolla is yellow and 0.6 cm long.

■ **CS 2.20** IIm

■ **FIGURE 2.3** *Duhaldea cappa* (Buch.Ham. ex D. Don) Pruski & Anderb.

The achenes are cylindrical, minute, hairy and whitish. The pappus consists of one row of bristles (Figure 2.3).

Medicinal Uses In China the plant is used to treat rheumatism, sore throats, malaria, dysentery and hepatitis.

Phytopharmacology The plant accumulates a plethora of germacranolide sesquiterpenes, including: 2,3-dihydroxy-9-angeloxygermacra-4-en-6,12-olide[42] and inulacappolide;[43] the triterpenes friedelin, epifriedelanol, α-amyrin, β-amyrin, oleanolic acid and ursolic acid;[43] the sterols stigmast-4-en-3-one, stigmasta-4,22-dien-3-one, β-sitosterol, stigmasterol, 7-oxo-β-sitosterol, stigmast-5-ene-3β,7β-diol, stigmasta-5,22-diene-3β,7β-diol, stigmast-5-ene-3β,7α-diol, stigmasta-5,22-diene-3β,7α-diol and daucosterol;[43] and the flavonoids apigenin and luteolin-3′-*O*-β -D-glucopyranoside.[44] The anti-inflammatory property of the plant is most probably owed to germanacrolides.[45]

Proposed Research Pharmacological study of inulacappolide and derivatives for the treatment of cancer.

■ **CS 2.21** Inulacappolide

■ **CS 2.22** Arucanolide

■ **CS 2.23** Parthenolide

Rationale The germacranolide sesquiterpene inulacappolide (CS 2.21) isolated from *Duhaldea cappa* (Buch.Ham. ex D. Don) Pruski & Anderb was cytotoxic against human epitheloid cervix carcinoma (Hela), human erythromyeloblastoid leukemia (K562) and human nasopharyngeal carcinoma (KB) cells, with IC_{50} values equal to $1.2\,\mu M$, $3.8\,\mu M$ and $5.3\,\mu M$, respectively.[43] Current evidence indicates that germacranolide sesquiterpenes induce apoptosis via the disruption of the NF-κB-induced transcription of the anti-apoptotic protein Bcl-2 and hence an increase in pro-apoptotic Bax and caspase 3 activation through mitochondrial insult. The germacranolides arucanolide (CS 2.22) and parthenolide (CS 2.23) isolated from *Calea urticifolia* (Mill.) DC. (family Asteraceae Bercht. & J. Presl) induced human colon adenocarcinoma (SW480) and human promyelocytic leukemia (HL-60) cell death, with DNA fragmentation, a fall in mitochondrial membrane potential and the release of AIF from the mitochondria.[46] Parthenolide, which is abundant in feverfew or *Tanacetum parthenium* (L.) Sch. Bip. (family Asteraceae Bercht. & J. Presl) abrogated the survival of human hepatoma (SH-J1) cells, with an IC_{50} value of $7.5\,\mu M$, by causing a decrease in cellular glutathione and hence an increase in intracellular ROS, a fall in mitochondrial membrane potential, release of cytochrome c and activation of caspase 9 via Apaf-1.[47] In human pancreas adenocarcinoma (PC-3) cells, parthenolide precipitated apoptosis (EC_{50} value of $14.5\,\mu M$), with mitochondrial insult, cytochrome c release and caspases 9 and 3 activation on account of a decreased level of anti-apoptotic Bcl-2 protein and an increased level of pro-apoptotic Bax and the consequent release of cytochrome c from mitochondria.[48] The inhibition of NF-κB by parthenolide may explain the decreased level of anti-apoptotic Bcl-2 protein.[49] In the cytoplasm (NF-κB) (p50/p65(RelA)) is attached to, sequestered and inhibited by nuclear factor of kappa-light-polypeptide-gene-enhancer in B-cells inhibitor α (IκBα). Once the cell is exposed to tumor necrosis factor-α (TNF-α), interleukin 1 (IL-1) or ROS, IκBα is phosphorylated byIκB kinase (IκBα kinase (IKKα) and IκBβ kinase (IKKβ)) and ubiquinated and destroyed by 26S proteasome, allowing NF-κB to enter the nucleus, bind to its DNA site and induce the synthesis of proteins in charge of cell survival, notably the anti-apoptotic Bcl-2 protein.[50] Bork et al. (1997) first observed that parthenolide inhibited the transcription induced by NF-κB at a dose of 5 μM.[51] Additionally, TNF-α or H_2O_2-induced DNA binding of NF-κB was abolished by 5 μM of parthenolide in mouse fibroblast (L929) cells by blocking the proteolysis of IκBα by the proteasome.[52] In fact, IκB kinase phosphorylation of IκBα is abrogated by parthenolide in human epitheloid cervix carcinoma (Hela) cells[53] on account of the alkylation of cysteine 179 in the activation loop of IKKβ.[54] This alkylation occurs as a result of the lactone ring of parthenolide,[45] which forms covalent modification of the oxomethylen group.[51] Parthenolide alkylates cysteine 38 residues in the DNA-binding domain of p65 and annihilates the binding of NF-κ to the κB-DNA motif.[55,56] The fall in glutathione mentioned earlier might very well involve alkylation of thiol groups. Note that the depletion of IKKβ activity leads to spindle abnormalities.[57]

2.1.4 *Ixeris Chinensis* (Thunb.) Kitag.

History This plant was first described by Masao Kitagawa in *Botanical Magazine*, published in 1934.

Synonym *Prenanthes chinensis* Thunb.

Family Asteraceae Bercht. & J. Presl, 1820

Common Name Zhong hua ku mai cai (Chinese)

Habitat and Description *Ixeris chinensis* is a perennial herb that grows to 45 cm tall in the wastelands of China, Vietnam, Mongolia, Russia, Korea, Japan, Thailand, Cambodia and Laos. The leaves are simple, arranged into a rosette or amplexicaul, elliptic, 5–20 cm × 1–2 cm, entire or pinnatifid and dull light green. The inflorescence is a lax corymb of conspicuous capitula. The involucre is cylindrical, 1 cm long and covered with acute phyllaries. The capitulum bears 20 yellow florets, which are yellow and produce a limb minutely dentate at apex. The achenes are brown, fusiform, 0.5 cm long, have 10 ribs and are beaked at apex. The pappus is white, 0.5 cm long and presents scabrid bristles (Figure 2.4).

Medicinal Uses In Cambodia, Laos and Vietnam the plant is used to make a cooling drink and affords a laxative remedy. In China, the plant is

■ **FIGURE 2.4** *Ixeris chinensis* (Thunb.) Kitag.

used to treat dysentery, infection of the lungs and throat, to break fever and to counteract poisoning.

Phytopharmacology The plant is known to accumulate triterpenes including: 17-epilupenyl acetate,[58] ixerenol,[59] 17-epilupenyl acetate and ixerenyl acetate,[60] 3β,21α-dihydroxylupen-18(19)-en, 3β,25-dihydroxytirucalla-7,23(24)-dien, and 21 α -hydroxy-19 α -hydrogentaraxasterol-20(30)-en,[61] the guaiane sesquiterpene lactones 8-epicrepioside G, 8-epidesacylcynaropicrin glucoside and ixerin D,[62] chinensiolides A, B, C,[63] D and E,[64] ixerochinolide, ixerochinoside, 8β-hydroxydehydrozaluzanin, lactucin, 10α -hydroxy-10,14-dihydro-desacylcynaropicrin,[65] (11*S*)-10-α−hydroxy-3-oxoguaia-4-eno-12,6-α-lactone, and (11*S*)-10-α-hydroxy-3-oxo-4β*H*-guaiano-12,6-α-lactone.[64]

The antidotal property of the plant may involve some hepatoprotective effects which are owed to luteolin-7-glucoside.[66]

Proposed Research Pharmacological study of chinensiolide B and derivatives for the treatment of cancer.

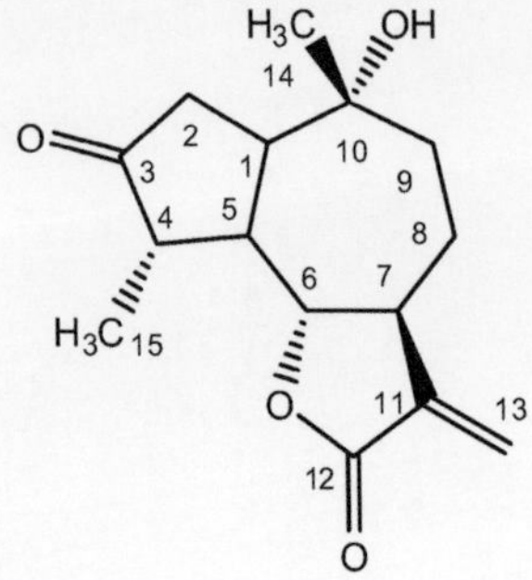

■ **CS 2.24** Chinensiolide B

Rationale There is a massive body of evidence that points to the fact that guaianolides are of chemotherapeutic value against a broad array of malignancies. Chinensiolide B (CS 2.24) from *Ixeris chinensis* (Thunb.) Kitag. and the semisynthetic derivatives 3β,10α-dihydroxy-1α,5α,15α-*H*-guaia-11(13)-en-6α,12-olide (CS 2.25), 3α, 10α-dihydroxy-1α,5α,15α-*H*-guaia-11(13)-en-6α,12-olide (CS 2.26) and 3β,10α-dihydroxy-1α,5α,11β,15α-*H*-guaia-6α,12-olide (CS 2.27) abrogated the survival of human normal lung fibroblast (WI-38) cells, with IC_{50} values of 0.3 μM, 1.6 μM, 0.3 μM and 1.9 μM, respectively,[67] suggesting that the substitution of C11 by a methylene is important for cytotoxicity. Chinenciolide E (CS 2.28) from *Ixeris chinensis* (Thunb.) Kitag. was cytocidal towards human normal lung fibroblast (WI-38), human lung tumor (VA-13) and human hepatocellular liver carcinoma (HepG2) cells, with IC_{50} values of 23 μM, 0.72 μM and 140 μM, respectively,[64] indicating that glycosylation in C3 reduces cytotoxic potencies against normal cells. Eupatorin (CS 2.29)

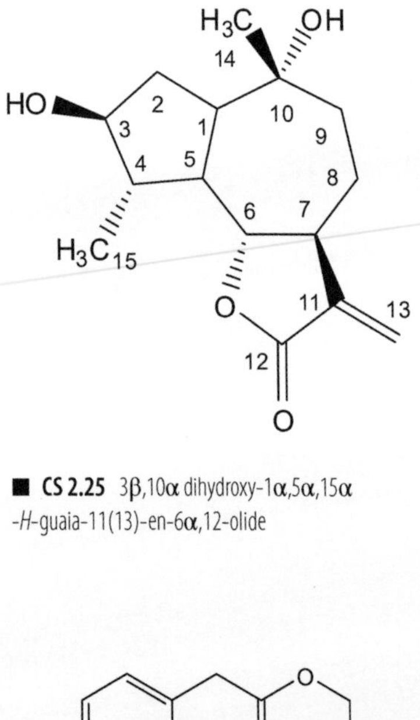

■ **CS 2.25** 3β,10α dihydroxy-1α,5α,15α -*H*-guaia-11(13)-en-6α,12-olide

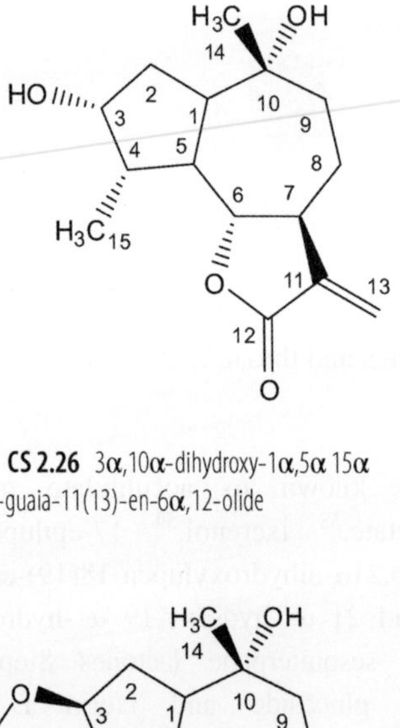

■ **CS 2.26** 3α,10α-dihydroxy-1α,5α 15α -*H*-guaia-11(13)-en-6α,12-olide

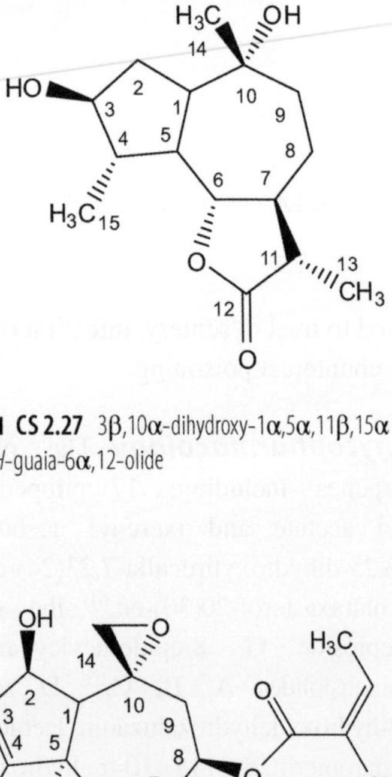

■ **CS 2.27** 3β,10α-dihydroxy-1α,5α,11β,15α -*H*-guaia-6α,12-olide

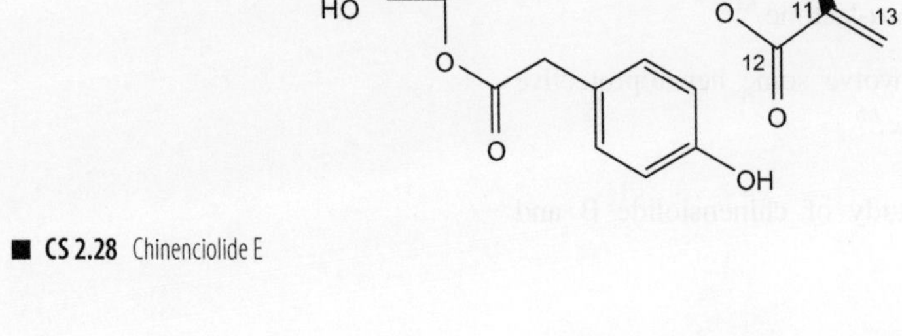

■ **CS 2.28** Chinenciolide E

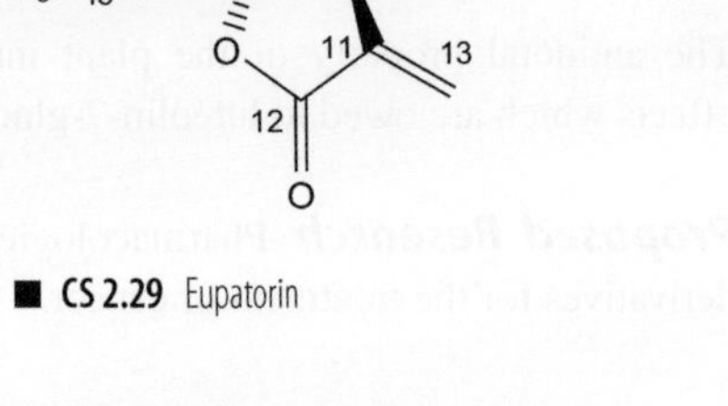

■ **CS 2.29** Eupatorin

and eupatorin acetate (CS 2.30) isolated from *Eupatorium rotundifolium* L. (family Asteraceae Bercht. & J. Presl) killed human nasopharyngeal carcinoma (KB) cells at a dose of $0.21\,\mu g/mL$,[68] suggesting that the presence of an epoxy group in C10 and acetylations of hydroxyl groups are beneficial for cytotoxic activity. Ixerin Y (CS 2.31) from *Ixeris denticulata* (Houtt.) Stebbins (family Asteraceae Bercht. & J. Presl) abated the proliferation of human breast adenocarcinoma (MCF-7) cells and human breast cancer (MDA-468) cells, with IC_{50} values of $6.3\,\mu g/mL$ and $11.8\,\mu g/mL$, respectively,[69] showing that a glycoside in C15 is not favorable for activity. 3β-O-(2-Methylbutyryl)-moroccolide A (CS 2.32), moroccolide A (CS 2.33), 8-desoxy-$3\alpha,4\alpha$-epoxyrupiculin-A and -B (CS 2.34 and CS 2.35), saharanolides A and B (CS 2.36 and CS 2.37) from *Warionia saharae* Bentham ex Coss. (family Asteraceae Bercht. & J. Presl) inhibited the growth of human nasopharyngeal carcinoma (KB) cells, with IC_{50} values of

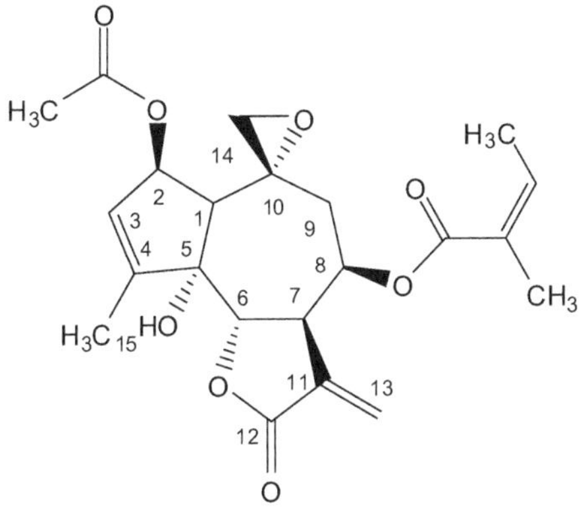

■ **CS 2.30** Eupatorin acetate

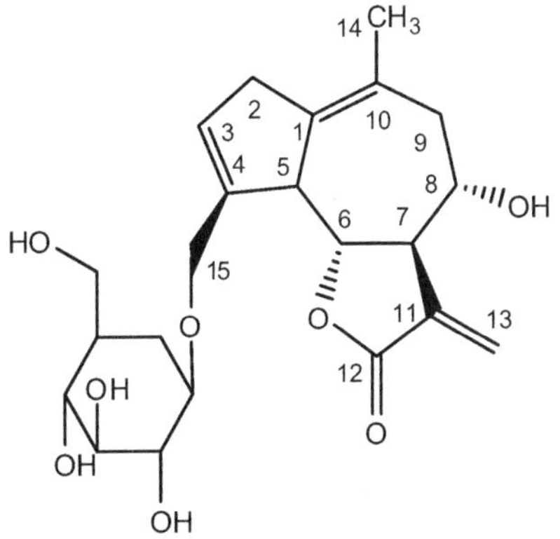

■ **CS 2.31** Ixerin Y

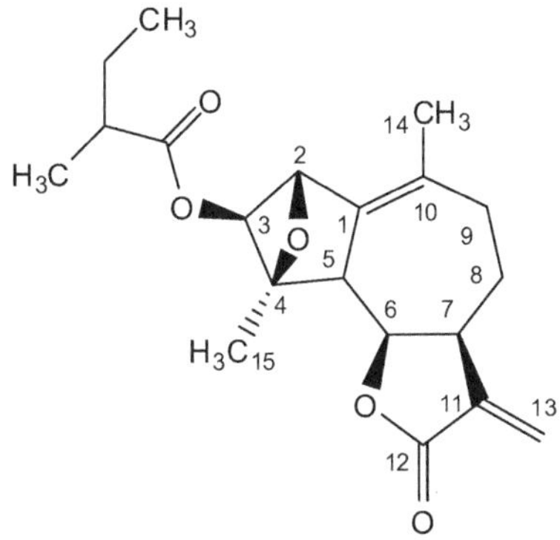

■ **CS 2.32** 3β-O-(2-Methylbutyryl)-moroccolide A

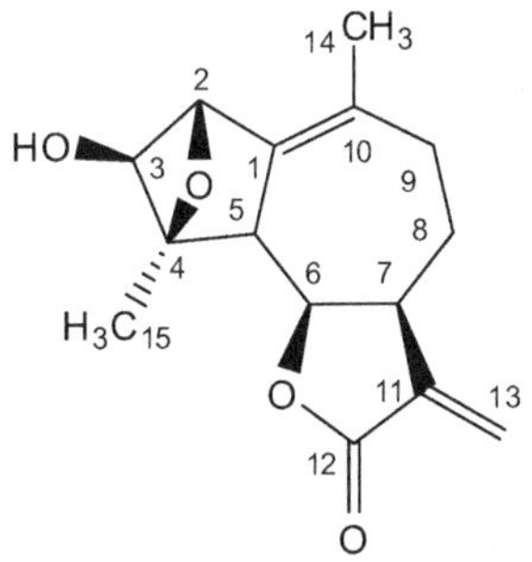

■ **CS 2.33** Moroccolide A

■ **CS 2.34** 8-Desoxy-3α,4α-epoxyrupiculin-A

■ **CS 2.35** 8-Desoxy-3α,4α-epoxyrupiculin-B

■ **CS 2.36** Saharanolide A

■ **CS 2.37** Saharanolide B

■ **CS 2.38** Chlorojanerin

■ **CS 2.39** Cynaropicrin

■ **CS 2.40** Janerin

1 μg/mL, 4.5 μg/mL, 1.7 μg/mL, 2 μg/mL, 3.3 μg/mL and 5.5 μg/mL, respectively,[70] suggesting that the presence of an ester in C3 or a C3-C4 α epoxyde are favorable for activity. Chlorojanerin (CS 2.38), cynaropicrin (CS 2.39) and janerin (CS 2.40) from *Centaurothamnus maximus* Wagenitz & Dittrich

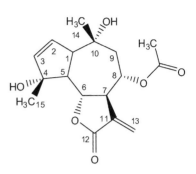

■ **CS 2.41** Dehydrocostuslactone

■ **CS 2.42** (+)-8α-Acetoxy-4α-hydroxyguaia-1(10), 2,11(13)-trien-12,6 α-olide

■ **CS 2.43** 8-Acetoxy-4,10-dihydroxy-2,11(13)-guaiadiene-12,6-olide

(family Asteraceae Bercht. & J. Presl) were cytotoxic against human skin melanoma (SK-MEL) cells, with IC_{50} values equal to 2.3 μg/mL, 2 μg/mL and 2.7 μg/mL, respectively,[71] implying that the C4 substitution by a methylene is beneficial. This is confirmed by dehydrocostuslactone (CS 2.41) from *Saussurea lappa* (Decne.) Sch. Bip. (family Asteraceae Bercht. & J. Presl), which was cytotoxic against human hepatocellular liver carcinoma (HepG2), human epitheloid cervix carcinoma (Hela) and human ovarian carcinoma (OVCAR-3) cells, with IC_{50} values equal to 3.5 μg/mL, 3.5 μg/mL and 2.5 μg/mL, respectively.[72] (+)-8α-Acetoxy-4α-hydroxyguaia-1(10),2,11(13)-trien-12,6 α-olide (CS 2.42) from a member of the genus *Anthemis* L. (family Asteraceae Bercht. & J. Presl) inhibited the growth of Hela and mouse melanoma (B16) cells, with IC_{50} values of 21 μM and 20.7 μM, respectively,[73] suggesting that multiple double bonds of the guaiane framework are not favorable for cytotoxic activity.

8-Acetoxy-4,10-dihydroxy-2,11(13)-guaiadiene-12,6-olide (CS 2.43) isolated from *Chrysanthemum boreale* Makino (family Asteraceae Bercht. & J. Presl) inhibited the growth of human renal carcinoma (ACHN), human melanoma (LOX-IMVI), human colon adenocarcinoma (SW 620), human prostate cancer (PC-3) and human lung adenocarcinoma epithelial (A549) cells, with IC_{50} values equal to 3.2 μg/mL, 4.1 μg/mL, 2.9 μg/mL, 1 μg/mL and 3.6 μg/mL, respectively.[74] Repin (CS 2.44) and chlorohyssopifolin C (CS 2.45) isolated from a member of the genus from *Centaurea* L. (family Asteraceae Bercht. & J. Presl) were active against the following cell lines (respective IC_{50} values in parentheses): A549 (2.5 μM, 9.3 μM), MCF-7 (1.1 μM, 1.5 μM), human ovarian carcinoma (1A9; 0.3 μM, 0.8 μM), human nasopharyngeal carcinoma (KB; 1.4 μM, 2 μM), human vincristine-resistant human nasopharyngeal carcinoma (KB-V; 0.8 μM, 1.5 μM) and human ileocecal adenocarcinoma (HCT-8; 0.8 μM, 1.6 μM).[75]

■ **CS 2.44** Repin

■ **CS 2.45** Chlorohyssopifolin C

■ **CS 2.46** Repin derivative with C13 paclixatel moiety

■ **CS 2.47** 3α,4α-Epoxyrupicolin C

Structure–activity relationship studies showed that esterifying repin in C3 with the C13 side chain of paclitaxel boosted the cytotoxic potencies (CS 2.46). In fact esterifications of guaianolides seem to be beneficial for activity. Presence of hydroxy groups on C4 and C15 and/or C3′ or C4′ abolished the activity of repin.[75] In regard to the cytotoxic mode of action of guaianolides, one can reasonably infer that the lactone ring alkylates cysteine residues and therefore impairs the NF-κB pathway.[45,54,55] This is exemplified with 3α,4α-epoxyrupicolin C (CS 2.47) from *Artemisia sylvatica* Maxim. (family Asteraceae Bercht. & J. Presl), which inhibited NF-κB activation, with IC_{50} values equal to $0.8\,\mu M$.[76] Thapsigargin (CS 2.48) from *Thapsia garganica* L. (family Apiaceae Lindl.) stands apart in the guaianolides by its structure, activity and botanical origin and one might wonder if it is actually an artifact. This highly oxygenated and esterified umbelliferous guaianolide sesquiterpene is a sarco-endoplasmic reticulum Ca^{2+}-ATPase (SERCA) inhibitor which boosts the cytoplasmic concentration of Ca^{2+}[77] and therefore cancer cell death by apoptosis.[78] Inhibition of SERCA is followed by Ca^{2+} leaks

■ **CS 2.48** Thapsigargin

■ **CS 2.49** 10-*epi*-8-deoxycumambrinB

■ **CS 2.50** Dehydroleucodin

■ **CS 2.51** Ludartin

from the endoplasmic reticulum that rends the cytoplasmic membrane permeable to extracellular Ca^{2+}.[79] The resulting tremendous cytoplasmic accumulation of Ca^{2+} leads to saturation of calmodulin. This in turn stimulates calcineurin, which dephosphorylates and therefore releases Bcl-2-associated death promoter (Bad), a pro-apoptotic protein from the 14-3-3 protein.[80]

The pro-apoptotic protein Bax interferes with the mitochondrial membrane voltage-activated channel and forms some pores through which cytochrome c and apoptosis inducing factor (AIF) are released.[81] Another astonishing finding in regard to the biomolecular events triggered by guaianolides is the fact that 10-*epi*-8-deoxycumambrinB (CS 2.49) from *Stevia yaconensis* Hieron (family Asteraceae Bercht. & J. Presl) and dehydroleucodin (CS 2.50) and ludartin (CS 2.51) from *Artemisia douglasiana* Besser (family Asteraceae Bercht. & J. Presl) inhibited the enzymatic activity of aromatase in human placental microsomes, with IC_{50} values equal to 7 μM, 15 μM and 55 μM, respectively.[82] Note that the aromatase inhibitor letrozole at a dose of 1 μM induced apoptosis in estrogen-dependent human breast cancer cells transfected with aromatase (MCF-7Ca) via a decreased level of anti-apoptotic Bcl-2 protein, an increased level of pro-apoptotic protein Bax and caspase 9 activation.[83]

2.1.5 *Parasenecio Delphiniphyllus* (H. Lév.) Y.L. Chen

History The plant was first described by Yi Lin Cheng in *Flora Republicae Popularis Sinicae*, published in 1999.

Synonyms *Cacalia delphiniifolia* Siebold & Zucc., Cacalia delphiniphylla (H. Lév.) Hand.-Mazz., Cacalia pilgeriana subsp. delphiniphylla (H. Lév.) Koyama, Parasenecio delphiniphyllus (H. Lév.) Y.L. Chen, Parasenecio tongchuanensis Y.L. Chen, Senecio delphiniphyllus H. Lév., Senecio zuccarinii Maxim.

Family Asteraceae Bercht. & J. Presl, 1820

Common Names Momijigasa (Japanese), cui que xie jia cao (Chinese)

Habitat and Description This is a perennial herb that grows to 1 m high in the forests of China, Korea and Japan. The rhizome is robust. The leaves are simple. The petiole is 4–6.5 cm long, and straight. The blade is 9–15 cm × 10–18 cm, palmately lobed and serrate. The inflorescence is a terminal raceme of cylindrical capitula. The involucre is 0.5 cm long and presents five phyllaries which are lanceolate and minute. The receptacle supports five florets which are purplish, 1 cm long and five-lobed. The anthers are caudate. The style protrudes from the corolla and produces a pair of conspicuously recurved stigmas. The achenes are brown, cylindrical, ribbed and 0.5 cm long. The pappus consists of capillary-like bristles that are brownish and 0.5 cm long (Figure 2.5).

Medicinal Uses In China the plant is a vegetable used to treat cancer.

Phytopharmacology The plant contains a rare type of sesquiterpenes known as eremophilanes including: cacalol,[84] 1β-hydroxy-2β-methylsenecioyloxyeremophil-7-(11)-en-8β-(12)-olide, 1β-hydroxy-2β-methylsenecioyloxy-8α-methoxyeremophil-7-(11)-en-8β-(12)-olide,[85] 2β-hydroxy-3β-methylsenecioyloxyeremophil-7(11)-en-8α (12)-olide, 2β,8β-dihydroxy-3β-methylsenecioyloxyeremophil-7(11)-en-8α (12)-olide and 1β,8β-dihydroxy-2β,3α-diangeloyloxyeremophil-7(11)-en-8α (12)-olide;[86]

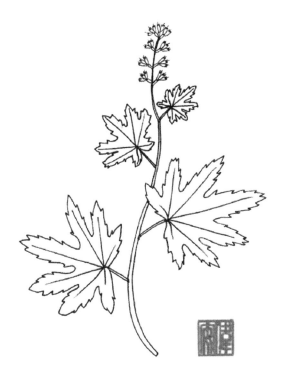

■ **FIGURE 2.5** *Parasenecio delphiniphyllus* (H. Lév.) Y.L. Chen

the caryolane sesquiterpene caryolane-1,9β-diol;[86] the bisabolane sesquiterpene 3,6-epidioxy-1,10-bisaboladiene;[87] and the ent-kaurane diterpenes 19-acetoxyl-*ent*-3β,17-dihydroxykaur-15-ene, 19-acetoxyl-*ent*-3β-hydroxykaur-15-en-17-al, *ent*-kaur-16-en-19-al, *ent*-kaur-16-en-19-oic acid, *ent*-kauran-16β,17-diol, *ent*-15β,16β-epoxy-17-hydroxykauran-19-oic acid,19-acetyl-*ent*-3β-hydroxyl-kaur-16-ene, *ent*-3 β,19-dihydroxykaur-16-ene, *ent*-17-hydroxykaur-15-ene,[88] 19-acetyl-*ent*-3-β,17-dihydroxykaur-15-ene and 19-acetyl-*ent*-3β-hydroxykaur-15-en-17-al.[89]

The anticancer property of the plant results most probably from the synergistic action of its terpenic constituents, such as cacalol[90] and 3,6-epidioxy-1,10-bisaboladiene.[87]

Proposed Research Pharmacological study of cacalol and derivatives for the treatment of cancer.

Rationale Some lines of evidence indicate that eremophilane and biosynthetic congeners such as cacalol are of chemotherapeutic value. Cacalol (CS 2.52) isolated from *Parasenecio delphiniphyllus* (H. Lév.) Y.L. Chen

■ **CS 2.52** Cacalol

■ **CS 2.53** 1-α-Hydroxy-9-deoxycacalol

■ **CS 2.54** 14-Angeloyloxy-2α,3α-epoxy-1-oxo-
O-methylcacalol

abated the survival of human breast adenocarcinoma (MDA-MB-231 and MCF-7) cells at doses of 70 μM via inactivation of Akt, which resulted in inhibition of sterol regulatory element binding protein1 (SREBP1)-induced fatty acid synthetase (FAS) expression[90] and therefore apoptosis.[91] Inhibition of FAS negates mammalian target of rapamycin (mTOR) and therefore the production of the anti-apoptotic FLICE-like inhibitory protein (FLIP), which normally blocks the activation of caspase 8.[92] One could reasonably infer that cacalol, being a strong antioxidant, lowers the cytoplasmic levels of ROS, which are known to activate phosphoinositide 3-kinase (PI3K) and therefore Akt.[93,94]

1-α-Hydroxy-9-deoxycacalol (CS 2.53) isolated from *Ligularia macrophylla* (Ledeb.) DC. (family Asteraceae Bercht. & J. Presl) killed human lung adenocarcinoma epithelial (A549) and MCF-7 cells, with IC_{50} values equal to 13.8 μg/mL and 15.3 μg/mL, respectively.[95] 14-Angeloyloxy-2α,3α-epoxy-1-oxo-*O*-methylcacalol (CS 2.54) from *Parasenecio deltophyllus* (Maxim.) Y.L. Chen (family Asteraceae Bercht. & J. Presl) was active against human hepatocellular liver carcinoma (HepG2) and human promyelocytic leukemia (HL-60) cells, with IC_{50} values equal to 27.8 μM and 49.8 μM, respectively,[96] suggesting that the presence of a ketone or a hydroxyl at C1 might be favorable for cytotoxic activity. This is confirmed with the C1 esterified eremophilane sesquiterpene 1,3-diangeloyloxyeremophila-9,7(11)-dien-8-one (CS 2.55) from *Robinsonecio gerberifolius* (Sch. Bip. ex Hemsl.) T.M. Barkley & Janovec (family Asteraceae Bercht. & J. Presl), which was lethal to human colorectal adenocarcinoma (HCT-15), MCF-7, human glioblastoma (U251), human prostate cancer (PC-3) and human erythromyeloblastoid leukemia (K562) cells, with IC_{50} values equal to 12.8 μM, 21 μM, 17.6 μM, 10.7 μM and 52.2 μM, respectively.[97]

3β-(Angeloyloxy)-eremophil-7(11)-en-12,8 β-olid-14-oic (CS 2.56) acid from *Ligularia atroviolacea* (Franch.) Hand.-Mazz. (family Asteraceae Bercht. & J. Presl), which is not oxygenated nor esterified at C1, mildly abated the proliferation of human nasopharyngeal carcinoma (KB), HL-60,

■ **CS 2.55** 1,3-Diangeloyloxyeremophila-9,7(11)-dien-8-one

■ **CS 2.56** 3β-(Angeloyloxy)eremophil-7(11)-en-12,8 β-olid-14-oic

human epitheloid cervix carcinoma (Hela), human nasopharyngeal carcinoma (CNE), and murine macrophage-like (P388D1) cells, with IC_{50} values of 99 μM, 208 μM, 89.3 μM, 167 μM and 62.3 μM, respectively.[98] Eremophilone (CS 2.57), 8-hydroxy-1,11-eremophiladien-9-one (CS 2.58), santalcamphor (CS 2.59) and 9-hydroxy-7(11),9-eremophiladien-8-one (CS 2.60) from *Eremophila mitchelli* Benth. (family Scrofulariaceae Juss.) inhibited the growth of macrophage-like (P388D1) cells, with IC_{50} values equal to 42 μg/mL, 76 μg/mL, 95 μg/mL, and 105 μg/mL, respectively,[99] suggesting that a double bond between C7 and C11 might not be a favorable chemical feature for cytotoxic activity.

The eremophilane 07H239-A (CS 2.61) isolated from a marine fungus and its amino derivative (CS 2.62) potently inhibited the growth of a number of cell lines (respective IC_{50} values in parentheses): Hela (9.9 μM, 3.3 μM), MCF-7 (7.1 μM, 3.4 μM), PC-3 (10 μM, 1 μM) A549 (6.3 μM, 6.5 μM) and human breast carcinoma (MDA-MB-435; 22.5 μM, 0.9 μM).[100] Cryptosphaerolide (CS 2.63) from a marine fungus belonging to the genus *Cryptosphaeria* was cytotoxic against human colorectal carcinoma (HCT-116) cells, with an IC_{50} value of 4.5 μM, and inhibited anti-apoptotic

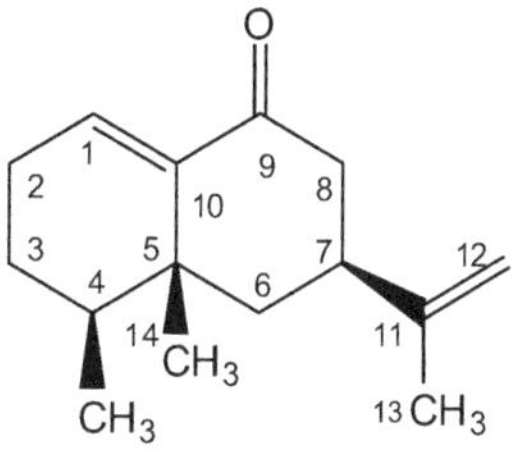

■ **CS 2.57** Eremophilone

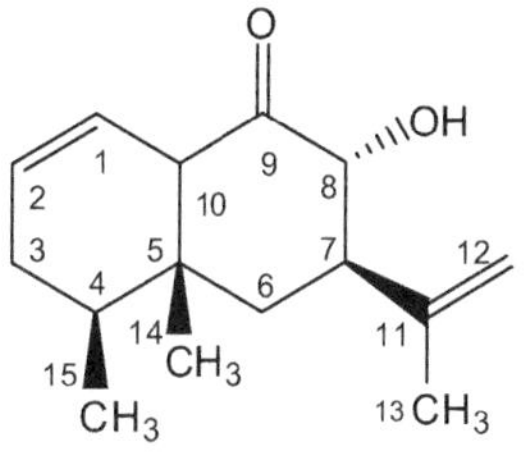

■ **CS 2.58** 8-Hydroxy-1,11-eremophiladien-9-one

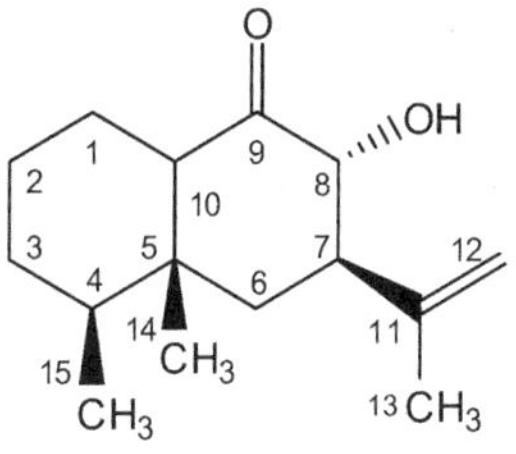

■ **CS 2.59** Santalcamphor

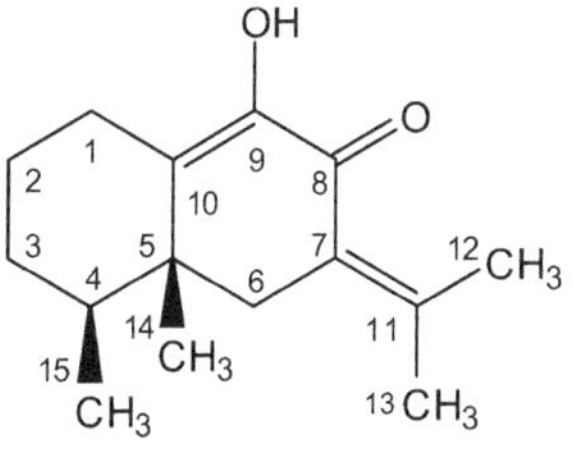

■ **CS 2.60** 9-Hydroxy-7(11), 9-eremophiladien-8-one

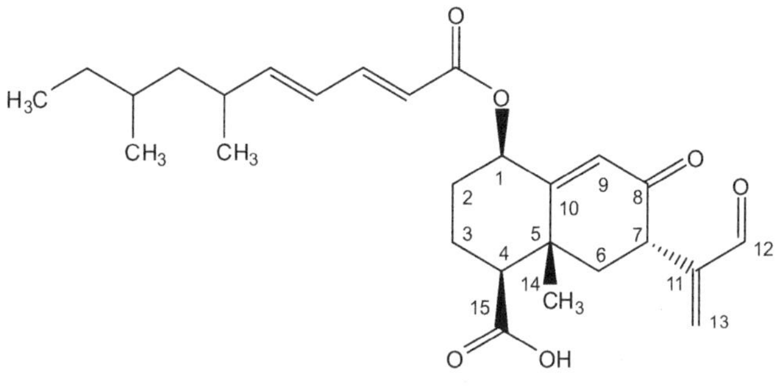

■ **CS 2.61** 07H239-A

■ **CS 2.62** 07H239-A amino derivative

■ **CS 2.63** Cryptosphaerolide

induced myeloid leukemia cell differentiation protein (Mcl-1), with an IC_{50} of 11.4 μM, and activities were mitigated by the replacement of the C1 ester by a hydroxyl group.[101]

REFERENCES

[1] Lin Y, Jin T, Wu X, Huang Z, Fan J, Chan WL. A novel bisesquiterpenoid, biatractylolide, from the Chinese herbal plant *Atractylodes macrocephala*. J Natural Prod 1997;60(1):27–8.

[2] Li CQ, He LC, Jin JQ. Atractylenolide I and atractylenolide III inhibit lipopolysaccharide-induced TNF-α and NO production in macrophages. Phytother Res 2007;21(4):347–53.

[3] Li CQ, He LC, Dong HY, Jin JQ. Screening for the anti-inflammatory activity of fractions and compounds from *Atractylodes macrocephala* koidz. J Ethnopharmacol 2007;114(2):212–7.

[4] Ding HY, Liu MY, Chang WL, Lin HC. New sesquiterpenoids from the rhizomes of *Atractylodes macrocephala*. Chinese Pharm J 2005;57(1):37–42.

[5] Chen ZL. The acetylenes from *Atractylodes macrocephala*. Planta Med 1987;53:493–4.

[6] Dong H, He L, Huang M, Dong Y. Anti-inflammatory components isolated from *Atractylodes macrocephala* Koidz. Nat Prod Res 2008;22(16):1418–27.

[7] Peng W, Han T, Wang Y, Xin WB, Zheng CJ, Qin LP. Chemical constituents of the aerial part of *Atractylodes macrocephala*. Chem Nat Comp 2011;46(6):959–60.

[8] Hsieh TJ, Chang FR, Chia YC, Chen CY, Chiu HF, Wu YC. Cytotoxic constituents of the fruits of *Cananga odorata*. J Nat Prod 2001;64(5):616–9.

[9] Tsuneki H, Ma EL, Kobayashi S, Sekizaki N, Maekawa K, Sasaoka T, et al. Antiangiogenic activity of β-eudesmol in vitro and in vivo. Eur J Pharmacol 2005;512(2–3):105–15.

[10] Ma EL, Li YC, Tsuneki H, Xiao JF, Xia MY, Wang MW, et al. β-Eudesmol suppresses tumour growth through inhibition of tumour neovascularisation and tumour cell proliferation. J Asian Nat Prod Res 2008;10(2):159–67.

[11] Xu X, Xie H, Hao J, Jiang Y, Wei X. Eudesmane sesquiterpene glucosides from lychee seed and their cytotoxic activity. Food Chem 2010;123(4):1123–6.

[12] El-Gamal AA. Sesquiterpene lactones from *Smyrnium olusatrum*. Phytochem 2001;57(8):1197–200.

[13] Wang CC, Chen LG, Yang LL. Cytotoxic activity of sesquiterpenoids from *Atractylodes ovata* on leukemia cell lines. Planta Med 2002;68(3):204–8.

[14] Wang CC, Lin SY, Cheng HC, Hou WC. Pro-oxidant and cytotoxic activities of atractylenolide I in human promyeloleukemic HL-60 cells. Food Chem Toxicol 2006;44(8):1308–15.

[15] Kang TH, Bang JY, Kim MH, Kang IC, Kim HM, Jeong HJ. Atractylenolide III, a sesquiterpenoid, induces apoptosis in human lung carcinoma A549 cells via mitochondria-mediated death pathway. Food Chem Toxicol 2011;49(2):514–9.

[16] Ye Y, Chou GX, Wang H, Chu JH, Fong WF, Yu ZL. Effects of sesquiterpenes isolated from largehead atractylodes rhizome on growth, migration, and differentiation of B16 melanoma cells. Integr Cancer Ther 2011;10(1):92–100.

[17] Melinda MM, Joseph MG, Michael GS, Richard JB. AKT: a novel target in pancreatic cancer therapy. Cancer Ther 2004;2:227–38.

[18] Coqueret O. New roles for p21 and p27 cell-cycle inhibitors: a function for each cell compartment?. Trends Cell Biol 2003;13:65–70.

[19] Ye Y, Wang H, Chu JH, Chou GX, Chen SB, Mo H, et al. Atractylenolide II induces G1 cell-cycle arrest and apoptosis in B16 melanoma cells. J Ethnopharmacol 2011;136(1):279–82.

[20] Yang ZD, Gao K, Jia ZJ. Eudesmane derivatives and other constituents from *Saussurea parviflora*. Phytochem 2003;62(8):1195–9.

[21] Jeong SH, Koo SJ, Ha JH, Ryu SY, Park HJ, Lee KT. Induction of apoptosis by yomogin in human promyelocytic leukemic HL-60 cells. Biol Pharm Bull 2004;27(7):1106–11.

[22] Wang XR, Wu QX, Shi YP. Terpenoids and sterols from *Saussurea cauloptera*. Chem Biodiver 2008;5(2):279–89.

[23] Harimaya K, Gao JF, Ohkura T, Kawamata T, Iitaka Y, Guo YT, et al. A series of sesquiterpenes with a 7α-isopropyl side chain and related compounds isolated from *Curcuma wenyujin*. Chem Pharm Bull 1991;39(4):843–53.

[24] Gao JF, Xie JH, Harimaya K, Kawamata T, Iitaka Y, Inayama S. The absolute structure and synthesis of wenjine isolated from *Curcuma wenyujin*. Chem Pharm Bull 1991;39(4):854–6.

[25] Lou Y, Zhao F, Wu Z, Peng KF, Wei XC, Chen LX, et al. Germacrane-type sesqui-terpenes from *Curcuma wenyujin*. Helvetica Chimica Acta 2009;92(8):1665–72.

[26] Cao J, Qi M, Zhang Y, Zhou S, Shao Q, Fu R. Analysis of volatile compounds in *Curcuma wenyujin* Y.H. Chen et C. Ling by headspace solvent microextraction-gas chromatography-mass spectrometry. Analytica Chimica Acta 2006;561(1–2):88–95.

[27] Lou Y, Zhao F, He H, Peng KF, Chen LX, Qiu F. Four new sesquiterpenes from *Curcuma wenyujin* and their inhibitory effects on nitric-oxide production. Chem Biodivers 2010;7(5):1245–53.

[28] Inayama S, Gao JF, Harimaya K. The absolute stereostructure of curcumol iso-lated from *Curcuma wenyujin*. Chem Pharm Bull 1984;32(9):3783–6.

[29] Wang D, Huang W, Shi Q, Hong C, Cheng Y, Ma Z, et al. Isolation and cyto-toxic activity of compounds from the root tuber of *Curcuma wenyujin*. Nat Prod Commun 2008;3(6):861–4.

[30] Huang W, Zhang P, Jin YC, Shi Q, Cheng YY, Qu HB, et al. Cytotoxic diter-penes from the root tuber of *Curcuma wenyujin*. Helvetica Chimica Acta 2008;91(5):944–50.

[31] Ma E, Wang X, Li Y, Sun X, Tai W, Li T, et al. Induction of apoptosis by fura-nodiene in HL60 leukemia cells through activation of TNFR1 signaling pathway. Cancer Lett 2008;271(1):158–66.

[32] Li QQ, Wang G, Huang F, Banda M, Reed E. Antineoplastic effect of β-elemene on prostate cancer cells and other types of solid tumour cells. J Pharm Pharmacol 2010;62(8):1018–27.

[33] Wang G, Li X, Huang F, Zhao J, Ding H, Cunningham C, et al. Antitumor effect of β-elemene in non-small-cell lung cancer cells is mediated via induction of cell cycle arrest and apoptotic cell death. Cell Mol Life Sci 2005;62(7–8):881–93.

[34] Yu Z, Wang R, Xu L, Dong J, Jing Y. N-(β-Elemene-13-yl)tryptophan methyl ester induces apoptosis in human leukemia cells and synergizes with arsenic trioxide through a hydrogen peroxide dependent pathway. Cancer Lett 2008;269(1):165–73.

[35] Tee AR, Proud CG. DNA-damaging agents cause inactivation of translational regulators linked to mTOR signalling. Oncogene 2000;19(26):3021–31.

[36] Liu J, Zhang Y, Qu J, Xu L, Hou K, Zhang J, et al. β-Elemene-induced autoph-agy protects human gastric cancer cells from undergoing apoptosis. BMC Cancer 2011;11:183.

[37] Xie CY, Yang W, Li M, Ying J, Tao SJ, Li K, et al. Cell apoptosis induced by δ-elemene in colorectal adenocarcinoma cells via a mitochondrial-mediated path-way. Yakugaku Zasshi 2009;129(11):1403–13.

[38] Wang XS, Yang W, Tao SJ, Li K, Li M, Dong JH, et al. The effect of δ-elemene on hela cell lines by apoptosis induction. Yakugaku Zasshi 2006;126(10):979–90.

[39] Xie CY, Yang W, Ying J, Ni QC, Pan XD, Dong JH, et al. B-cell lymphoma-2 over-expression protects δ-elemene-induced apoptosis in human lung carcinoma mucoepidermoid cells via a nuclear factor kappa B-related pathway. Biol Pharm Bull 2011;34(8):1279–86.

[40] Zhu T, Zhao Y, Zhang J, Li L, Zou L, Yao Y, et al. β-Elemene inhibits prolifer-ation of human glioblastoma cells and causes cell-cycle G0/G1 arrest via mutually compensatory activation of MKK3 and MKK6. Int J Oncol 2011;38(2):419–26.

[41] Xu L, Tao S, Wang X, Yu Z, Wang M, Chen D, et al. The synthesis and anti-proliferative effects of β-elemene derivatives with mTOR inhibition activity. Bioorg Med Chem 2006;14(15):5351–6.

[42] Goswami AC, Baruah RN, Sharma RP, Baruah JN, Kulanthaivel P, Herz W. Germacranolides from *Inula cappa*. Phytochem 1984;23(2):367–72.

[43] Xie HG, Chen H, Cao B, Zhang HW, Zou ZM. Cytotoxic germacranolide sesquiterpene from *Inula cappa*. Chem Pharm Bull 2007;55(8):1258–60.

[44] Wu ZJ, Shan L, Lu M, Shen YH, Tang J, Zhang WD. Chemical constituents from *Inula cappa*. Chem Nat Comp 2010;46(2):298–300.

[45] Kwok BHB, Koh B, Ndubuisi MI, Elofsson M, Crews CM. The anti-inflammatory natural product parthenolide from the medicinal herb feverfew directly binds to and inhibits IκB kinase. Chem Biol 2001;8(8):759–66.

[46] Nakagawa Y, Iinuma M, Matsuura N, Yi K, Naoi M, Nakayama T, et al. A potent apoptosis-inducing activity of a sesquiterpene lactone, arucanolide, in HL60 cells: a crucial role of apoptosis-inducing factor. J Pharmacol Sci 2005;97(2):242–52.

[47] Wen J, You KR, Lee SY, Song CH, Kim DG. Oxidative stress-mediated apoptosis: the anticancer effect of the sesquiterpene lactone parthenolide. J Biol Chem 2002;277(41):38954–38964.

[48] Liu JW, Cai MX, Xin Y, Wu QS, Ma J, Yang P, et al. Parthenolide induces proliferation inhibition and apoptosis of pancreatic cancer cells in vitro. J Exp Clin Cancer Res 2010;29(1):108.

[49] Fahy BN, Schlieman MG, Mortenson MM, Virudachalam S, Bold RJ. Targeting BCL-2 overexpression in various human malignancies through Nf-kappaB inhibition by the proteasome inhibitor bortezomib. CancerChemother Pharmacol 2005;56:46–54.

[50] Karin M. How NF-kappaB is activated: the role of the IkappaB kinase (IKK) complex. Oncogene 1999;18(49):6867–74.

[51] Bork PM, Schmitz ML, Kuhnt M, Escher C, Heinrich M. Sesquiterpene lactone containing Mexican Indian medicinal plants and pure sesquiterpene lactones as potent inhibitors of transcription factor NF-κB. FEBS Lett 1997;402:85–90.

[52] Hehner SP, Heinrich M, Bork PM, Vogt M, Ratter F, Lehmann V, et al. Sesquiterpene lactones specifically inhibit activation of NF-κB by preventing the degradation of IκB-α and IκB-β. J Biol Chem 1998;273:1288–97.

[53] Hehner SP, Hofmann TG, Droge W, Schmitz ML. The antiinflammatory sesquiterpene lactone parthenolide inhibits NF-κB by targeting the IκB kinase complex. J Immunol 1999;163:5617–23.

[54] Hehner SP, Heinrich M, Bork PM, Vogt M, Ratter F, Lehmann V, et al. Sesquiterpene lactones specifically inhibit activation of NF-kappa B by preventing the degradation of I kappa B-alpha and I kappa B-beta. J Biol Chem 1998;273:1288–97.

[55] García-Pineres AJ, Castro V, Mora G, Schmidt TJ, Strunck E, Pahl HL, et al. Cysteine 38 in p65/NFkappaB plays a crucial role in DNA binding inhibition by sesquiterpene lactones. J Biol Chem 2001;276:39713–39720.

[56] García-Pineres AJ, Lindenmeyer MT, Merfort I. Role of cysteine residues of p65/NF-kappaB on the inhibition by the sesquiterpene lactone parthenolide and N-ethyl maleimide, and on its transactivating potential. Life Sci 2004;75(7):841–56.

[57] Irelan JT, Murphy TJ, DeJesus PD, Teo H, Xu D, Gomez-Ferreria MA, et al. A role for IκB kinase 2 in bipolar spindle assembly. Proc Nat Acad Sci USA 2007;104(43):16940–16945.

[58] Shiojima K, Suzuki H, Kodera N, Kubota K, Tsushima S, Ageta H, et al. Composite constituent: novel triterpenoid, 17-epilupenyl acetate, from aerial parts of *Ixeris chinensis*. Chem Pharm Bull 1994;42(10):2193–5.

[59] Shiojima K, Suzuki H, Kodera N, Ageta H, Chang HC, Chen YP. Composite constituent: novel triterpenoid, ixerenol, from aerial parts of *Ixeris chinensis*. Chem Pharm Bull 1995;43(1):180–2.

[60] Shiojima K, Suzuki H, Kodera N, Ageta H, Chang HC, Chen YP. Composite constituents: thirty-nine triterpenoids including two novel compounds from *Ixeris chinensis*. Chem Pharm Bull 1996;44(3):509–14.

[61] Zhang SJ, Wang JL, Deng QG, Ando M. New triterpenes from Siyekucai (*Ixeris chinensis*). Chin Chem Letters 2006;17(2):195–7.

[62] Lee SW, Chen ZT, Chen CM. A new sesquiterpene lactone glucoside of *Ixeris chinensis*. Heterocycles 1994;38(8):1933–6.

[63] Zhang S, Wang J, Xue H, Deng Q, Xing F, Ando M. Three new guaianolides from siyekucai (*Ixeris chinensis*). J Nat Prod 2002;65(12):1927–9.

[64] Zhang S, Zhao M, Bai L, Hasegawa T, Wang J, Wang L, et al. Bioactive guaianolides from siyekucai (*Ixeris chinensis*). J Nat Prod 2006;69(10):1425–8.

[65] Khalil AT, Shen YC, Guh JH, Cheng SY. Two new sesquiterpene lactones from *Ixeris chinensis*. Chem Pharm Bull 2005;53(1):15–17.

[66] Qiusheng Z, Xiling S, Xubo GL, Meng S, Changhai W. Protective effects of luteolin-7-glucoside against liver injury caused by carbon tetrachloride in rats. Pharmazie 2004;59(4):286–9.

[67] Dai J, Yang L, Sakai JI, Ando M. Biotransformation of chinensiolide B and the cytotoxic activities of the transformed products. J Mol Catalysis B: Enzym 2005;33(3–6):87–91.

[68] Kupchan SM, Hemingway JC, Cassady JM, Knox JR, McPhail AT, Sim GA. The isolation and structural elucidation of euparotin acetate, a novel guaianolide tumor inhibitor from *Eupatorium rotundifolium* J Am Chem Soc 1967;89(2):465–6.

[69] Ma JY, Wang ZT, Xu LS, Xu GJ. A sesquiterpene lactone glucoside from *Ixeris denticulata* f. *pinnatipartita*. Phytochem 1999;50(1):113–5.

[70] Hilmi F, Sticher O, Heilmann J. New cytotoxic 6,7-*cis* and 6,7-*trans* configured guaianolides from *Warionia saharae*. J Nat Prod 2002;65(4):523–6.

[71] Muhammad I, Takamatsu S, Mossa JS, El-Feraly FS, Walker LA, Clark AM. Cytotoxic sesquiterpene lactones from *Centaurothamnus maximus* and *Vicoa pentanema*. Phytother Res 2003;17(2):168–73.

[72] Sun CM, Syu WJ, Don MJ, Lu JJ, Lee GH. Cytotoxic sesquiterpene lactones from the root of *Saussurea lappa*. J Nat Prod 2003;66(9):1175–80.

[73] Vučković I, Vujisić L, Stešević D, Radulović S, Lazić M, Milosavljević S. Cytotoxic guaianolide from *Anthemis segetalis* (Asteraceae). Phytother Res 2010;24(2):225–7.

[74] Park KH, Yang MS, Park MK, Kim SC, Yang CH, Park SJ, et al. A new cytotoxic guaianolide from *Chrysanthemum boreale*. Fitoterapia 2009;80(1):54–6.

[75] Bruno M, Rosselli S, Maggio A, Raccuglia RA, Bastow KF, Lee KH. Cytotoxic activity of some natural and synthetic guaianolides. J Nat Prod 2005;68(7):1042–6.

[76] Jin HZ, Lee JH, Lee D, Hong YS, Kim YH, Lee JJ. Inhibitors of the LPS-induced NF-κB activation from *Artemisia sylvatica*. Phytochem 2004;65(15):2247–53.

[77] Treiman M, Caspersen C, Christensen SB. A tool coming of age: thapsigargin as an inhibitor of sarco-endoplasmic reticulum Ca2+-ATPases. Trends Pharmacol Sci 1998;19(4):131–5.

[78] Furuya Y, Lundmo P, Short AD, Gill DL, IsaaCS JT. The role of calcium, pH, and cell proliferation in the programmed (apoptotic) death of androgen-independent prostatic cancer cells induced by thapsigargin. Cancer Res 1994;54(23):6167–75.

[79] Randriamampita C, Tsien RY. Emptying of intracellular Ca2+ stores releases a novel small messenger that stimulates Ca2+ influx. Nature 1993;364(6440):809–14.

[80] Denmeade SR, IsaaCS JT. The SERCA pump as a therapeutic target. Making a smart bomb for prostate cancer. Cancer Biol Ther 2005;4(1):69–77.

[81] Christensen SB, Mondrup Skytte D, Denmeade SR, Dionne C, Møller JV, Nissen P, et al. A trojan horse in drug development: targeting of thapsigargins towards prostate cancer cells. Anti-Cancer Agents Med Chem 2009;9(3):276–94.

[82] Blanco JG, Gil RR, Alvarez CI, Patrito LC, Genti-Raimondi S, Flury A. A novel activity for a group of sesquiterpene lactones: inhibition of aromatase. FEBS Letts 1997;409(3):396–400.

[83] Thiantanawat A, Long BJ, Brodie AM. Signaling pathways of apoptosis activated by aromatase inhibitors and antiestrogens. Cancer Res 2003;63(22):8037–50.

[84] Shindo K, Kimura M, Iga M. Potent antioxidative activity of cacalol, a sesquiterpene contained in *Cacalia delphiniifolia* Sleb et Zucc. Biosci Biotechnol Biochem 2004;68(6):1393–4.

[85] Li EW, Gao K, Jia ZJ. Two new eremophilenolides from Cacalia pilgeriana. Chinese Chem Lett 2005;16(9):1230–2.

[86] Li EW, Pan J, Gao K, Jia ZJ. New eremophilenolides from Cacalia pilgeriana. Planta Med 2005;71(12):1140–4.

[87] Nishikawa K, Aburai N, Yamada K, Koshino H, Tsuchiya E, Kimura KI. The bisabolane sesquiterpenoid endoperoxide, 3,6-epidioxy-1,10-bisaboladiene, isolated from *Cacalia delphiniifolia* inhibits the growth of human cancer cells and induces apoptosis. Biosci Biotechnol Biochem 2008;72(9):2463–6.

[88] Li EW, Gao K, Jia ZJ. *Ent*-Kaurenoids from the roots of *Cacalia pilgeriana*. J Asian Nat Prod Res 2007;9(3):191–5.

[89] Li EW, Gao K, Jia ZJ, Zhao X. Two new *ent*-kaurenoids from *Cacalia pilgeriana*. Chinese Chem Lett 2005;16(5):625–6.

[90] Liu W, Furuta E, Shindo K, Watabe M, Xing F, Pandey PR, et al. Cacalol, a natural sesquiterpene, induces apoptosis in breast cancer cells by modulating Akt-SREBP-FAS signaling pathway. Breast Cancer Res Treat 2011;128(1):57–68.

[91] Kuhajda FP. Fatty-acid synthase and human cancer: new perspectives on its role in tumor biology. Nutr 2000;16:202–8.

[92] Knowles LM, Yang C, Osterman A, Smith JW. Inhibition of fatty-acid synthase induces caspase-8-mediated tumor cell apoptosis by up-regulating DDIT4. J Biol Chem 2008;283(46):31378–31384.

[93] Shindo K, Kimura M, Iga M. Potent antioxidative activity of cacalol, a sesquiterpene contained in *Cacalia delphiniifolia* Sleb et Zucc. Biosci Biotechnol Biochem 2004;68(6):1393–4.

[94] Cruz CM, Rinna A, Forman HJ, Ventura ALM, Persechini PM, Ojcius DM. ATP activates a reactive oxygen species-dependent oxidative stress response and secretion of proinflammatory cytokines in macrophages. J Biol Chem 2007;282(5):2871–9.

[95] Wang Q, Mu Q, Shibano M, Morris-Natschke SL, Lee KH, Chen DF. Eremophilane sesquiterpenes from *Ligularia macrophylla*. J Nat Prod 2007;70(8):1259–62.

[96]   Huang GD, Yang YJ, Wu WS, Zhu Y. Terpenoids from the aerial parts of *Parasenecio deltophylla*. J Nat Prod 2010;73(11):1954–7.

[97]   Arciniegas A, Pérez-Castorena AL, Reyes S, Contreras JL, De Vivar AR. New oplopane and eremophilane derivatives from *Robinsonecio gerberifolius*. J Nat Prod 2003;66(2):225–9.

[98]   Wang X, Sun L, Huang K, Shi S, Zhang L, Xu J, et al. Phytochemical investigation and cytotoxic evaluation of the components of the medicinal plant *Ligularia atroviolacea*. Chem Biodivers 2009;6(7):1053–65.

[99]   Beattie KD, Waterman PG, Forster PI, Thompson DR, Leach DN. Chemical composition and cytotoxicity of oils and eremophilanes derived from various parts of *Eremophila mitchellii* Benth. (Myoporaceae). Phytochem 2011;72(4–5):400–8.

[100]   Song YX, Cheng B, Zhu X, Qiao LT, Wang JJ, Gu YC, et al. Synthesis and cytotoxic evaluation of eremophilane sesquiterpene 07H239-A derivatives. Chem Pharm Bull 2011;59(9):1186–9.

[101]   Oh H, Jensen PR, Murphy BT, Fiorilla C, Sullivan JF, Ramsey T, et al. Cryptosphaerolide, a cytotoxic Mcl-1 inhibitor from a marine-derived ascomycete related to the genus *Cryptosphaeria*. J Nat Prod 2010;73(5):998–1001.

Topic **2.2**

Diterpenes

2.2.1 *Clerodendrum Kaichianum* P.S. Hsu

History   The plant was first described by Ping Sheng Hsu in *Observationes ad Florulam Hwangshanicum*, published in 1965.

Family   Lamiaceae Martinov, 1820

Common Name   Zhe jiang da qing (Chinese)

Habitat and Description   It is a treelet which grows to 5 m tall in China. The stems are pitted, quadrangular and hairy at apex. The leaves are simple. The petiole is 2.5–5 cm long and straight. The blade is ovate oblong, dark green above, light green beneath, papery, round to wedge-shaped at base, acuminate at apex and presents five to six pairs of secondary nerves.

■ **FIGURE 2.6** *Clerodendrum kaichianum* P.S. Hsu

The inflorescence is a lax and terminal cyme. The calyx is cupular, some-what pink and develops five acute tiny lobes. The corolla is salver-shaped and white, with a slender 1-cm-long tube that develops five lobes which are oblong and 0.5 cm long. The androecium consists of four filamentous sta-mens which are conspicuously exerted. The style is filiform. The fruit is a blue drupe which is globose and seated in the calyx (Figure 2.6).

Medicinal Uses In China the plant is used to treat hypertension.

Phytopharmacology Phytochemical analyses of the plant are quite recent and resulted in the isolation of series of abietane diterpenes, such as 11-methoxyl-12,14-dihydroxy-13-(2-hydroxypropyl)-3,5,8,11,13-abietapentaen-7-one,[102] (3*S*,16*R*)-12,16-epoxy-3,6,11,14,17-pentahydroxy-17(15→16)-*abeo*-5,8,11,13-abietatetraen-7-one,[103] 17-hydroxyteuvincenone G, 17-hydroxyteuvincen-5(6)-enone G, teuvincenone A, 11,14-dihydroxy-abieta-8,11,13-trien-7-one, dehydroabietan-7-one, sugiol,[104] 5,6-dehydro-sugiol, villosin A and salvinolone.[105] To date the medicinal property of the plant remains unconfirmed. It might be interesting to look for vasorelaxant

■ **CS 2.64** 17-Hydroxyteuvincenone G

■ **CS 2.65** 17-Hydroxyteuvincen-5(6)-enone G

■ **CS 2.66** Carnosol

abietane diterpenes. Note that other types of diterpene, like forskolin, are well-known hypotensive agents.[106]

Proposed Research Pharmacological study of tanshinone IIA for the treatment of cancer.

Rationale Current evidence indicates that abietane diterpenes induce apoptosis in cancer cells mainly via mitochondrial insult, release of cyto-chrome c, caspase 3 activation and a decreased level of anti-apoptotic Bcl-2 protein. The abietane diterpenes 17-hydroxyteuvincenone G (CS 2.64) and 17-hydroxyteuvincen-5(6)-enone G (CS 2.65) isolated from *Clerodendrum kaichianum* P.S. Hsu inhibited the growth of human promyelocytic leukemia (HL-60) and human lung adenocarcinoma epithelial (A549) cells, with IC_{50} values equal to 5.9 µM, 9.3 µM and 15.91 µM and 10.35 µM, respectively.[104] Carnosol (CS 2.66) from *Rosmarinus officinalis* L. (family Lamiaceae Martinov) is a polyoxygenated abietane diterpene which abated the growth of human lymphoblastoid leukaemia (REH) cells by 60% at a dose of 9 µg/mL via mitochondrial membrane depolarization, a decrease in the level of anti-apoptotic Bcl-2 protein of 42% and apoptosis.[107] In fact, the anti-apoptotic Bcl-2 protein is necessary for maintaining normal mitochondrial membrane potential and mitigating release of cytochrome c and the resulting activa-tion of caspase 3.[108] Kellner et al. (2004) reported that carnosol, which is a strong antioxidant, increased the amount of nitric oxide in the cytoplasm of REH cells resulting in cytochrome c oxidase inhibition and therefore in a fall in mitochondrial membrane potential.[109] Additionally, human prostate cancer (PC-3) cell growth was lowered by 50% when 34 µmol/L of carnosol was administered, and this was associated with increases in cyclin-dependent kinase inhibitors p21 and p27 and pro-apoptotic Bcl-2-associated X protein (Bax) and decreases in cyclin-dependent kinases 2 and 6, cyclins A, D1 and D2, and in phosphoinositide 3-kinase (PI3K).[110]

Ferruginol from *Persea nubigena* L.O. Williams (family Lauraceae Juss.) differs from carnosol by the absence of oxygen atoms at C7, C11 and C20 and it inhibited the proliferation of PC-3 cells, with an IC_{50} value of 55 µM.[111] Like carnosol, ferruginol increased cyclin-dependent kinase inhibitor p21 and therefore decreased cyclin D1 and PI3K and activated caspase 3. Moreover, ferruginol was able to activate mitogen-activated pro-tein kinase (MAPK) p38, to induce the release of apoptosis inducing factor (AIF) from mitochondria and therefore chromatin condensation, to inhibit protein kinase B (Akt), to decrease of low molecular weight protein tyros-ine phosphatase (LMWPTP) and to increase nuclear factor kappa-light-chain-enhancer of activated B cells (NF-κB).[111]

■ **CS 2.67** Ferruginol

■ **CS 2.68** 6-Hydroxy-5,6-dehydrosugiol

■ **CS 2.69** Kayadiol

Ferruginol (CS 2.67) is structurally close to 6-hydroxy-5,6-dehydrosugiol (CS 2.68) from *Cryptomeria japonica* (Thunb. ex L. f.) D. Don (family Taxodiaceae Saporta) and induced: apoptosis in adenocarcinoma (LNCaP) cells at a dose of $10\,\mu$mol/L via caspase 3 activation; an increase in pro-apoptotic protein (p53); and a decrease in the anti-apoptotic Bcl-xL.[112] Kayadiol (CS 2.69) from *Torreya nucifera* (L.) Siebold & Zucc. (family Taxaceae Gray) was cytocidal to human epitheloid cervix carcinoma (Hela) cells, with an IC_{50} value of $30\,\mu$M, and was associated with DNA fragmentation, and, like carnosol, with depolarization of mitochondrial membrane potential, activation of caspase 3 and with an increase in pro-apoptotic Bax protein.[113]

Tanshinone IIA is an abietane diterpene from *Salvia miltiorrhiza* Bunge (family Lamiaceae Martinov) which activated caspase 3 in HL-60 cells and resulted in apoptosis with a dose of $3\,\mu$g/mL.[114] In human hepatocellular liver carcinoma (HepG2) cells, tanshinone IIA abrogated cell survival, with an IC_{50} value equal to $4.2\,\mu$M, via non-related glutathione apoptosis.[115] In addition, this compound repressed the growth of human hepatoma (SMMC-7721) cells (wild-type p53) by 85% at a dose of 1 mg/mL via an increase in Fas, pro-apoptotic protein (p53), pro-apoptotic Bax and decreased level of anti-apoptotic Bcl-2 protein.[116] Tanshinone IIA inhibited the growth of human promyelocytic leukemia (NB4) cells with an IC_{50} value of $19.3\,\mu$mol/L with, like carnosol, a decreased level of anti-apoptotic Bcl-2 protein, a fall in mitochondrial membrane potential and activation of caspase 3.[117] Tanshinone IIA (CS 2.70), like ferruginol, activated caspase 3 and decreased PI3K and Akt.[118] Wu et al. (1991) studied the structure–activity relationship of several tanshinone derivatives and framed the hypothesis that the planar aromatic rings allow DNA intercalation of the abietane diterpenes whereas the furano-*o*-quinone moiety could account for the generation of DNA-damaging reactive oxygen species (ROS).[119]

■ **CS 2.70** Tanshinone IIA

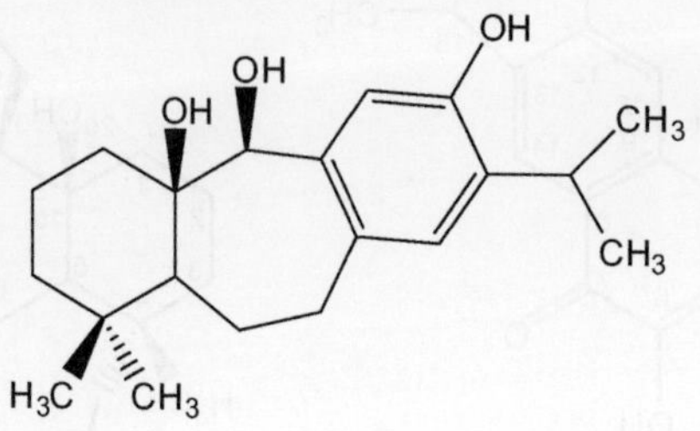

■ **CS 2.71** Pisiferdiol

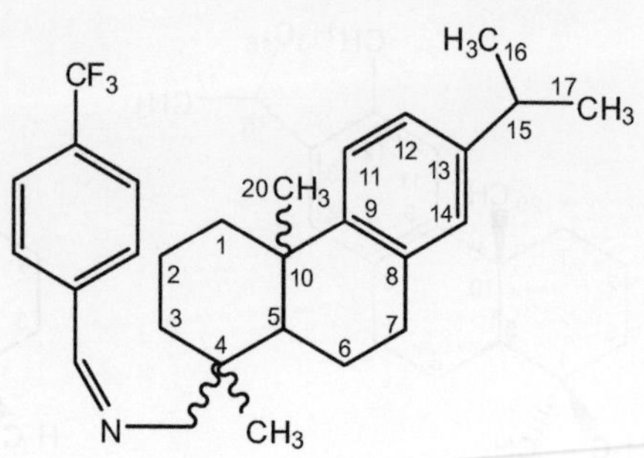

■ **CS 2.72** TBIDOM

Pisiferdiol (CS 2.71) from *Chamaecyparis pisifera* (Siebold & Zucc.) Endl. (family Cupressaceae Gray) is a cycloheptane-bearing abietane diterpene which induced apoptosis in HL-60 cells, with an IC_{50} value of 18.3 μM, via DNA fragmentation, chromatin condensation, caspase 3 activation, protein phosphatase 2C (PPC2) activation and proapoptotic protein Bad dephosphorylation.[120] The survival of SMMC-7721 cells exposed to 250 μM of *N*-(4-(2,2-trifluoroethyl)benzylidene)(7-isopropyl-1,4a-dimethyl-1,2,3,4,4a,9,10,10a-octahydrophenanthren-1-yl)meth-anamine (TBIDOM) (CS 2.72) was reduced by 93%, which, as with carnosol, was associated with a fall in the mitochondrial membrane potential followed by the release of cytochrome c, a decrease in the level of anti-apoptotic Bcl-2 protein and induction of caspase 3.[121]

2.2.2 *Euphorbia Fischeriana* Steud.

History This plant was first described by Ernst Gottlieb von Steudel in *Nomenclator Botanicus*, published in 1840.

Synonyms *Euphorbia fischeriana* var. *pilosa* (Regel) Kitag., *Euphorbia pallasii* Turcz., *Euphorbia pallasii* Turcz. ex Ledeb., *Euphorbia pallasii* var. *pilosa* Regel, *Tithymalus fischeranus* (Steud.) Soják, *Euphorbia verticillata* Fischer.

Family Euphorbiaceae Juss., 1789

Common Name Lang du (Chinese)

Habitat and Description *Euphorbia fischeriana* is a perennial herb that grows to 30 cm tall in the dry grasslands of Siberia, Mongolia, China,

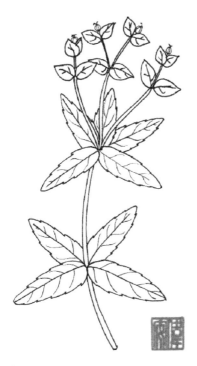

■ **FIGURE 2.7** *Euphorbia fischeriana* Steud.

Korea and Japan. The root is massive and somewhat anthropomorphic. The stem is fleshy, terete, dichotomously branched, reddish and laticiferous. The latex is pure white. The leaves are simple, opposite and sessile. The blade is light green, oblong and 5–7 cm × 1–2 cm in size. The inflorescence is a cyathium, which is sessile and seated in a campanulate involucre which is 0.5 cm long and develops five rounded lobes. The cyathium includes a female flower reduced to an ovary and stigma on a 0.4 cm pedicel which is surrounded by five male flowers reduced to a single stamen. The cyathophylls are triangular, 2 cm long and arranged in pairs. The fruit is a dehiscent explosive capsule which is ovoid, 0.6 cm across and pubescent. Each locule includes a seed which is 0.4 cm long (Figure 2.7).

Medicinal Uses In Korea the plant is used to relieve the bowels from constipation. In China, the plant is used to expel worms from the intestine and to treat skin diseases, edema, ascites and cancer.

Phytopharmacology The plant is known to contain a series of campesterol, stigmasterol and sitosterol derivatives,[122] triterpenes such as

■ **CS 2.73** Jolkinolide B

■ **CS 2.74** Jolkinolide D

■ **CS 2.75** 17-Acetoxyjolkinolide B

lupeol and β-amyrin-3-acetate,[123] the tigliane-type diterpenes 12-deoxy phorbol-13-hexadecanoate, 12-deoxyphorbol-13-acetate,[124] langduin A, prostratin,[125] 12-deoxyphorbaldehyde-13-acetate, 12-deoxyphorbaldehyde-13-hexadecacetate, 12-deoxyphorbol-13-(9Z)-octadecanoate-20-acetate, 12-deoxyphorbol-13-decanoate,[126] fischerosides A-C,[127] the *ent*-abietane diterpenes langduin B, 17-acetoxyjolkinolide A and B, jolkinolide A and B, 17-hydroxyjolkinolide A and B,[128] 7β,11β,12β-trihydroxy-ent-abieta-8(14), 13(15)-dien-16,12-olide, 17-acetoxyjolkinolide B, 13-hydroxy-ent-abiet-8(14)-en-7-one,[126] the pimarane diterpene 3R,17-dihydroxy-ent-pimara-8(14), 15-diene[126] unusual diterpenes of which langduin C[129] and D[130] and some phenolics such as scopoletin, physcion and 2,4-dihydroxy-6-methoxy-1-acetophenone.[123]

The anticancer property of the plant is owed to the *ent*-abietane 17-acetoxyjolkinolide A and B.[131,132]

Proposed Research Pharmacological study of 17-acetoxyjolkinolide B and derivatives for the treatment of cancer.

Rationale The *ent*-abietane diterpene jolkinolide B (CS 2.73) isolated from *Euphorbia fischeriana* Steud. abrogated the survival of adenocarcinoma (LNCaP), human prostate carcinoma (DU-145) and human prostate cancer (PC-3) cells, with IC_{50} values equal to $12.5\,\mu g/mL$, $45\,\mu g/mL$ and $75\,\mu g/mL$, respectively, by blocking thymidine incorporation into DNA and apoptosis.[133] Additionally, jolkinolide B inhibited the growth of human erythromyeloblastoid leukemia (K562) and human esophageal carcinoma (Eca-109) cells, with IC_{50} values equal to $12.1\,\mu g/mL$ and $23.7\,\mu g/mL$, respectively, accompanied by DNA fragmentation and apoptosis.[134] Several lines of evidence point to the fact that jolkinolide B and analogs have the ability to bring cancer cells to death via the formation of covalent bonds between the α,β-unsaturated γ lactone[135] and epoxydes[131] with the nucleophilic proteinic thiol group of cysteine and the amines of DNA. Sakakura et al. (2002) first identified the cytotoxic pharmacophore of jolkinolide D (CS 2.74) from *Euphorbia jolkini* Boiss. as the γ,δ-unsaturated β-hydroxy-α-methylene lactone group which alkylates thymidine as well as the thiol group of cysteine.[136,137] Yan et al. (2008) later observed that 17-acetoxyjolkinolide B (CS 2.75) inhibited tumor necrosis factor-α (TNF-α)-induced NF-κB signal transduction in human hepatocellular liver carcinoma (HepG2) cells by hampering p65 nuclear translocation and inhibitor of IκB kinase β (IKKβ) activity. The team proposed that the IKKβ activity resulted from a Michael addition between the α,β-unsaturated lactone with the thiols of cysteine.[131] Congruently, Wang et al. (2009) obtained evidence that 17-hydroxy-jolkinolide B (CS 2.76) inhibited interleukin-6 (IL-6)-induced signal transducer and activator of transcription 3 (STAT3) tyrosine phosphorylation by 100% at a dose of 10 μmol/L in HepG2 cells via cross-linking between the diterpene with the thiol group on cysteine residue of Janus kinase (JAK).[132]

2.2.3 *Vitex Rotundifolia* **L.f.**

History The plant was first described by Carl von Linnaeus filius in *Supplementum Plantarum*, published in 1782.

Synonyms *Vitex ovata* Thunb., *Vitex ovata* var. *subtrisecta* Kuntze, *Vitex trifolia* var. *ovata* (Thunb.) Makino, *Vitex trifolia* var. *simplicifolia* Cham., *Vitex trifolia* var. *unifoliolata* Schauer

Family Lamiaceae Martinov., 1820

Common Names Beach vitex, man hyung ja (Korea), hamagou (Japanese), dan ye man jing (Chinese)

Habitat and Description This invasive shrub, the seeds of which are dispersed by the sea, grows on the shores of the Mediterranean, Central Asia, India, Southeast Asia, China, Korea, Japan, Taiwan and the Pacific Islands. The stems are hairy at apex. The leaves are simple and sessile. The blade is ovate, 2.5–5 cm × 1.5–3 cm and hairy beneath. The inflorescence is a terminal thyrse. The calyx is cupular, 0.5 cm long and obscurely five-dentate. The corolla is purplish and develops five lobes. The four stamens

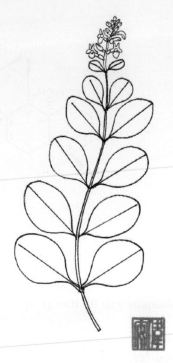

■ **FIGURE 2.8** *Vitex rotundifolia* L.f.

are exerted. The gynoecium includes a globular ovary. The fruit is a drupe which is dark brown and enclosed in the vestigial calyx (Figure 2.8).

Medicinal Uses In China the plant is used to treat asthma. In Japan, it is used to alleviate headaches.

Phytopharmacology Phytochemical analysis resulted in the identification of flavonoids such as vitexicarpin (casticin), luteolin,[138] artemetin,[139] lignans,[140] the iridoid agnuside, eurostoside,[141] eucommiol, 1-oxo-eucommiol, agnuside, VR-I, viteoid I-II, iridolactone, pedicularis-lactone,[142] the phenylpropanoid glycosides vitexfolin A,[143] cohorts of labdane diterpenes including rotundifuran, prerotundifuran, vitexilactone, previtexilactone,[138] viteoside A, vitexifolin A, D and E,[144–147] the abietane diterpenes ferruginol and abietatrien-3β-ol,[145] the clerodane diterpene vitexifolin B,[147] and the halimane diterpene vitetrifolin D.[147]

The analgesic and anti-inflammatory properties of the plant have been substantiated in vitro and are imparted to vitexfolin A,[143] vitexicarpin[148] and casticin.[149] 1H,8H-Pyrano[3,4-c]pyran-1,8-dione has been identified as the anti-asthmatic principle.[150]

Proposed Research Pharmacological study of rotundifuran and derivatives for the treatment of leukemia and/or myeloma.

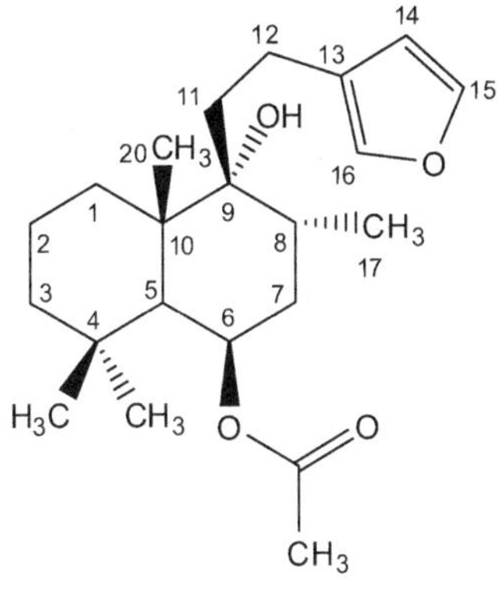

Rationale There is a growing body of evidence suggesting that labdane diterpenes induce apoptosis of leukemic cells via multiple pathways. One such labdane is rotundifuran (CS 2.77) isolated from *Vitex rotundifolia* L.f., which inhibited the growth of human promyelocytic leukemia (HL-60) cells, with an IC_{50} value of 22.5 μM, via DNA fragmentation and apoptosis.[151] The labdane diterpene sclareol (CS 2.78) from *Cistus incanus* L. (family Cistaceae, Juss.) elicited apoptotic effects against a broad spectrum of leukemic cell lines, especially human lymphoma HUT78 and H9 cells, with IC_{50} values equal to 9.5 μg/mL and 6 μg/mL, respectively,[152] by possible downregulation of c-Myc and therefore transcription.[153] In addition, sclareol at a dose of 100 μM induced apoptosis in the human breast adenocarcinoma (MCF-7) cell-line variants MN1 (wild-type p53) and MDD2 (p53 null) independently of p53.[154]

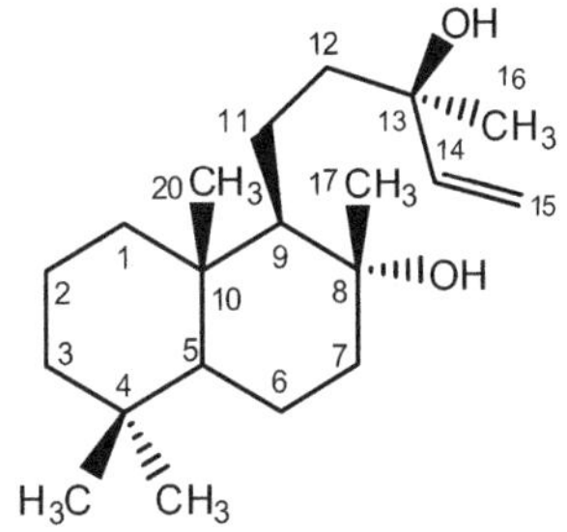

The labdane diterpene myriadenolide (CS 2.79) from *Alomia myriadenia* Sch. Bip. ex Baker (family Asteraceae, Bercht. & J. Presl) induced apoptosis in both Jurkat human T cell lymphoblast-like cells and human acute monocytic leukemia (THP-1) cells, with IC_{50} values equal to 6 μM, that was associated with DNA fragmentation, a fall in mitochondrial membrane potential, activation of caspases 3, 8 and 9 and cleavage of the pro-apoptotic BH3 interacting-domain death agonist (Bid),[155] which interacts with the Bax protein. In human chronic myeloid leukemia (KBM-5) cells, coronarin D (CS 2.80) suppressed TNF-α-induced NF-κB activation at a dose of 10 μmol/L and inhibited NF-κB activation by abating IKKβ activation,

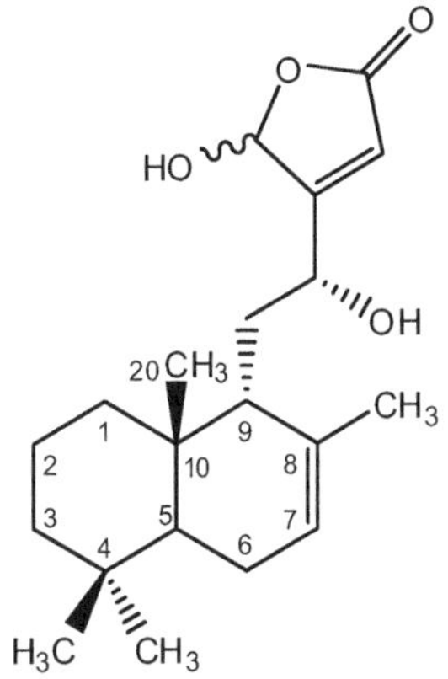

■ **CS 2.77** Rotundifuran

■ **CS 2.78** Sclareol

■ **CS 2.79** Myriadenolide

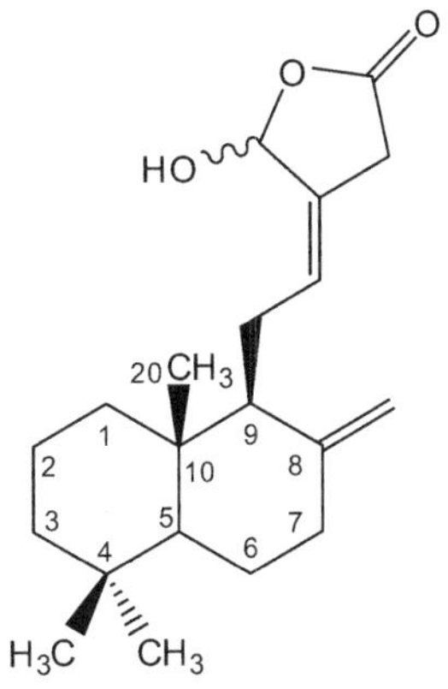

■ **CS 2.80** Coronarin D

■ **CS 2.81** Forskolin

resulting in the suppression of IκBα phosphorylation, p65 phosphorylation and nuclear translocation.[156] The survival of multiple mouse myeloma (MOPC315) cells was severely compromised by the polyoxygenated labdane diterpene forskolin (CS 2.81) from *Coleus forskohlii* (Willd.) Briq. (family Lamiaceae, Martinov) at a dose of 50 μM, owing to cAMP-induced apoptosis with an accompanying fall in mitochondrial membrane potential, release of cytochrome c, activation of caspases 9 and 3 and cleavage of PARP (Poly [ADP-ribose] polymerase).[157]

2.2.4 *Isodon Rubescens* (Hemsl.) H. Hara

History This plant was first described by Hiroshi Hara in *Journal of Japanese Botany*, published in 1985.

Synonyms *Isodon henryi* var. *dichromophyllus* (Diels) Kudô, *Isodon ricinispermus* (Pamp.) Kudô, *Plectranthus dichromophyllus* Diels, *Plectranthus ricinispermus* Pamp., *Plectranthus rubescens* Hemsl., *Rabdosia dichromophylla* (Diels) H. Hara, *Rabdosia ricinisperma* (Pamp.) H. Hara, *Rabdosia rubescens* (Hemsl.) H. Hara, *Rabdosia rubescens* fo. *lushanensis* Z.Y. Gao & Y.R. Li, Rabdosia rubescens var. lushiensis Z.Y. Gao & Y.R. Li.

Family Lamiaceae Martinov., 1820

Common Name Sui mi ya (Chinese)

Habitat and Description *Isodon rubescens* is a Chinese shrub which grows to 1 m. The stems are hairy at apex. The petiole is 1–3.5 cm long. The blade is ovate, 2–5 cm × 1.5–3 cm, thin, papery, cuneate at base, serrate, acuminate at apex, and with four or five pairs of secondary nerves. The inflorescence is a terminal panicle of cymes which is 5–15 cm long and hairy. The calyx is campanulate, 0.3 cm long, hairy, reddish and with two lips. The corolla is 1 cm long, light purple and has two lips, the upper lip having four clefts that are recurved, the lower lip being entire and concave. The androecium includes four exerted stamens. The style is exerted. Nutlets are brown, ovoid and minute (Figure 2.9).

Medicinal Uses In China the plant is used to treat cancer of the liver, pancreas, esophagus, breast and colon.

Phytopharmacology The plant produces an exceptional array of *ent*-kaurane diterpenes taibairubescensin A and B,[158] taibairubescensin C,[159] lushanrubescensins A–E, lasiodonin, oridonin, ponicidin, isodonoiol, isodonal,

■ **FIGURE 2.9** *Isodon rubescens* (Hemsl.) H. Hara

rabdosin B, enmenol, epinodosin, inflexusin,[160] rabdoternins A–G[160–162], rubescensins C and I–Q,[161–163] hebeirubescensins A–L, lasiokaurin, macrocalin B and xerophilusin B,[162] guidongnins B–H, ludongnin A and B, guidongnin, angustifolin and 6-epiangustifolin,[163] and bisrubescensins A-C.[164]

The anticancer property of the plant is due to oridonin and ponicidin which are the main constituents.[162]

Proposed Research Pharmacological study of oridonin and derivatives for the treatment of cancers.

Rationale Several lines of evidence demonstrate the *ent*-abietane diterpene oridonin precipitates apoptosis in a broad array of malignancies via manifold apoptotic pathways. The apoptotic mechanism of oridonin in 'wild-type' p53 cancer cells involves inactivation of Akt, activation of pro-apoptotic protein (p53), translocation of pro-apoptotic Bcl-2-associated X protein (Bax), a reduction in the extracellular-signal release of cytochrome c and subsequent activation of caspase 3. Indeed, Zhang et al. (2004)

■ **CS 2.82** Oridonin

observed that oridonin induced DNA fragmentation, caspase 3 activation, an increase in the protein level of pro-apoptotic Bax and a decreased level of anti-apoptotic Bcl-2 protein in human epitheloid cervix carcinoma (Hela) cells 'wild-type' p53.[166] Hela cells exposed to 25 µmol/L of oridonin (CS 2.82) became apoptotic, with inactivation of Akt and dephosphorylation of glycogen synthetase kinase 3 (GSK3) followed by the release of cytochrome c, activation of caspase 3, PARP cleavage and downregulation of the anti-apoptotic proteins cIAP1, X-linked inhibitor of apoptosis protein (XIAP) and survivin.[167] Of compelling oncological interest is the fact that oridonin inhibited the growth of human lung adenocarcinoma (SPC-A-1) p53 mutant cells, with an IC_{50} value equal to 27.8 µmol/L, involving DNA fragmentation, a decreased level of anti-apoptotic Bcl-2 protein and increased Bax.[168] Furthermore, oridonin induced apoptosis in human hepatocellular liver carcinoma (HepG2) cells (wild-type p53), with an IC_{50} value equal to 30 µM, via pro-apoptotic protein (p53)-induced generation of ROS, which induced, simultaneously, MAPK p38 activation and a fall in mitochondrial membrane potential, release of cytochrome c and activation of caspases 9 and 3.[169] Oridonin at a dose of 20 µM induced apoptosis in human epithelial carcinoma (A-431) cells, with activation of caspase 3, reduced epidermal growth factor (EGFR) tyrosine phosphorylation and therefore a decrease in growth factor receptor-bound protein 2 (Grb2), G protein Ras, Raf and extracellular-signal-regulated kinases (ERK).[170] Furthermore, oridonin abrogated the survival of human malignant melanoma (A375-S2) cells (wild-type p53), with an IC_{50} value of 15.1 µmol/L, via DNA fragmentation, release of cytochrome c, activation of caspases 9, 8 and 3, an increase in the level of pro-apoptotic Bax,[171] which induced the activation of extracellular-signal-regulated kinases (ERK) and therefore the activation of pro-apoptotic protein (p53).[172]

Onirodin induced apoptosis in human laryngeal carcinoma (Hep2) cells, with an IC_{50} value equal to 37.1 µM, through an increase in ROS and pro-apoptotic Bax protein, which both altered the mitochondrial membrane potential and compelled the release of cytochrome c and AIF. This resulted in caspase 3 activation and therefore in cleavage of inhibitor of caspase-activated DNase (ICAD) into caspase-activated DNase (CAD), thereby bringing about fragmentation of DNA.[173]

In addition, onirodin induced the cleavage of cyclin-dependent kinase inhibitor p21[173] and inhibited the growth both of human colorectal carcinoma (HCT-116) and human colon adenocarcinoma (SW1116) cells, with IC_{50} values equal to 12.5 µM, by causing a decrease in cyclin B and c-Myc and elevation of cyclin-dependent kinase inhibitor p21.[174] Oridonin prompted apoptosis in human promyelocytic leukemia (HL-60) cells at a dose of 8 μmol/L with a decrease in human telomerase reverse transcriptase (hTERT) mRNA and telomerase activity.[175] Additionally, oridonin provoked apoptosis in Jurkat (human T cell lymphoblast-like) cells via caspase 3 activation at a dose of 32 μmol/L.[176] Oridonin impaired the growth of human leukemia (HPB-ALL) cells, with an IC_{50} value of 23.7 μmol/L, via DNA fragmentation, a fall of mitochondrial membrane potential, activation of caspase 3, a decrease in the levels of anti-apoptotic Bcl-2 and Bcl-xL proteins, and an increase in the levels of pro-apoptotic Bax and the death agonist Bid.[177] Moreover, oridonin abated the proliferation of human T-lymphotropic virus Type I-infected T cells (human lymphocyte MT1 and MT2), multiple myeloma (human multiple myeloma U266 and RPMI8226) and acute lymphoblastic T-cell leukemia (Jurkat human T cell lymphoblast-like cells), with ED_{50} values equal to 1.2 μg/mL, 1.2 μg/mL, 0.8 μg/mL, 1.7 μg/mL and 2.7 μg/mL, respectively, which was mediated by falls in anti-apoptotic myeloid leukemia cell differentiation protein (Mcl-1) and in anti-apoptotic protein Bcl-xL levels and by blockade of NF-κB binding to DNA.[178] Oridonin abrogated the multiplication of human osteosarcoma (U2OS) and (MG63) and human osteogenic sarcoma (SaOS-2) cells, with an IC_{50} of 35 μM in each case, via a fall of mitochondrial membrane potential, release of cytochrome c, binding of cytochrome c to apoptotic protease activating factor 1 (Apaf-1) (and consequent caspases 9 and 3 activation and PARP cleavage), inactivation of Akt, activation of MAPK p38 and pro-apoptotic c-Jun N-terminal kinases (JNK) and downregulation of the anti-apoptotic proteins cIAP1, XIAP and survivin.[179] Oridonin induced the death of healthy mouse fibroblast (L929) cells (wild-type p53), with an IC_{50} value of 9.1 μM, through pro-apoptotic protein (p53) activation and a consequent increase in the level of pro-apoptotic Bax.[180] The activation of p53 by oridonin was promulgated by decreases of cyclin B, G protein Ras and Raf and the subsequent activation of extracellular-signal-regulated kinases (ERK).[181]

In human premalignant breast epithelial (MCF-10A) cells oridonin at 5 μM reduced the levels of cyclin B1 and transcription factor p65, activated retinoblastoma tumor suppressor protein (Rb), diminished the levels of the transcription factor E2F and suppressed NF-κB.[182]

REFERENCES

[102]   Xu M, Shen L, Wang K, Du Q. A new abietane diterpenoid from *Clerodendrum kaichianum* Hsu. J Chem Res 2010;34(12):722–3.

[103]   Xu MF, Shen LQ, Wang KW, Du QZ, Wang N. A new rearranged abietane diterpenoid from *Clerodendrum kaichianum* Hsu. J Asian Nat Prod Res 2011;13(3):260–4.

[104]   Xu M, Shen L, Wang K, Du Q. Two new abietane diterpenoids from the stems of *Clerodendrum kaichianum* P. S. Hsu. Helvetica Chimica Acta 2011;94(3):539–44.

[105]   Xu M, Shen L, Wang K, Du Q, Wang N. Bioactive diterpenes from *Clerodendrum kaichianum*. Nat Prod Commun 2011;6(1):3–5.

[106]   Tirapelli CR, Ambrosio SR, De Oliveira AM, Tostes RC. Hypotensive action of naturally occurring diterpenes: a therapeutic promise for the treatment of hypertension. Fitoterapia 2010;81(7):690–702.

[107]   Dörrie J, Sapala K, Zunino SJ. Carnosol-induced apoptosis and downregulation of Bcl-2 in B-lineage leukemia cells. Cancer Lett 2001;170(1):33–9.

[108]   Vander Heiden MG, Thompson CB. Bcl-2 proteins: regulators of apoptosis or of mitochondrial homeostasis. Nat Cell Biol 1999;1:209–16.

[109]   Kellner C, Zunino SJ. Nitric oxide is synthesized in acute leukemia cells after exposure to phenolic antioxidants and initially protects against mitochondrial membrane depolarization. Cancer Lett 2004;215(1):43–52.

[110]   Johnson JJ, Syed DN, Heren CR, Suh Y, Adhami VM, Mukhtar H. Carnosol, a dietary diterpene, displays growth inhibitory effects in human prostate cancer PC3 cells leading to G2-phase cell cycle arrest and targets the 5'-AMP-activated protein kinase (AMPK) pathway. Pharmaceutical Research 2008;25(9):2125–34.

[111]   Bispo de Jesus M, Zambuzzi WF, Ruela de Sousa RR, Areche C, Santos de Souza AC, Aoyama H, et al. Ferruginol suppresses survival signaling pathways in androgen-independent human prostate cancer cells. Biochimie 2008;90(6):843–54.

[112]   Lin FM, Tsai CH, Yang YC, Tu WC, Chen LR, Liang YS, et al. A novel diterpene suppresses CWR22Rv1 tumor growth in vivo through antiproliferation and proapoptosis. Cancer Res 2008;68(16):6634–42.

[113]   Chen SP, Dong M, Kita K, Shi QW, Cong B, Gou WZ, et al. Anti-proliferative and apoptosis-inducible activity of labdane and abietane diterpenoids from the pulp of *Torreya nucifera* in HeLa cells. Mol Med Rep 2010;3(4):673–8.

[114]   Sung HJ, Choi SM, Yoon Y, An KS. Tanshinone IIA, an ingredient of *Salvia miltiorrhiza* BUNGE, induces apoptosis in human leukemia cell lines through the activation of caspase-3. Exp Mol Med 1999;31(4):174–8.

[115]   Lee WYW, Chiu LCM, Yeung JHK. Cytotoxicity of major tanshinones isolated from Danshen (*Salvia miltiorrhiza*) on HepG2 cells in relation to glutathione perturbation. Food Chem Toxicol 2008;46(1):328–38.

[116]   Yuan SL, Wei YQ, Wang XJ, Xiao F, Li SF, Zhang J. Growth inhibition and apoptosis induction of tanshinone IIA on human hepatocellular carcinoma cells. World J. Gastroenterol. 2004;10(14):2024–8.

[117]   Liu JJ, Lin DJ, Liu PQ, Huang M, Li XD, Huang RW. Induction of apoptosis and inhibition of cell adhesive and invasive effects by tanshinone IIA in acute promyelocytic leukemia cells in vitro. J Biomed Sci 2006;13(6):813–23.

[118]   Won SH, Lee HJ, Jeong SJ, Lee HJ, Lee EO, Jung DB, et al. Tanshinone IIa induces mitochondria dependent apoptosis in prostate cancer cells in association with an inhibition of phosphoinositide 3-kinase/AKT pathway. Biol Pharm Bull 2010;33(11):1828–34.

[119]   Wu WL, Chang WL, Chen CF. Cytotoxic activities of tanshinones against human carcinoma cell lines. Am. J. Chin. Med 1991;19(3–4):207–16.

[120]   Aburai N, Yoshida M, Ohnishi M, Kimura K. Pisiferdiol and pisiferic acid isolated from *Chamaecyparis pisifera* activate protein phosphatase 2C in vitro and

induce caspase-3/7-dependent apoptosis via dephosphorylation of Bad in HL60 cells. Phytomed 2010;17(10):782–8.

[121]   Li F, He L, Song ZQ, Yao JC, Rao XP, Li HT. Cytotoxic effects and pro-apoptotic mechanism of TBIDOM, a novel dehydroabietylamine derivative, on human hepatocellular carcinoma SMMC-7721 cells. J Pharm Pharmacol 2006;60(2):205–11.

[122]   Schroeder G, Rohmer M, Beck JP, Anton R. 7-Oxo-, 7α-hydroxy- and 7β-hydroxysterols from Euphorbia fischeriana. Phytochem 1980;19(10):2213–5.

[123]   Sun YX, Liu JC. Chemical constituents and biological activities of Euphorbia fischeriana Steud. Chem Biodivers 2011;8(7):1205–14.

[124]   Liu WZ, Wu XY, Yang GJ, Ma QG, Zhou TX, Tang XC, et al. 12-Deoxyphorbolesters from *Euphorbia fischeriana*. Chin Chem Letts 1996;7(10):917–8.

[125]   Ma QG, Liu WZ, Wu XY, Zhou TX, Qin GW. Diterpenoids from *Euphorbia fischeriana*. Phytochem 1997;44(4):663–6.

[126]   Wang YB, Huang R, Wang HB, Jin HZ, Lou LG, Qin GW. Diterpenoids from the roots of *Euphorbia fischeriana*. J Nat Prod 2006;69(6):967–70.

[127]   Pan LL, Fang PL, Zhang XJ, Ni W, Li L, Yang LM, et al. Tigliane-type diterpenoid glycosides from *Euphorbia fischeriana*. J Nat Prod 2011;74(6):1508–12.

[128]   Che CT, Zhou TX, Ma QG, Qin GW, Williams ID, Wu HM, et al. Diterpenes and aromatic compounds from *Euphorbia fischeriana*. Phytochem 1999;52(1):117–21.

[129]   Zhou TX, Bao GH, Ma QG, Qin GW, Che CT, Lv Y, et al. Langduin C, a novel dimeric diterpenoid from the roots of *Euphorbia fischeriana*. Tetrahedron Lett 2003;44(1):135–7.

[130]   Wang YB, Yao GM, Wang HB, Qin GW. A novel diterpenoid from *Euphorbia fischeriana*. Chem Lett 2005;34(8):1160–1.

[131]   Yan SS, Li Y, Wang Y, Shen S, Gu Y, Wang HB, et al. 17-Acetoxyjolkinolide B irreversibly inhibits IκB kinase and induces apoptosis of tumor cells. Mol Cancer Ther 2008;7(6):1523–32.

[132]   Wang Y, Ma X, Yan S, Shen S, Zhu H, Gu Y, et al. 17-Hydroxy-jolkinolide B inhibits signal transducers and activators of transcription 3 signaling by covalently cross-linking Janus kinases and induces apoptosis of human cancer cells. Cancer Res 2009;69(18):7302–10.

[133]   Liu WK, Ho JCK, Qin G, Che CT. Jolkinolide B induces neuroendocrine differentiation of human prostate LNCaP cancer cell line. Biochem Pharmacol 2002;63(5):951–7.

[134]   Luo H, Wang A. Induction of apoptosis in K562 cells by jolkinolide B. Canadian J Physiol Pharmacol 2006;84(10):959–65.

[135]   Kupchan SM, Giacobbe TJ, Krull IS, Thomas SM, Eakin MA, Fessler DC. Tumor inhibitors. LVII. Reaction of endocyclic .alpha.,.beta.unsaturated gamma. lactones with thiols. J Org Chem 1970;35(10):3539–43.

[136]   Sakakura A, Takayanagi Y, Kigoshi H. Jolkinolide D pharmacophore: synthesis and reaction with amino acids, nucleosides, and DNA. Tetrahedron Lett 2002;43(34):6055–8.

[137]   Sakakura A, Takayanagi Y, Shimogawa H, Kigoshi H. Jolkinolide D pharmacophore: synthesis and reaction with biomolecules. Tetrahedron 2004;60(33):7067–75.

[138]   Kondo Y, Sugiyama K, Nozoe S. Studies on the constituents of *Vitex rotundifolia* L. fil. Chem Pharm Bull 1986;34(11):4829–32.

[139] Ko WG, Kang TH, Lee SJ, Kim NY, Kim YC, Sohn DH, et al. Polymethoxyflavonoids from *Vitex rotundifolia* inhibit proliferation by inducing apoptosis in human myeloid leukemia cells. Food ChemToxicol 2000;38(10):861–5.

[140] Kawazoe K, Yutani A, Takaishi Y. Aryl naphthalenes norlignans from *Vitex rotundifolia*. Phytochem 1999;52(8):1657–9.

[141] Kouno I, Inoue M, Onizuka Y, Fujisaki T, Kawano N. Iridoid and phenolic glucoside from *Vitex rotundifolia*. Phytochem 1988;27(2):611–2.

[142] Ono M, Ito Y, Kubo S, Nohara T. Two new iridoids from *viticis trifoliae fructus* (fruit of *Vitex rotundifolia* L.). Chem Pharm Bull 1997;45(6):1094–6.

[143] Okuyama E, Fujimori S, Yamazaki M, Deyama T. Pharmacologically active components of *Viticis fructus* (*Vitex rotundifolia*). II. The components having analgesic effects. Chem Pharm Bull 1998;46(4):655–62.

[144] Ono M, Ito Y, Nohara T. A labdane diterpene glycoside from fruit of *Vitex rotundifolia*. Phytochem 1998;48(1):207–9.

[145] Ono M, Yamamoto M, Masuoka C, Ito Y, Yamashita M, Nohara T. Diterpenes from the fruits of *Vitex rotundifolia*. J Nat Prod 1999;62(11):1532–7.

[146] Ono M, Yamamoto M, Yanaka T, Ito Y, Nohara T. Ten new labdane-type diterpenes from the fruit of *Vitex rotundifolia*. Chem Pharm Bull 2001;4(1):82–6.

[147] Ono M, Yanaka T, Yamamoto M, Ito Y, Nohara T. New diterpenes and norditerpenes from the fruits of *Vitex rotundifolia*. J Nat Prod 2002;65(4):537–41.

[147] You KM, Son KH, Chang HW, Kang SS, Kim HP. Vitexicarpin, a flavonoid from the fruits of *Vitex rotundifolia*, inhibits mouse lymphocyte proliferation and growth of cell lines in vitro. Planta Med 1998;64(6):546–50.

[149] Lin S, Zhang H, Han T, Wu JZ, Rahman K, Qin LP. In vivo effect of casticin on acute inflammation. J Chinese Integr Med 2007;5(5):573–6.

[150] Lee H, Han AR, Kim Y, Choi SH, Ko E, Lee NY, et al. A new compound, 1H,8H-Pyrano[3,4-c]pyran-1,8-dione, suppresses airway epithelial cell inflammatory responses in a murine model of asthma. Int J Immunopathol Pharmacol 2009;22(3):591–603.

[151] Ko WG, Kang TH, Lee SJ, Kim YC, Lee BH. Rotundifuran, a labdane type diterpene from *Vitex rotundifolia*, induces apoptosis in human myeloid leukaemia cells. Phytother Res 2001;15(6):535–7.

[152] Dimas K, Kokkinopoulos D, Demetzos C, Vaos B, Marselos M, Malamas M, et al. The effect of sclareol on growth and cell cycle progression of human leukemic cell lines. Leukemia Res 1999;23(3):217–34.

[153] Dimas K, Demetzos C, Vaos V, Ioannidis P, Trangas T. Labdane type diterpenes down-regulate the expression of c-myc protein, but not of bcl-2, in human leukemia T-cells undergoing apoptosis. Leukemia Res 2001;25(6):449–54.

[154] Dimas K, Papadaki M, Tsimplouli C, Hatziantoniou S, Alevizopoulos K, Pantazis P, et al. Labd-14-ene-8,13-diol (sclareol) induces cell cycle arrest and apoptosis in human breast cancer cells and enhances the activity of anticancer drugs. Biomed Pharmacother 2006;60(3):127–33.

[155] Souza-Fagundes EM, Brumatti G, Martins-Filho OA, Corrêa-Oliveira R, Zani CL, Amarante-Mendes GP. Myriadenolide, a labdane diterpene isolated from *Alomia myriadenia* (asteraceae) induces depolarization of mitochondrial

membranes and apoptosis associated with activation of caspases-8, -9, and -3 in Jurkat and THP-1 cells. Exp Cell Res 2003;290(2):420–6.

[156] Kunnumakkara AB, Ichikawa H, Anand P, Mohankumar CJ, Hema PS, Nair MS, et al. Coronarin D, a labdane diterpene, inhibits both constitutive and inducible nuclear factor-κB pathway activation, leading to potentiation of apoptosis, inhibition of invasion, and suppression of osteoclastogenesis. Mol Cancer Ther 2008;7(10):3306–17.

[157] Follin-Arbelet V, Hofgaard PO, Hauglin H, Naderi S, Sundan A, Blomhoff R, et al. Cyclic AMP induces apoptosis in multiple myeloma cells and inhibits tumor development in a mouse myeloma model. BMC Cancer 2011;11:301.

[158] Li BL, Chen SN, Shi ZX, Tian X, Chen YZ. Two new diterpenoids from *Isodon rubescens*. Chinese Chem Lett 2000;11(1):43–4.

[159] Li BL, Shi ZX, Pan YJ, Tian XH. A new diterpenoid, taibairubescensin C, from *Isodon rubescens*. Polish J Chem 2002;76(5):721–4.

[160] Han QB, Li ML, Li SH, Mou YK, Lin ZW, Sun HD. *Ent*-kaurane diterpenoids from *Isodon rubescens* var. lushanensis. Chem Pharm Bull 2003;51(7):790–3.

[161] Han QB, Jiang B, Zhang JX, Niu XM, Sun HD. Two novel ent-kaurene diterpenoids from *Isodon rubescens*. Helvetica Chimica Acta 2003;86(3):773–7.

[162] Han QB, Zhao QS, Li SH, Peng LY, Sun HD. ent-Kaurane diterpenoids from *Isodon rubescens* collected in Guizhou Province. Acta Chimica Sinica 2003;61(7):1077–82.

[163] Han QB, Li RT, Zhang JX, Sun HD. New *ent*-Abietanoids from *Isodon rubescens*. Helvetica Chimica Acta 2004;87(4):1007–15.

[164] Huang SX, Xiao WL, Li LM, Li SH, Zhou Y, Ding LS, et al. Bisrubescensins A–C: three new dimeric *ent*-kauranoids isolated from *Isodon rubescens*. Org Lett 2006;8(6):1157–60.

[165] Huang SX, Zhou Y, Pu JX, Li RT, Li X, Xiao WL, et al. Cytotoxic *ent*-kauranoid derivatives from *Isodon rubescens*. Tetrahedron 2006;62(20):4941–7.

[166] Zhang CL, Wu LJ, Tashiro SI, Onodera S, Ikejima T. Oridonin induces apoptosis of HeLa cells via altering expression of Bcl-2/Bax and activating caspase-3/ICAD pathway. Acta Pharmacologica Sinica 2004;25(5):691–8.

[167] Hu HZ, Yang YB, Xu XD, Shen HW, Shu YM, Ren Z, et al. Oridonin induces apoptosis via PI3K/Akt pathway in cervical carcinoma HeLa cell line. Acta Pharmacologica Sinica 2007;28(11):1819–26.

[168] Liu JJ, Huang RW, Lin DJ, Peng J, Wu XY, Pan XL, et al. Anti-proliferative effects of oridonin on SPC-A-1 cells and its mechanism of action. J Int Med Res 2004;32(6):617–25.

[169] Huang J, Wu L, Tashiro SI, Onodera S, Ikejima T. Reactive oxygen species mediate oridonin-induced HepG2 apoptosis through p53, MAPK, and mitochondrial signaling pathways. J Pharmacol Sci 2008;107(4):370–9.

[170] Li D, Wu LJ, Tashiro SI, Onodera S, Ikejima T. Oridonin-induced A431 cell apoptosis partially through blockage of the Ras/Raf/ERK signal pathway. J Pharmacol Sci 2007;103(1):56–66.

[171] Zhang CL, Wu LJ, Tashiro SI, Onodera S, Ikejima T. Oridonin induced A375-S2 cell apoptosis via bax-regulated caspase pathway activation, dependent on the cytochrome C/caspase-9 apoptosome. J Asian Nat Prod Res 2004;6(2):127–38.

[172]   Zhang CL, Wu LJ, Zuo HJ, Tashiro SI, Onodera S, Ikejima T. Cytochrome c release from oridonin-treated apoptotic A375-S2 cells is dependent on p53 and extracellular signal-regulated kinase activation. J Pharmacol Sci 2004;96(2):155–63.

[173]   Kang N, Zhang JH, Qiu F, Chen S, Tashiro SI, Onodera S, et al. Induction of G2/M phase arrest and apoptosis by oridonin in human laryngeal carcinoma cells. J Nat Prod 2010;73(6):1058–63.

[174]   Gao FH, Hu XH, Li W, Liu H, Zhang YJ, Guo ZY, et al. Oridonin induces apoptosis and senescence in colorectal cancer cells by increasing histone hyperacetylation and regulation of p16, p21, p27 and c-myc. BMC Cancer 2010;10:610.

[175]   Liu JJ, Wu XY, Peng J, Pan XL, Lu HL. Antiproliferation effects of oridonin on HL-60 cells. Annals of Hematology 2004;83(11):691–5.

[176]   Liu J, Huang R, Lin D, Wu X, Wu X, Lu H, et al. Oridonin-induced apoptosis of Jurkat cells and its mechanism. Comparative Clinical Pathology 2004; 13(2):65–9.

[177]   Liu JJ, Huang RW, Lin DJ, Wu XY, Peng J, Pan XL, et al. Antiproliferation effects of oridonin on HPB-ALL cells and its mechanisms of action. American J Hematol 2006;81(2):86–94.

[178]   Ikezoe T, Yang Y, Bandobashi K, Saito T, Takemoto S, Machida H, et al. Oridonin, a diterpenoid purified from Rabdosia rubescens, inhibits the proliferation of cells from lymphoid malignancies in association with blockade of the NF-κB signal pathways. Mol Cancer Ther 2005;4(4):578–86.

[179]   Jin S, Shen JN, Wang J, Huang G, Zhou JG. Oridonin induced apoptosis through Akt and MAPKs signaling pathways in human osteosarcoma cells. Cancer Biol Ther 2007;6(2):261–8.

[180]   Huang J, Wu L, Tashiro SI, Onodera S, Ikejima T. Bcl-2 up-regulation and P–p53 down-regulation account for the low sensitivity of murine L929 fibrosarcoma cells to oridonin-induced apoptosis. Biol Pharm Bull 2005;28(11):2068–74.

[181]   Cheng Y, Qiu F, Ye Yc, Tashiro Si, Onodera S, Ikejima T. Oridonin induces G2/M arrest and apoptosis via activating ERK-p53 apoptotic pathway and inhibiting PTK-Ras-Raf-JNK survival pathway in murine fibrosarcoma L929 cells. Arch Biochem Biophys 2009;490(1):70–5.

[182]   Hsieh TC, Wijeratne EK, Liang JY, Gunatilaka AL, Wu JM. Differential control of growth, cell cycle progression, and expression of NF-κB in human breast cancer cells MCF-7, MCF-10A, and MDA-MB-231 by ponicidin and oridonin, diterpenoids from the Chinese herb Rabdosia rubescens. Biochem Biophys Res Commun 2005;337(1):224–31.

Topic **2.3**

Triterpenes

2.3.1 *Alisma Orientale* **(Sam.) Juz.**

History This plant was first described by Vladimir Leontjevich Komarov in *Flora URSS*, published in 1934.

Synonyms *Alisma plantago-aquatica* L. var. orientale Sam., *Alisma plantago-aquatica* subsp. *orientale* (Sam.) Sam.

Family Alismataceae Vent., 1799

Common Names Takusha (Japanese), dong fang ze xie (Chinese)

Habitat and Description This aquatic herb is found in the marshes, ponds, rivers and lakes of China, Korea, Japan, Taiwan, Mongolia, Russia, India, Nepal, Burma and Vietnam. The tubers are globose and 2 cm across. The leaves are aerial, simple and basal. The petiole is 5–35 cm long and fleshy. The blade is fleshy, broadly lanceolate, 3.5–10 cm × 1.5–7 cm, rounded at base, acuminate at apex and displays between five and seven longitudinal nerves. The inflorescence is a slender panicle of little white flowers. The calyx consists of three sepals. The corolla presents three petals which are white, yellow at base, membranaceous, marginally undulate and orbicular. The androecium consists of six stamens which are minute. The gynoecium comprises numerous carpels which are irregularly arranged. The achenes are elliptic and minute (Figure 2.10).

Medicinal Uses In Japan the plant is used as a diuretic and to treat inflammation.

Phytopharmacology The plant harbors a compelling array of protostane triterpenes: alisols A and B,[183] alisol A acetate, alisol B acetate, alisol C acetate, 16β-methoxyalisol B acetate, 16β-hydroxyalisol B acetate,[184] alisol E 23 acetate, alisols F and G, 13-17 epoxyalisol A,[185] 11-deoxyalisol B, 11-deoxyalisol B 23 acetate,[186] 13,17epoxyalisol B 23 acetate, alisol C,

■ FIGURE 2.10 *Alisma orientale* (Sam.) Juz.

16-oxoalisol A, 25-*O*-methyl-alisol A, 25-anhydroalisol A, 16,23-oxido-alisol B, 11-deoxyalisol A, 11-deoxyalisol B, 11-deoxyalisol B acetate, 11-deoxyalisol C acetate, 13β,17β-epoxyalisol B 23 acetate, 13β,17β-epoxyalisol B, 13β,17β-epoxyalisol A, 11-deoxy-13β,17β-epoxyalisol B 23 acetate, 11-deoxy-13β,17β-epoxyalisol A, 16,23-oxidoalisol B,[187] alis-maketone 23-acetate and alismaketone-A 23-acetate,[188] alisol H, I, alisol J 23 acetate, alisol K 23-acetate, alisol L 23 acetate, alisol M 23 acetate and alisol N 23-acetate,[189] alisolide, alisols O and P, 25-*O*-methylalisol A, 25-anhydroalisol A 24 acetate, 25-anhydroalisol, alisol E 23 acetate, 13β,17β-epoxyalisol A,[190] 11,25-anhydro-alisol F,[191] 25-anhydro-alisol F and 11-anhydro-alisol F.[192] In addition, the plant manufactures the guaiane sesquiterpene alismol, alismoxide,[193] orientalols A–C[194] and D,[185] alis-moxide 10-*O*-methyl ether,[187] orientalols E and F,[195] alismorientols A and B,[196] the germacrane sesquiterpenes germacrene C and D,[186] the eudes-mane sesquiterpene eudesma-4(14)-en-1β6α-diol[187] and the kaurane diter-pene 16R-(−)-kaurane-2,12-dione.[187]

■ **CS 2.83** Alisol B

■ **CS 2.84** Alisol B acetate

■ **CS 2.85** Alisol A 24-acetate

■ **CS 2.86** Alisol A

The diuretic and anti-inflammatory properties of the plant are due to alisols.[189,197]

Proposed Research Pharmacological study of alisol B and derivatives for the treatment of cancer.

Rationale Emerging evidence points to the fact that alisol-type protostane triterpenes are of chemotherapeutic value. Alisol B (CS 2.83) was cytotoxic to human ovary adenocarcinoma (SK-OV-3), murine melanoma (B16F10) and human fibrosarcoma (HT1080) cells, with IC_{50} values of 7.5 μg/mL, 7.5 μg/mL and 4.9 μg/mL, respectively.[198] Alisol B acetate (CS 2.84), alisol B and alisol A 24-acetate (CS 2.85) inhibited the survival of human breast adenocarcinoma (MCF-7) cells, with IC_{50} values equal to 28.9 μmol/L, 29.9 μmol/L and 38.2 μmol/L, whereas alisol A (CS 2.86) was

■ **CS 2.87** Alisol C 23-acetate

inactive, with an IC_{50} value of 229 μmol/L,[199] implying that the acetoxylation of C23 (CS 2.87) and/or C24 is crucial for activity of Alisol A and B and derivatives.

Mitochondrial insult and blockade of cellular ATP-dependent pumps have been identified as oncomolecular events triggered by alisols. Alisol B acetate inhibited the incorporation of [3H]thymidine into the DNA of murine vascular smooth muscle (A7r5) and human leukemia (CEM) cells at a concentration of 10^{-4} M and induced 78.3% and 69.7% apoptosis, respectively, with an increase of pro-apoptotic c-Myc, hence Bcl-2-associated X protein (Bax) activation and the consequent loss of mitochondrial membrane potential, release of cytochrome c from the mitochondria and DNA fragmentation.[200,201] Alisol B acetate inhibited the growth of human prostate cancer (PC-3) cells, with an IC_{50} value of 13.5 μM, via a fall of mitochondrial membrane potential induced by Bax, release of cytochrome c, which binds to apoptotic protease activating factor-1 (Apaf-1) and pro-caspase 9 in the presence of ATP to form the apoptosome that activates caspase 9 (and therefore caspase 3) and cleavage of PARP (Poly [ADP-ribose] polymerase).[202] Note that alisol B acetate is a strong antioxidant[196] that probably reduces the level of reactive oxygen species (ROS), which are chronically present in cancer cells[203] where they stimulate protein kinase B (Akt) phosphorylation.[204]

Alisol B acetate destroyed human gastric cancer (SGC-7901) cells, with an IC_{50} of 30 μmol/L, mediated by lowered phosphorylated Akt, resulting in an increase in the level of pro-apoptotic Bax and hence a fall in mitochondrial membrane potential, release of cytochrome c and activation of caspases 9 and 3.[205] Note that Akt inactivation may result in an increase of CD95 death machinery, which interacts with pro-apoptotic c-Myc to induce apoptosis.[206] Alisol B acetate at a dose of 10 μM reversed the resistance of multidrug-resistant forms of human hepatocellular liver carcinoma

(HepG2-DR) and human erythromyeloblastoid leukemia (K562-DR) cells to vinblastine, with cells arresting in G_2/M phase owing to blockage of the ATP-dependent efflux pump P-glycoprotein (P-gp).[207] Alisol B boosted cytoplasmic Ca^{2+} concentration in MCF-7 cells by blocking sarco-endoplasmic reticulum Ca^{2+}-ATPases (SERCA) inhibitor pumps, resulting in autophagy through the calmodulin-dependent protein kinase (CaMKK) phosphorylation of 5′-AMP-activated protein kinase (AMPK), which phosphorylates and binds to the mammalian autophagy-initiating kinase (Ulk1) released from mammalian target of rapamycin (mTOR).[199,208,209] In addition, alisol B induces apoptosis via endoplasmic reticulum stress, hence activation of protein kinase RNA-like endoplasmic reticulum kinase (PERK) and activating transcription factor 6 (ATF6) of the unfolded protein response (UPR) pathway.[199] Lee et al. (2002) synthesized series of alisol B acetate derivatives and found that a hydroxy imino group at C3 enhanced the cytotoxic activity, such as is found in 23S-acetoxy-24R(25)-epoxy-11β,23S-dihydroxyprotost-13(17)-en-3-hydroxy-imine which was lethal to human lung adenocarcinoma epithelial (A549), SK-OV-3, B16F10, and HT1080 cells, with IC_{50} values of $10\,\mu g/mL$, $8.7\,\mu g/mL$, $5.2\,\mu g/mL$, and $3.1\,\mu g/mL$, respectively.[210] In another structure–activity relationship study, alisol C 23-acetate inhibited the enzymatic activity of farnesyl-protein transferase (FPTase),[211] which is a compelling oncological target.[212]

2.3.2 *Schisandra Henryi* **C.B. Clarke**

History This plant was first described by Charles Baron Clarke in *The Gardeners' Chronicle*, published in 1905.

Family Schisandraceae Blume, 1830

Common Name Yi geng wu wei zi (Chinese)

Habitat and Description This is a woody climber that grows in the forests of China. The stems are winged. The leaves are simple and originate in groups from some buds along the stems. The petiole is 1–4 cm long. The blade is ovate, cuneate at base, acuminate at apex, serrate, 7–10 cm × 3–9 cm, thinly coriaceous, glaucous beneath and presents three to five pairs of secondary nerves. The inflorescence is simple and axillary. The flower pedicel is 1.5–4.5 cm long. The flowers are male or female. The perianth includes 6–10 yellow tepals which are 0.5 cm × 1.5 cm. The androecium present numerous stamens spirally arranged along a floral axis. The gynaecium includes several carpels which are free. The fruit is

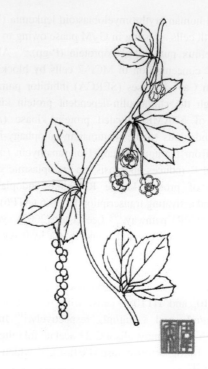

■ **FIGURE 2.11** *Schisandra henryi* C.B. Clarke

a 1.5–10 cm-long aggregate of globose apocarps which are red, and 1 cm across (Figure 2.11).

Medicinal Uses In China the plant is used to invigorate, to promote blood circulation, to regulate menses, to heal broken bones and putrefied ulcers, to treat rheumatism and to alleviate stomachache.

Phytopharmacology The plant contains several lignans, including enshicine,[213] henricine[214] (+)-anwulignan,[215] schisanhenrin, schisanhenol, deoxyschizandrin, schisantherin B, epienshicine methyl ether, epischisandrone, schisandrone, epiwulignan A_1, wulignans A_1 and A_2, enshicine, enshizhisu,[216] gomisin G, schisantherin A, benzoylgomisin Q, isoanwulignan,[217] henricines A and B, ganshisandrine and schisandrol A.[218] In addition, the plant produces the lanostane triterpene kadsuric acid,[219] anwuweizic acid,[220] a series of lanostane derivatives including schiprolactone A, nigranoic acid, schisanlactone B, schisandronic acid,[216] schisanlactone A,[219] nigranoic acid 3-ethyl ester and isoschizandronic acid,[220]

■ **CS 2.88** Daedaleaside B

■ **CS 2.89** Dehydrotrametenolic acid

henrischinins A–C,[221] juncoside I,[219] the nortriterpenoid henridilactones A–D,[219] changnic acid[219] and schisanhenric acid.[216]

Proposed Research Pharmacological study of schisandronic acid and derivatives for the treatment of cancer.

Rationale Lanostane triterpenes *sensu stricto* are mainly found in mushrooms. One such compound, daedaleaside B (CS 2.88), isolated from *Daedalea dickinsii* Yasuda (family Fomitopsidaceae Jülich), was lethal to human promyelocytic leukemia (HL-60) and human colorectal adeno-carcinoma (HCT-15) cells, with IC_{50} values equal to 9.4 µM and 55 µM, respectively, via apoptosis.[222] Dehydrotrametenolic acid (CS 2.89) from *Poria cocos* F.A. Wolf (family Polyporaceae Corda) abrogated the survival of Ras-transformed rat embryo fibroblast-like (Rat2) cells, with an IC_{50} value of 39.2 µM, via apoptosis with decreased G protein Ras and there-fore attenuation of the Ras/Raf/MEK/ERK and Ras/PI3K/PTEN/Akt/

■ **CS 2.90** Polyporenic acid C

■ **CS 2.91** Poricotriol A

mTOR pathways.[223] Polyporenic acid C (CS 2.90) from the same mushroom inhibited the growth of human lung adenocarcinoma epithelial (A549) cells (wild-type p53) by 50% at a dose of $60\,\mu M$, with hypophosphorylation of Akt and hence activation of the pro-apoptotic protein p53 and caspases 3 and 8.[224] Poricotriol A (CS 2.91) isolated from the same species killed cells from the following cell lines (IC$_{50}$values in parentheses): HL-60 ($1.9\,\mu M$), A549 ($4.2\,\mu M$), human melanoma (CRL1579; $5.5\,\mu M$), human ovarian carcinoma (OVCAR-3; $2.2\,\mu M$), human breast cancer (SK-BR-3; $1.2\,\mu M$), human prostate carcinoma (DU-145; $14.5\,\mu M$), human gastric cancer (AZ521; $1.4\,\mu M$) and human pancreatic carcinoma (PANC-1; $26.3\,\mu M$),[225] suggesting that the presence of hydroxyl groups at C3, C16 and C21 on a seco-lanostane are favorable for cytotoxic activity. Poricotriol A induces apoptosis with activation of caspases 3, 8, and 9, a decreased level of anti-apoptotic Bcl-2 protein and an increase in the level of pro-apoptotic Bax, hence the release of apoptosis inducing factor (AIF) from mitochondria.[225] Ganoderic acid T (CS 2.92) from *Ganoderma*

■ **CS 2.92** Ganoderic acid T

■ **CS 2.93** Ganoderic acid Me

lucidum (Curtis) P. Karst. (Ganodermataceae (Donk) Donk) reduced the viability of human lung cancer (95-D) cells (wild-type p53) by 70% at 50 μg/mL via a mechanism implying that stimulation of pro-apoptotic protein (p53) (and perhaps DNA damage) increases the protein level of pro-apoptotic protein Bax and therefore triggers the release of cytochrome c from mitochondria that then activates caspase 3.[226] Ganoderic acid Me (CS 2.93) from the same species negated the growth of a broad array of cancer cells including 95-D cells (wild-type p53), with an IC_{50} value of 55.1 μM, by a mechanism dependent on pro-apoptotic protein (p53).[227] In addition, ganoderic acid Me inhibited the multiplication of human colorectal carcinoma (HCT-116) cells (p53 wild-type), with an IC_{50} of 36.9 μM, via apoptosis with upregulation of pro-apoptotic protein p53, pro-apoptotic Bax translocation, a fall in mitochondrial membrane potential, release of cytochrome c, caspase 3 activation and DNA fragmentation.[228]

■ **CS 2.94** Lucidumol A

■ **CS 2.95** Ganoderiol F

■ **CS 2.96** Lucialdehyde C

Lucidumol A (CS 2.94) and ganoderiol F (CS 2.95) from the same species were active against mouse sarcoma (Meth-A) cells and Lewis lung carcinoma (LLC) cells, with IC_{50} values equal to 4.2 μg/mL, 2.3 μg/mL and 4.4 μg/mL and 6 μg/mL, respectively.[229] Lucialdehyde C (CS 2.96) annihilated LLC, human breast carcinoma (T-47D) and mouse sarcoma (S-180 and Meth-A) cells, with IC_{50} values equal to 10.7 μg/mL, 4.7 μg/mL,

■ **CS 2.97** Inotodiol

■ **CS 2.98** Ganoderic acid X

$7.1\,\mu g/mL$ and $3.8\,\mu g/mL$, respectively,[230] suggesting that ketone moieties at both C3 and C7 are favorable for activity of lanostanes. Inotodiol (CS 2.97) from *Inonotus obliquus* (Ach. ex Pers.) Pilát (Hymenochaetaceae Imazeki & Toki) induced apoptosis in mouse leukemia (P388) cells, with an IC_{50} value of $13.9\,\mu M$, via caspase 3 activation,[231] suggesting that esterification of a hydroxyl moiety at C3 is not favorable for cytotoxicity.

The precise cellular mechanism by which pro-apoptotic protein p53 is stimulated in wild-type p53 cancer cells by lanostanes is as yet elusive, but one could reasonably think of a DNA damage-induced mechanism since lanostanes are somewhat planar and might inhibit the enzymatic activity of topoisomerase II. One such compound is ganoderic acid X (CS 2.98), which inhibited the growth of human hepatocellular carcinoma (HuH-7; mutant p53), HCT-116 (wild-type p53), human B lymphocyte Burkitt's lymphoma (Raji; mutant p53) and HL-60 cells (p53 null), with IC_{50} values equal to $20.3\,\mu g/mL$, $38.3\,\mu g/mL$, $39.2\,\mu g/mL$ and $26.5\,\mu g/mL$, respectively, via topoisomerase inhibition.[232] In addition, decreased G protein Ras may

■ **CS 2.99** Methyl quadrangularate D

■ **CS 2.100** Acetyl methyl quadrangularate D

stimulate the pro-apoptotic protein p53,[233] implying a p53-independent cell death, which would therefore be of tremendous oncological interest.

In flowering plants, lanostane triterpenes present of a covalent bond between C9 and C19, the consequence of which is the formation of a cyclopropane ring and the impossibility of forming Δ_{8-9} or Δ_{9-11} double bonds and therefore reduced planarity of the cyclo-lanostane or cyclo-artanes, which may be less able to inhibit topoisomerase and induce apoptosis via different routes. An example of cycloartane is methyl quadrangularate D (CS 2.99) from *Combretum quadrangulare* Kurz (family Combretaceae R. Br.), which was lethal to murine colon carcinoma (26-L5) cells, with an IC_{50} value of 5.4 μM, whereas its acetyl derivative (CS 2.100) was inactive, suggesting that acetylation of the C3 hydroxyl group is deleterious for activity.[234]

■ **CS 2.101** 9,19-Cycloart-25-ene-3β,24-diol

■ **CS 2.102** Cycloartane-23-ene-3β,25-diol

■ **CS 2.103** Cycloartane-3β,24,25-triol
(P-glycoprotein)

Several cytotoxic cycloartanes have been isolated from members of the family Euphorbiaceae Juss. For instance, cycloartane-25-ene-3β,24-diol, (CS 2.101) and cycloartane-23-ene-3β,25-diol (CS 2.102) from *Euphorbia pulcherrima* Willd. ex Klotzsch (family Euphorbiaceae Juss.) inactivated Ehrlich's ascites carcinoma cells, with an IC_{50} value of 7.5 μM and 15 μM.[235] Note that cycloartane-23-ene-3β,25-diol and cycloartane-3β,24,25-triol (CS 2.103) isolated from *Euphorbia portlandica*

■ **CS 2.104** 21α-Hydroxytaraxasterol (P-glycoprotein)

■ **CS 2.105** Macrostachyoside A

■ **CS 2.106** Macrostachyoside B

L. (family Euphorbiaceae Juss.) potently inhibited the P-gp efflux pump in multidrug-resistant L5178Y mouse T-cell lymphoma at a dose of 40 µg/mL,[236] confirming the potential of euphorbiaceous triterpenes like 21α-hydroxytaraxasterol (CS 2.104) to mitigate multidrug resistance.[237] Macrostachyosides A and B (CS 2.105 and CS 2.106) from *Mallotus macrostachyus* (Miq.) Müll. Arg. (family Euphorbiaceae Juss.) killed human nasopharyngeal carcinoma (KB) and human lung adenocarcinoma (Lu-1) cells, with IC_{50} values equal to 7.1 µg/mL (A v KB), 6.4 µg/mL (A v Lu-1), 5.1 µg/mL (B v KB) and 4.3 µg/mL (B v Lu-1),[238] showing that an olefin moiety is not necessary in the side chain for cytotoxicity but moderate hydroxylation is favorable. 3β,21,22,23-Tetrahydroxy-cycloart-24(31),25(26)-diene (CS 2.107) isolated from a member of the genus *Amberboa* (Pers.) Less. (family Asteraceae Bercht. & J. Presl) inhibited the enzymatic activity of tyrosinase, with an IC_{50} value equal to 1.32 µM,[239] and is most probably cytotoxic. Note the triterpenes are known for inhibiting tyrosinase[240] but, as with P-gp inhibition, there is a dearth

■ **CS 2.107** 3β,21,22,23-Tetrahydroxy-cycloart-24(31),25(26)-diene

■ **CS 2.108** Secaubryolide

■ **CS 2.109** Sootepin A

of information about their structure–activity relationship. The flavonoid derivative haginine from *Lespedeza cyrtobotrya* Miq. (family Fabaceae Lindl.) is a potent tyrosinase inhibitor which induces the phosphorylation and therefore the activation of ERK and hence c-Myc activation, which has the potential to be pro-apoptotic.[241] It is interesting to note that increased resistance to drugs is mediated by the overexpression of tyrosinase.[242]

The seco-cycloartane secaubryolide (CS 2.108) from *Gardenia aubryi* Vieill. (family Rubiaceae Juss.) was active against human breast adenocarcinoma (MCF-7), human breast adenocarcinoma (MDA-MB-231), human prostate cancer (PC-3) and human epitheloid cervix carcinoma (Hela) cells, with IC_{50} values of 52 μM, 21 μM, 35 μM and 40 μM, respectively,[243] and it is probable that the alkylating property of the methylene furanolactone ring is responsible for this. The same pharmacophore is present in the seco-cycloartane sootepin A (CS 2.109) from *Gardenia sootepensis* Hutch.

(family Rubiaceae Juss.), which was active (IC_{50} values in parentheses) against human ductal breast carcinoma (BT-474; 5.9 µg/mL), human stomach adenocarcinoma (KATO-3; 2.1 µg/mL), human lung tumor (CHAGO; 3.9 µg/mL), human colon cancer (SW620; 1.8 µg/mL) and human hepatocellular liver carcinoma (HepG2; 2.9 µg/mL) cells.[244]

Gardenoin E (CS 2.110) from *Gardenia obtusifolia* Roxb. ex Hook. f. (family Rubiaceae Juss.) was cytotoxic against SW620, KATO-3 and CHAGO cells, with IC_{50} values of 5.5 µM, 7.2 µM and 5.5 µM, respectively,[245] showing that the opening of the A ring into a seco-cycloartane is not favorable for activity. Cycloartane-24-en-1α,2α,3β-triol (CS 2.111) isolated from *Commiphora opobalsamum* (L.) Engl. (family Burseraceae Kunth) mitigated the multiplication of PC-3 and human prostate adenocarcinoma (LNCaP) cells, with IC_{50} values of 5.7 µM and 22.1 µM, respectively.[246] 3β-Acetoxycycloartan-24-ene-1α,2α-diol (CS 2.112) from the same plant terminated PC-3 and DU-145 cells, with IC_{50} values equal to 15.4 µM and 20.7 µM,[247] respectively, confirming that acetylation of the C3 hydroxyl mitigates the cytocidal properties of cycloartanes. 24-Methylenecycloartane-3β,16β,23β-triol or longitriol (CS 2.113) isolated from *Polyalthia longifolia* (Sonn.) Thwaites (family Annonaceae Juss.) was lethal to KB, MCF-7, A549) and human

■ **CS 2.110** Gardenoin E

■ **CS 2.111** Cycloartane-24-en-1α,2α,3β-triol

cervix carcinoma (C33A) cells, with IC_{50} values equal to 24.1 µg/mL, 30.8 µg/mL, 13.1 µg/mL and 10 µg/mL, respectively,[248] implying that a hydroxyl moiety at C16 is not detrimental for activity. The cyclization of the side chain of cycloartanes is not detrimental to cytotoxic properties. For instance, argenteanones A and E (CS 2.114 and CS 2.115) isolated from

■ **CS 2.112** 3β-Acetoxycycloartan-24-ene-1α,2α-diol

■ **CS 2.113** Longitriol

■ **CS 2.114** Argenteanone A

■ **CS 2.115** Argenteanone B

■ **CS 2.116** Argentatin A

■ **CS 2.117** 2α-Bromo-argentatin A

Aglaia argentea Bl. (family Meliaceae Juss.) were cytotoxic against KB cells, with IC_{50} values of 7.5 μg/mL and 3.7 μg/mL, respectively,[249,250] implying that the replacement of a hydroxyl moiety by a ketone at C21 of the tetrahydro-furan side chain promotes the cytotoxic activity. Parra-Delgado et al. (2005) synthesized series of cycloartane derivatives with cyclic side chains from argentatin A (CS 2.116) isolated from *Parthenium argentatum* A. Gray (family Asteraceae Bercht. & J. Presl.) and inferred that the presence of a double bond between C1 and C2, or a bromine atom at C2 or a formyl moiety at C2, enhanced cytotoxicity, as did an isopropyl moiety at C24. Argentatin A and its bromo derivative (CS 2.117) abated the proliferation of the following cell lines

■ **CS 2.118** Schiprolactone A

■ **CS 2.119** Schisanlactone B

(IC_{50} values given in parentheses for the two respective compounds): human colorectal adenocarcinoma (HCT-15; 31.7 μM, 3.2 μM), human erythromyeloblastoid leukemia (K562; 38.6 μM, 4.3 μM), human prostate cancer (PC-3; 20.2 μM, 11.0 μM) and human glioblastoma (U251; 27.3 μM, 13.8 μM) cells.[251]

The lanostane derivatives schiprolactone A (CS 2.118), schisanlactone B (CS 2.119), nigranoic acid (CS 2.120) and schisandronic acid (CS 2.121) isolated from *Schisandra henryi* C.B. Clarke inhibited the growth of human T cell lymphoblast-like cell (CCRF-CEM), with IC_{50} values equal to 5 μg/mL, 4.6 μg/mL, 45.6 μg/mL and 4.5 μg/mL, respectively. The respective values obtained in the inhibition of Hela cells by the same compounds were 50.6 μg/mL, 46.6 μg/mL, 45.6 μg/mL and 45 μg/mL,[216]

■ **CS 2.120** Nigranoic acid

■ **CS 2.121** Schisandronic acid

confirming that the opening of ring A mitigates the cytotoxic activity of the resulting seco-cycloartane. Tian et al. (2007) studied the apoptotic effect of schisandrolic acid from *Schisandra propinqua* (Wall.) Baill (Schisandraceae) against HepG2 cells and observed the activation of caspase 3 and cleavage of PARP (Poly [ADP-ribose] polymerase),[252] suggesting mitochondrial insult.

2.3.3 *Cucurbita Pepo* L.

History This plant was first described by Carl von Linnaeus in *Species Plantarum*, published in 1753.

Synonyms *Cucurbita aurantia* Willd., *Cucurbita courgero* Ser., *Cucurbita elongata* Bean ex Schrad., *Cucurbita esculenta* Gray, *Cucurbita melopepo* L., *Cucurbita ovifera* L., *Cucurbita pepo* subsp. *ovifera* (L.) D.S. Decker, *Cucurbita pepo* var. condensa L.H. Bailey, *Cucurbita pepo* var. melopepo (L.) Alef., *Cucurbita pepo* var. ovifera (L.) Alef., *Cucurbita pepo* var. torticollis Alef., *Cucurbita subverrucosa* Willd., *Cucurbita verrucosa* L.

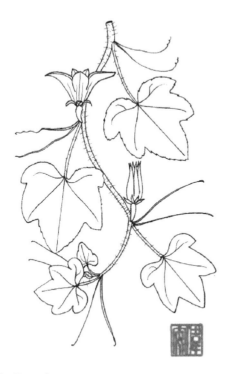

■ **FIGURE 2.12** *Cucurbita pepo* L.

Family Cucurbitaceae Juss., 1799

Common Name Pumpkin, xi hu lu (Chinese)

Habitat and Description *Cucurbita pepo* is an annual, monoecious climber native to Mexico and cultivated worldwide for its edible fruits. The stems are fleshy, stout, setose and develop series of multifid tendrils. The leaves are simple. The petiole is stout, fleshy, setose and up to 10 cm long. The blade is triangular, 20–30 cm across, irregularly five-lobed, setose, cordate at base, dentate at base and acute at apex. The flowers are massive and solitary on a 2–20-cm-long, setose pedicel. The corolla is orangish, membranaceous, infundibuliform and five-lobed and can grow up to 10 cm long. In male flowers the calyx is campanulate and develops five linear segments which are 1–2.5 cm long. The androecium includes three stamens with 1.5-cm-long filaments. Female flowers present a reduced calyx and an ovary which is ovoid and unilocular. The fruiting pedicel is robust and pentagonal. The fruit is a berry which has a multitude of shapes and colors according to the varieties. The seeds are ovoid, flattened, 1–1.5 cm × 0.5–1 cm, white and smooth (Figure 2.12).

Medicinal Uses In Cambodia, Laos and Vietnam the plant is used to treat gastritis and constipation, to expel worms from the intestines and to heal abscesses. In Malaysia, the plant is used to mitigate pregnancy discomfort and to expel worms from the intestines. In India, the plant is used to treat tuberculosis and to induce urination. In China, the plant is used to treat colds and to alleviate aches.

Phytopharmacology The plant contains some flavonoids such as *O*-glycosides: isoquercitrin, astragalin, rhamnocitrin 3-*O*-glucoside, isorhamnetin 3-*O*-glucoside, narcissin, nicotiflorin, rutin and rhamnocitrin 3-*O*-rutinoside;[253,254] the lignans secoisolariciresinol and lariciresinol;[255] phenolic glycosides phenolic cucurbitosides F–M;[256] a series of sterols – 24β-ethyl-5α-cholesta-7,25(27)dien-3 β-ol, 24 β-ethyl-5 α-cholesta-7, trans-22,25(27)-trien-3 β-ol, spinasterol,[257] codisterol, 25(27)-dehydroporiferasterol, clerosterol, isofucosterol, stigmasterol, campesterol, sitosterol, 25(27)-dehydrofungisterol, 25(27)-dehydrochondrillasterol, 24b-ethyl25(27)-dehydrolathosterol, avenasterol and 22-dihydrospinasterol;[258] multiflorane triterpenes such as 7-epizucchini factor A, 24-dihydrocucurbitacin D, cucurbitacin B and cucurbitacin E;[259–262] cucurbitane glycosides cucurbitacin L 2-*O*-β-D-glucopyranoside, cucurbitacin K 2-*O*-β-D-glucopyranoside, 2,16-dihydroxy-22,23,24,25,26,27-hexanorcucurbit-5-en-11,20-dione 2-*O*-β-D-glucopyranoside and 16-hydroxy-22,23,24,25,26,27-hexanorcucurbit-5-en-11,20-dione 3-*O*-α-l-rhamnopyranosyl-(1→2)- β-D-glucopyranoside,[263] and cucurbitaglycosides A and B.[264]

Proposed Research Pharmacological study of cucurbitacin E and derivatives for the treatment of cancer.

Rationale There is a massive body of evidence pointing to the fact that cucurbitacins possess very high orders of cytotoxicity against a vast array of malignancies. Cucurbitacins are cucurbitane triterpenes (19-(10→9β)-abeo-10α-lanost-5-ene) with double bonds between C4 and C5, a hydroxyl at C16, C20 and C25 and a ketone at C11 and C22.[265] Isocucurbitacin B (CS 2.122), cucurbitacin B, 22,23 dihydrocucurbitacin B (CS 2.123) and cucurbitacin E isolated from *Marah oregana* (Torr. & A. Gray) Howell (family Cucurbitaceae Juss.) were cytotoxic against human nasopharyngeal carcinoma (KB) cells, with IC_{50} values equal to $4 \times 10^{-1} \mu g/mL$, $2 \cdot 5 \times 10^{-6} \mu g/mL$, $1 \cdot 7 \times 10^{-3} \mu g/mL$ and $4.5 \times 10^{-7} \mu g/mL$, respectively.[266] Cucurbitacin P (CS 2.124) and Q from *Brandegea bigelovii* (S. Watson) Cogn. (family Cucurbitaceae Juss.) showed significant activity against KB cells, with IC_{50} values equal to $0.5 \mu g/mL$ and $0.03 \mu g/mL$, respectively.[267]

■ **CS 2.122** Isocucurbitacin B

■ **CS 2.123** 23,24-Dihydrocurbitacin B

■ **CS 2.124** Cucurbitacin P

Cucurbitacin B, D and E from *Cucurbita andreana* Naudin (family Cucurbitaceae Naudin) compromised the proliferation of human colorectal carcinoma (HCT-116) cells by 81.5%, 80.4% and 77%, respectively, at 0.4 μM, abrogated the growth of human breast adenocarcinoma (MCF-7)

■ **CS 2.125** 2-Deoxycucurbitacin D

■ **CS 2.126** 25-Acetylcucurbitacin F

cells by 87%, 78% and 66.5%, respectively, at 0.4 μM, abated the proliferation of human large-cell lung cancer (NCI-H460) cells by 96%, 43% and 37%, respectively, at 0.1 μM and reduced the number of human glioblastoma (SF268) cells by 92%, 25% and 27%, respectively, at 0.05 μM. In contrast, cucurbitacin I was not active.[268] 2-Deoxycucurbitacin D (CS 2.125), 25-acetylcucurbitacin F (CS 2.126) and cucurbitacin D from *Sloanea zuliaensis* Pittier (family Elaeocarpaceae Juss.) were active against the following cell lines (respective IC$_{50}$ values in parentheses): MCF-7 (0.04 μg/mL, 0.03 μg/mL, 0.2 μg/mL), NCI-H460 (0.1 μg/mL, 0.06 μg/mL, 0.08 μg/mL) and SF268 (0.02 μg/mL, 0.01 μg/mL, 0.02 μg/mL),[269] confirming the beneficial effect of C2 hydroxylation on cucurbitacin cytotoxicity.

Cucurbitacin F (CS 2.127) from *Elaeocarpus dolichostylus* Schltr. (family Elaeocarpaceae Juss.) displayed significant cytotoxic activity against KB and mouse leukemia (P388) cells, with IC$_{50}$ values equal to 0.07 μg/mL and 0.04 μg/mL, respectively, whereas 23,24-dihydrocucurbitacin F was

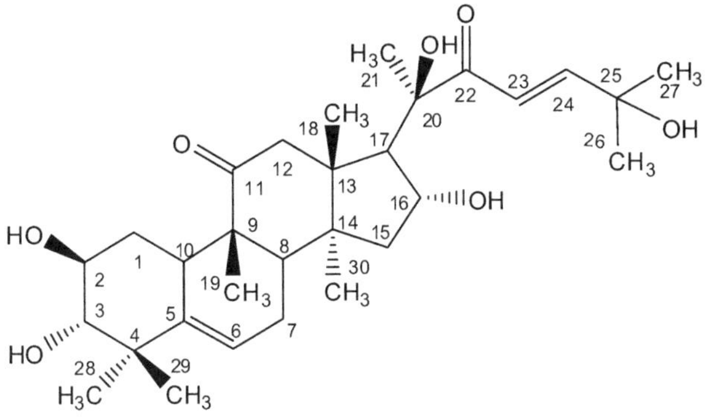

■ **CS 2.127** Cucurbitacin F

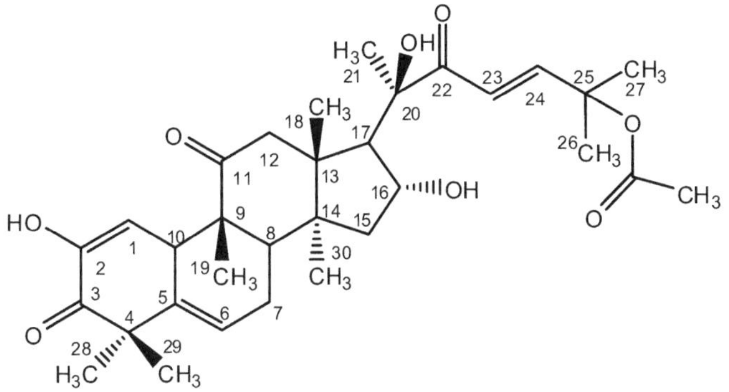

■ **CS 2.128** Cucurbitacin E

inactive, showing the Δ_{23-24} is necessary for activity.[270] Other chemical features requested for the cytotoxicity of cucurbitacins are: an A ring with hydroxyl moieties at C2 and C3 and/or a ketone at C3, a 16α-hydroxyl group, a 20β-hydroxyl group, an α,β-unsaturated ketone in the side chain and a 25-acetyl group.[270,271]

In addition, the hydroxyl group at C16 forms a hydrogen bond with the C22 ketone, which activates the α,β-unsaturated ketone toward nucleophilic attack and hence alkylation of thiol groups.[272] Such chemical features are perfectly exemplified with cucurbitacin B and E, which are among the most cytotoxic cucurbitacins. Cucurbitacin E (CS 2.128) inhibited the growth of an impressive array of cancer cells (IC$_{50}$ values in parentheses): SF268 (27 nM), human leukemic lymphoblast (CCRF-CEM; 32 nM), human non-small-cell lung carcinoma (H460; 26 nM), human colon cancer (HT-29; 23 nM), human skin melanoma (SK-MEL-28; 21 nM), human ovarian adenocarcinoma (IGROV1; 29 nM), human kidney carcinoma (Caki-1; 22 nM) and human prostate cancer (PC-3; 50 nM).[273] In contrast, cucurbitacins B (CS 2.129), I (CS 2.130), D (CS 2.131) and Q (CS 2.132) killed human prostate cancer (PC-3) cells, with IC_{50} values equal to 35 nM, 223 nM, 367 nM and 455 nM, respectively.[273]

■ **CS 2.129** Cucurbitacin B

■ **CS 2.130** Cucurbitacin I

■ **CS 2.131** Cucurbitacin D

Some of the earliest cellular effects of cucurbitacins observed were the inhibition of protein synthesis and disruption of actin skeleton. It was found that 1 μM of cucurbitacin I from *Bryonia alba* L. (family Cucurbitaceae Juss.) abated the incorporation of [³H]thymidine into DNA, RNA and [³H]leucine into proteins in human epithelial cancer (Hela S3) cells.[274] Cucurbitacin E disrupted actin cytoskeleton in PC-3 cells.[273] A turning point in the understanding of cucurbitacin cellular activities has been provided by Blaskovich et al. (2003) who noted that 10 μM of cucurbitacin I reduced the levels of phosphorylated Janus kinase 2 (JAK2), phosphorylated signal transducer and activator of transcription 3 (STAT3), inhibited STAT3 signaling and also hampered STAT3 DNA binding and therefore STAT3-mediated gene expression in human lung adenocarcinoma epithelial (A549) and human breast cancer (MDA-468) cells.[275] In normal conditions cytokines binding to a membrane receptor induce the phosphorylation of JAK2 followed by the phosphorylation and therefore activation of STAT3, which dimerises and translocates into the nucleus. Inside the nucleus it induces the synthesis of the anti-apoptotic protein B-cell lymphoma-extra large (Bcl-xL), myeloid leukemia cell differentiation protein (Mcl-1) and survivin, and proliferation of the inducing protein cyclin D1.[276] In several malignancies, such as multiple myeloma, glioblastoma, colorectal and hepatocellular carcinoma, STAT3 is constitutively activated and compels the survival and proliferation of cancer cells; therefore STAT3 inhibitors, like cucurbitacins, are indeed of considerable interest for the treatment of cancers.[277]

Sun et al. (2005) studied the effects of several cucurbitacins on STAT3 and JAK2 levels and concluded that hydroxyl at C3 lowers anti-JAK2 activity. Cucurbitacin Q inhibited STAT3 activation, with IC_{50} values of 3.7 μM, 1.4 μM and 0.9 μM in A549, MDA-468 and human breast carcinoma

■ **CS 2.132** Cucurbitacin Q

■ **CS 2.133** Cucurbitacin A

(MDA-MB-435) cells, respectively. Cucurbitacin A (CS 2.133) specifically inhibited JAK2 activity, with IC_{50} values of 1.5 μM, 0.86 μM, and 0.6 μM in A549, MDA-468 and MDA-MB-435 cells, respectively. Cucurbitacin I inhibited the activation of both STAT3 and anti-JAK2 in A549, MDA-468 and MDA-MB-435 cells.[278]

In fact, the ability of the cucurbitacins to induce apoptosis is mainly related to their ability to suppress the activation of STAT3 and, to a lesser extent, anti-JAK2 levels in cancer cells.

The inhibition of STAT3 results in manifold downstream effects. For instance, cucurbitacin E inhibited the growth of human pancreatic carcinoma (PANC-1) cells (p53 mutant), with an IC_{50} value of 1 μM, via decreased levels of phosphorylated STAT3 and p53-independent apoptosis.[279] Pro-apoptotic protein (p53) inhibition is mediated by activated STAT3.[280]

The striking ability of cucurbitacin E to mediate apoptosis without having to stimulate a wild-type p53 in cancer cells has been further confirmed as it has also been shown to induce apoptosis in human bladder cancer (T24) cells (mutant p53), with an IC_{50} value equal to 1012 nM, via inhibition of STAT3, C95 and caspase 8 activation, pro-apoptotic BH3 interacting-domain death agonist (Bid) cleavage and Bax translocation, mitochondrial insult, release of cytochrome c and activation of caspase 3.[281] The decrease in cyclin-B1 induced by 23,24-dihydrocucurbitacin B in murine melanoma (B16F10) cells hindered the cyclin-B/CDK1 complex, and hence actin cytoskeleton disorganization.[282] Another effect of STAT3 inhibition is the downregulation of Bcl-xL, Bcl-2, induced myeloid leukemia cell differentiation protein (Mcl-1), cyclin-dependent kinase 1 (CDK1), anti-apoptotic FLICE-like inhibitory protein (FLIP) and X-linked inhibitor of apoptosis protein (XIAP), the latter being a critical downstream target of STAT3 activation leading to cell survival.[283]

In human hepatocellular liver carcinoma (HepG2) cells cucurbitacin B inhibited STAT3 phosphorylation and decreased the level of anti-apoptotic Bcl-2 protein.[284] Cucurbitacin B killed PANC-1 cells (mutant p53) via apoptosis, STAT3 phosphorylation, activation of caspase 3 and cyclin-dependent kinase inhibitor p21, and decreased levels of anti-apoptotic Bcl-2 protein and survivin.[285] Note that activation of the Fas/CD95 receptor in Jurkat human T cell lymphoblast-like cells increases the phosphorylation of eukaryotic translation initiation factor 2 subunit kinase (eIF2K) and therefore progression of cell death.[286] Cucurbitacin E at a dose of 50 nmol/L inhibited the growth of HL-60 cells (p53 null), via induction of eukaryotic translation initiation factor 2 subunit (eIF2), decreased the levels of the anti-apoptotic proteins XIAP and survivin, induced Mcl-1, increased the levels of pro-apoptotic Bax and p21, and activated caspases 8, caspase 9 and 3.[287] Treatment of human chronic B-cell leukemia (I83), human B-lymphoma (BJAB) and human B-lymphoma (NALM-6) cells with 1 μmol/L of cucurbitacin I inhibited STAT3, released cytochrome c and decreased the expression of XIAP as well as death receptor 4 (DR4), suggesting the involvement of the tumor necrosis factor-related apoptosis-inducing ligand (TRAIL) apoptotic pathway.[283] Cucurbitacin E at a dose of 0.5 nM inhibited the migration and tubulogenesis of human umbilical vein endothelial (HUVEC) cells and reduced vascular endothelial growth factor (VEGF)-induced STAT3 nuclear translocation, hence bringing about a reduction in Bcl-xL, anti-apoptotic Bcl-2 protein, survivin and cyclin D1.[288]

Cucurbitacin B abrogated the survival of the following cell lines (IC$_{50}$ values in parentheses): CCRF-CEM (20 nM), K562 (15 nM), human malignant T-lymphoblastic (MOLT-4; 35 nM), human multiple myeloma (RPMI8226; 30 nM) and human large-cell immunoblastic lymphoma (SR; 20 nM) cells. In K562 cells cucurbitacin B inhibited induced STAT3 activation at a dose of 50 μM with inhibition of c-Raf, MEK and ERK, suggesting that cucurbitacin B inhibits the Raf/MEK/ERK pathway.[289] Note that the phosphorylation of STAT3 is abrogated by inhibition of ERK activity.[290] Duangmano et al. (2010) observed that cucurbitacin B induced apoptosis in human breast adenocarcinoma (SKBR-3), human breast adenocarcinoma (MCF-7), human ductal breast epithelial tumor (T47D) and human breast epithelial (HBL100) cells, with IC$_{50}$ values equal to 3.2 μg/mL, 63 μg/mL, 88.4 μg/mL and 32.7 μg/mL, respectively, with reduction of c-Myc and telomerase reverse transcriptase (hTERT) and therefore inhibition of telomerase activity[291] via possible STAT3 inhibition since c-Myc is a downstream effector of STAT3 signaling.[292] Cucurbitacin E and cucurbitacin I both hindered the proliferation of human histiocytic lymphoma (U937), human promyelocytic leukemia (HL-60) and human fibrosarcoma (HT1080) cells, with IC_{50} values for cucurbitacin E equal to 16 nM, 18 nM and 40 nM, respectively, the values for cucurbitacin I being 0.3 μM, 0.1 μM and 0.4 μM. In U937 cells inhibition of STAT3 was observed at high doses whereas at low doses cofilin severing was stimulated, hence the depolymerization of actin in HT1080 cells.[293] 23,24-Dihydrocucurbitacin B from *Trichosanthes kirilowii* Maxim. (family Cucurbitaceae Juss.) inhibited the growth of human breast cancer (BCap-37) cells, with an IC_{50} value of 1 μM, via apoptosis with DNA fragmentation, release of cytochrome c, caspases 9 and 3 activation and PARP cleavage, suggesting mitochondrial insult,[294] which could very well result from STAT3 inhibition.

Although STAT3 suppression is now broadly recognized as the main target for cucurbitacins, some scholars have highlighted the involvement of other targets such as nuclear factor kappa-light-chain-enhancer of activated B cells (NF-κB). For instance, cucurbitacin B from *Trichosanthes kirilowii* Maxim. (family Cucurbitaceae Juss.) inhibited the tumor necrosis factor-α (TNF-α)-induced expression of NF-κB in human epitheloid cervix carcinoma (Hela) cells, with an IC_{50} value of 14.7 nM, by hindering phosphorylation of Ser536 in RelA/p65 without interfering with nuclear factor of kappa-light-polypeptide-gene-enhancer in B-cells inhibitor (IκB)[295] on probable account of unphosphorylated STAT3 (U-STAT3). Ding et al. (2011) reported that cucurbitacin D at 1 μg/mL lowered the levels of Bcl-xL and Bcl-2 proteins in human T-cell lymphoma (HUT102) cells by inhibiting NF-κB activity by inhibition of IκB phosphorylation and therefore degradation by the proteasome,[296] which might be caused by U-STAT3. The latter interacts with the p65 subunit of NF-κB to inhibit the binding IκB.[297] It should also be noted that 23,24-dihycrocucurbitacin D (CS 2.134) from *Bryonia alba* L. (family Cucurbitaceae Juss.) abated the production of nitric oxide (NO) in peritoneal macrophages at

■ **CS 2.134** 23,24-dihycrocucurbitacin D

a dose of 80 μM by inhibiting the mRNA level and expression of the inducible nitric oxide synthetase (iNOS) protein by blocking the activation of NF-κB.[298] Cucurbitacin D from *Trichosanthes kirilowii* Maxim. (family Cucurbitaceae Juss.) induced apoptosis in human hepatocellular carcinoma (Hep3B) cells, with an IC_{50} value of 1 μg/mL, effected by induction of pro-apoptotic c-Jun N-terminal kinase- (JNK) mediated c-Jun phosphorylation and its transcriptional activation independently of STAT3.[299] Note that JNK, which is inhibited by NF-κB, induces pro-apoptotic Bax translocation to mitochondria and therefore apoptosis via release of cytochrome c and activation of caspase 3.[300]

In some instances the cytotoxic effects of cucurbitacins is clearly independent of STAT3. Cucurbitacin B destroyed human hepatocellular carcinoma (BEL-7402) cells, with an IC_{50} of 0.3 μM, via inhibition of cyclin D1, CDK1, c-Raf activation and an increase in ERK phosphorylation, suggesting the Raf/MEK/ERK pathway, without affecting STAT3 phosphorylation.[301] Activation of c-Raf directly promotes cell growth and survival.[302]

Several lines of evidence suggest that actin is a target for cucurbitacins. The viability of human colon adenocarcinoma (SW480) cells was lowered to 65.3% by treatment with 40 nM of cucurbitacin B that resulted in apoptosis with DNA fragmentation, reduced levels of cyclin B1 and cdc25C, activation of caspases 3, 7, 8, and 9 and accumulation of reactive oxygen species (ROS) independently of JAK2/STAT3.[303] Note that cucurbitacin B induced ROS-dependent actin aggregation.[304] Cucurbitacins are known to disrupt actin polymerization,[273,282,293,305] the consequences of which are the inhibition of the JAK2/STAT3 signaling pathway in cells harboring activated STAT3 and the direct inhibition of survivin in other cells.[306] Cucurbitacin IIa (CS 2.135) curbed the

■ **CS 2.135** Cucurbitacin IIa

■ **CS 2.136** Karavelagenin C

proliferation of human prostate cancer (CWR22rv1) cells by 43% at a dose of 100 µg/mL with aggregation of actin and inhibition of survivin independently of JAK2/STAT3.[306]

Other miscellaneous oncocellular mechanisms provoked by cucurbitacins have been reported. Human breast adenocarcinoma (SKBR-3) cells exposed to 10 µg/mL of cucurbitacin B became apoptotic via inhibition of galectin-3 resulting in a decrease in phosphorylated glycogen synthetase kinase 3β (GSK-3β), an increase in β-catenin degradation reduction of β-catenin/TCF4 in the nucleus and impaired transcription of cyclin D1 and c-Myc.[307] Cucurbitacin D and 23,24-dihydro-cucurbitacin D from *Trichosanthes kirilowii* Maxim. (family Cucurbitaceae Juss.) inhibited tyrosinase activity, with IC_{50} values equal to 0.18 µM and 6.7 µM, respectively, and the synthesis of melanin, with IC_{50} values of 0.16 µM and 7.5 µM, respectively, in mouse melanoma (B16) cells.[308] Karavelagenin C (CS 2.136) from *Momordica balsamina* L. (family Cucurbitaceae Juss.) at a dose of 20 µM inhibited P-glycoprotein-induced multidrug resistance in mouse lymphoma (L15178) cells.[309]

2.3.4 *Brucea Javanica* (L.) Merr.

History This plant was first described by Elmer Drew Merrill in *Journal of the Arnold Arboretum*, published in 1928.

Synonyms *Brucea sumatrana* Roxb., *Gonus amarissimus* Lour., *Rhus javanica* L.

■ **FIGURE 2.13** *Brucea javanica* (L.) Merr.

Family Simaroubaceae DC., 1811

Common Name Ya dan zi (Chinese)

Habitat and Description *Brucea javanica* is a treelet which grows in the forests of India, Sri Lanka, Burma, China, India, Malaysia, Indonesia, the Philippines and Australia. The wood is extremely bitter. The leaves are compound and 20–40 cm long and display 3–15 leaflets. The leaflets are lanceolate, 5–10 cm × 2.5–5 cm, rounded at base, asymmetrical, serrate, acuminate at apex, and present conspicuous nervations. The inflorescence is an axillary panicle which is 7.5–25 cm long. The flowers are minute and purplish. The male flowers present four ovate sepals and four oblong petals, which are minute, and an androecium made of four stamens around a four-lobed disc. The female flowers present four ovate sepals and four oblong petals, which are minute, and a gynaecium made of four tiny carpels which are free. The fruit is a drupe which is 0.5 cm × 0.8 cm, black, glossy, *Rauvolfia*-like and oblong (Figure 2.13).

Medicinal Uses In China the plant is used to treat dysentery, malaria, diarrhea, hemorrhoids and cancer.

Phytopharmacology The bitterness of the plant is imparted by a series of quassinoid triterpenes such as bruceoside A,[310] bruceoside B, bruceins D and E,[311] bruceins K and L,[312] bruceoside C,[313] bruceosides D–F,[314] brusatol,[315] yadanziosides A–P,[316–320] yadanziolides A–D,[318,321] dehydrobrusatol and dehydrobruceantinol,[322] yandanziolide D,[323] javanicin,[324] bruceantin, bruceolide,[325] 15-O-benzoylbrucein D, bruceantarin, bruceantinol, bruceantinoside, A, brucein A-C, F, G and Q, brucein E 2-O-β-D-glucoside, dehydrobruceins A and B, dihydrobrucein A and yadanzigan,[321] desmethyl-brusatol, desmethyl-bruceantinoside A,[326] javanicolides A and B, and javanicoside A,[327] apotirucallane-type triterpenoids named bruceajavanin A, dihydrobruceajavanin A and bruceajavanin B,[328] javanicosides G and H,[329] javanicolides C and D and javanicosides B–F,[330] a pregnane glycosides,[331,332] β-carboline alkaloids such as canthin-6-one, 11-hydroxycanthin-6-one, 11-methoxycanthin-6-one, 5-methoxycanthin-6-one, 4-hydroxy-5-methoxycanthin-6-one and canthin-6-one-3*N*-oxide,[333] bruceacanthinoside,[328] and lignans such as pinoresinol and syringaresinol.[321,334] The antiplasmodial property of the plant is the result of the quassinoids bruceins A, B, D and brusatol.[335,336]

Proposed Research Pharmacological study of brucein D and derivatives for the treatment of pancreatic cancer.

Rationale There is a bewildering array of evidence indicating that quassinoids are powerful cytotoxic agents of chemotherapeutic value. Brucein D (CS 2.137) from *Brucea javanica* (L.) Merr. (family Simaroubaceae DC.) inhibited the growth of human pancreatic carcinoma (PANC-1), human pancreatic cancer (SW1990), and human pancreatic carcinoma (CAPAN-1) cells, with IC_{50} values equal to 2.5 μM, 5.2 μM and 1.3 μM, respectively.[337] Bruceantin (CS 2.138) from *Brucea javanica* (L.) Merr. (family Simaroubaceae DC.) was cytotoxic against human colon

■ **CS 2.137** Brucein D

■ **CS 2.138** Bruceantin

cancer (HT-29), human epitheloid cervix carcinoma (Hela) and human promyelocytic leukemia (HL-60) cells, with IC_{50} values equal to $0.01\,\mu g/mL$, $0.01\,\mu g/mL$ and $0.008\,\mu g/mL$, respectively.[338] Bruceantin was evaluated in Phase II trials in several malignancies, but was ineffective.[339,340] Note that brusatol (CS 2.139) from *Brucea javanica* (L.) Merr. (family Simaroubaceae DC.) abated the survival of mouse leukemia (P388) cells, with an IC_{50} value equal to $0.006\,\mu g/mL$.[341]

Bruceanols A (CS 2.140) and B (CS 2.141) from *Brucea antidysenterica* J.F. Mill. (family Simaroubaceae DC.) were cytotoxic in vivo against P388, at doses of $0.5\,mg/kg$ and $2\,mg/kg$, respectively.[342] Bruceanol C (CS 2.142) from *Brucea antidysenterica* J.F. destroyed human nasopharyngeal carcinoma (KB), human lung adenocarcinoma epithelial (A549), human intestinal adenocarcinoma (HCT-8) and mouse leukemia (P388) cells, with IC_{50} values less than $0.04\,\mu g/mL$, $0.4\,\mu g/mL$, $0.4\,\mu g/mL$, and $0.5\,\mu g/mL$, respectively.[343] Bruceanols D (CS 2.143), E (CS 2.144) and F (CS 2.145) from *Brucea antidysenterica* J.F. Mill. (family Simaroubaceae DC.) inhibited the growth of the following cell lines (respective IC_{50} values

■ **CS 2.139** Brusatol

■ **CS 2.140** Bruceanol A

■ **CS 2.141** Bruceanol B

■ **CS 2.142** Bruceanol C

■ **CS 2.143** Bruceanol D

■ **CS 2.144** Bruceanol E

in parentheses): KB (0.08 μg/mL, 0.5 μg/mL, 0.4 μg/mL), A549 (0.5 μg/mL, 3.7 μg/mL, 0.5 μg/mL), HCT-8 (0.09 μg/mL, 0.3 μg/mL, 0.1 μg/mL), P388 (0.09 μg/mL, 0.5 μg/mL, 0.3 μg/mL), human rhabdomyosarcoma (TE-671; 0.08 μg/mL, 0.1 μg/mL, 0.09 μg/mL) and human malignant melanoma (RPMI-7951; 0.09 μg/mL, 0.1 μg/mL, 0.09 μg/mL),[344] implying that $\Delta^{185-186}$ is beneficial for cytotoxic activity.[344]

■ **CS 2.145** Bruceanol F

■ **CS 2.146** Bruceanol G

■ **CS 2.147** 13,18-Dehydro-6α-senecioyloxychaparrin

Bruceanol G (CS 2.146) from *Brucea antidysenterica* J.F. Mill. (family Simaroubaceae DC.) exhibited significant cytotoxicity against the human colon adenocarcinoma (CoLo 205) and human oesophageal squamous cell (KE3) carcinoma, with IC_{50} values of 0.4 μM and 0.5 μM, respectively.[345] 13,18-Dehydro-6α-senecioyloxychaparrin (CS 2.147) isolated from *Simaba multiflora* A. Juss. (family Simaroubaceae DC.) abrogated the survival of P388 cells, with an IC_{50} value of 1.7 μg/mL, whereas 12-dehydro-6α-senecioyloxychaparrin (CS 2.148) was inactive,[346] suggesting that a ketone at C12 abrogates cytotoxic activity. 6α-Senecioyloxychaparrinone (CS 2.149), 6α-senecioyloxychaparrin (CS 2.150) and chaparrinone (CS 2.151) from *Simaba multiflora* A. Juss. (family Simaroubaceae DC.) negated the growth of P388, with IC_{50} values equal to 0.003 μg/mL, 0.03 μg/mL and 0.02 μg/mL, respectively.[347] 6α-Tigloyloxychaparrin (CS 2.152) and 6α-tigloyloxychaparrinone (CS 2.153) from *Simaba cuspidata* Spruce ex Engl. (family Simaroubaceae DC.) hindered the multiplication of P388 cells, with IC$_{50}$ values of 0.24 µg/mL and <0.01 µg/mL, respectively.[348] Cedronolactone A (CS 2.154) and simalikalactone D (CS 2.155) isolated from *Simaba cedron* Planch. (family Simaroubaceae DC.) expedited the death of P388, with IC$_{50}$ values of 0.007 µg/mL and 0.005 µg/mL, respectively.[349]

Eurycomanone (CS 2.156) from *Eurycoma longifolia* Jack (family Simaroubaceae Juss.) abated the proliferation of KB and vincristine-resistant KB-V cells, with IC$_{50}$ values equal to 1.9 µg/mL and 0.8 µg/mL, respectively.[350] 14,15β-Dihydroxyklaineanone (CS 2.157) and

■ **CS 2.148** 12-Dehydro-6α-senecioyloxychaparrin

■ **CS 2.149** 6α-Senecioyloxychaparrinone

■ **CS 2.150** 6α-Senecioyloxychaparrin

■ **CS 2.151** Chaparrinone

■ **CS 2.152** 6α-Tigloyloxychaparrin

■ **CS 2.153** 6α-Tigloyloxychaparrinone

■ **CS 2.154** Cedronolactone A

■ **CS 2.155** Simalikalactone D

■ **CS 2.156** Eurycomanone

■ **CS 2.157** 14,15β-Dihydroxyklaineanone

■ **CS 2.158** 11-Dehydroklaineanone

11-dehydroklaineanone (CS 2.158) isolated from the same plant were cytocidal against P388 (IC$_{50}$ values equal to 0.2 μg/mL and 0.3 μg/mL, respectively) and KB cells (1.8 μg/mL and 1.6 μg/mL, respectively).[351] A series of compounds isolated from *Eurycoma longifolia* Jack (family Simaroubaceae Juss.) displayed potent cytotoxic effects against P388 cells as follows: 13α (18)-epoxyeurycomanone (CS 2.159), 14 μg/mL; 15-acetyl-13α(18)-epoxyeurycomanone (CS 2.160), 6.6 μg/mL; 12,15-diacetyl-13α(18)-epoxyeurycomanone (CS 2.161), 7.2 μg/mL; 12-acetyl-13,18-dihydroeurycomanone (CS 2.162), 0.9 μg/mL; 15β-acetyl-14-hydroxy-klaineanone (CS 2.163), 7.8 μg/mL; 6α-acetoxy-14,15β-dihydroxyklaineanone (CS 2.164) 12 μg/mL, and 6α-acetoxy-15β-hydroxyklaineanone (CS 2.165), 15 μg/mL.[352]

■ **CS 2.159** 13α(18)-Epoxyeurycomanone

■ **CS 2.160** 15-Acetyl-13α(18)-epoxyeurycomanone

■ **CS 2.161** 12,15-Diacetyl-13α(18)-epoxy-eurycomanone

■ **CS 2.162** 12-Acetyl-13,18-dihydroeurycomanone

The quassinoids 13,18-dehydroglaucarubinone (CS 2.166) and glaucarubinone (CS 2.167) isolated from *Simarouba amara* Aubl. (family Simaroubaceae DC.) were cytotoxic against PS leukemia cells, with IC_{50} values equal to 0.9 µg/mL and 0.3 µg/mL, respectively.[353] Sergiolide (CS 2.168) and isobrucein B (CS 2.169) isolated from *Cedronia granatensis* Cuatrec. (family Simaroubaceae DC.) displayed potent cytotoxic activity against a vast panel of cancer cells at concentrations of 10^{-5}–10^{-8} M.[354] 2'-(R)-O-acetylglaucarubinone (CS 2.170) from *Quassia gabonensis* Pierre (family Simaroubaceae DC.) inhibited the survival of human prostate carcinoma (DU-145), A549, KB and KB-V cells, with IC_{50} values equal to 0.04 µM, 0.06 µM, 0.05 µM and 0.4 µM, respectively.[355] 15-Deacetylsergeolide (CS 2.171) from *Picrolemma pseudocoffea* Ducke (family Simaroubaceae DC.) inhibited

■ **CS 2.163** 15β-Acetyl-14-hydroxy-klaineanone

■ **CS 2.164** 6α-Acetoxy-14,15β-dihydroxyklaineanone

■ **CS 2.165** 6α-Acetoxy-15β-hydroxyklaineanone

■ **CS 2.166** 13,18-Dehydroglaucarubinone

■ **CS 2.167** Glaucarubinone

■ **CS 2.168** Sergiolide

■ **CS 2.169** Isobrucein B

■ **CS 2.170** 2′-(R)-O-Acetylglaucarubinone

■ **CS 2.171** 15-Deacetylsergeolide

the proliferation of P388 at 0.6 mg/kg.[356] Soularubinone (CS 2.172) from a member of the genus *Soulamea* Lam. (family Simaroubaceae DC.) displayed potent in vivo activity against P388 at a dose of 4 mg/kg.[357] 15-Desacetylundulatone (CS 2.173) from *Hannoa klaineana* Pierre ex Engl. (family Simaroubaceae DC.) increased the life span of rodents infested with P388 cells at a dose of 10 mg/kg.[358]

Structure–activity relationship studies demonstrated that the following chemical features were beneficial for the cytotoxic effect of quassinoids: the intracyclic oxo bridge (C11–C30 or C13–C30),[271,359] an ester group at C15, a hydroxyl group at C12,[357] a ketone at C2[341,358–361], a Δ^3-2-oxo moiety in ring A, and an ester group at C6.[362] The hydroxyl group at C1 may boost the cytotoxic effect of quassinoids due to intramolecular hydrogen bonding between the hydroxyl and the ketone in C2, thus further activating the enone,[363] which acts as a Michael acceptor (CS 2.174) for biological nucleophiles.[361]

Preliminary study of the cellular mechanism of quassinoids indicated that bruceantin inhibited protein synthesis[361] due to some impairment of pep-tidyltransferase[364] and inhibition of DNA/RNA synthesis via phosphoro-bosyl pyrophosphate aminotransferase.[365] Bruceantin, isolated from *Brucea antidysenterica* J.F. Mill. (family Simaroubaceae DC.) exhibited potent antileukemic in vitro and in vivo activity[366,367] via inhibition of peptidyl transferase of ribosomes and therefore peptide bond formation.[365] Morre et al. (1999) showed that glaucarubolone (CS 2.175) from *Castela polyandra* Moran & Felger (family Simaroubaceae Juss.) inhibited the enzymatic activity of NADH oxidase in Hela cell membrane vesicles, with an IC_{50} value equal to 0.05 nM, and blocked the proliferation of the cells at a dose of 1 μM.[368] Brusatol inhibited the growth of human pre-B leukemia (BV-173) cells, human Burkitt's lymphoma (Daudi) and human B-cell

■ **CS 2.172** Soularubinone

■ **CS 2.173** 15-Desacetylundulatone

■ **CS 2.174** Ring A: Michael acceptor

■ **CS 2.175** Glaucarubolone

lymphoma (DHL-6) cells, with an IC_{50} value of 5 ng/mL via apoptosis by downregulation of c-Myc,[369] which normally results in apoptotic blockage caused by an increase in anti-apoptotic Bcl-xL and anti-apoptotic Bcl-2. Brusatol may very well induce apoptosis by NF-κB inactivation followed by downregulation of c-Myc, hence increased levels of p27 followed by decreased cyclin-dependent kinase 2 (CDK2) activity, lymphoma proteins hypophosphorylation, G_1 arrest and apoptosis.[370] In fact, bruceantin exhibited modest NF-κB inhibitory activity, with IC_{50} values of 19.6 μg/mL.[338] Incongruously, brusatol at a dose of 25 μg/mL induced in HL-60 cells the phosphorylation of IκBα via the translocation of NF-κB, which normally regulates the transcription of c-Myc.[371] This suggests that NF-κB is not the pivotal target of quassinoids. Brucein D from *Brucea javanica* (L.) Merr. (family Simaroubaceae DC.) inhibited the growth of PANC-1, SW1990 and CAPAN-1 cells via apoptosis, with activation of caspases 3 and 8, increase of pro-apoptotic protein Bak (hence mitochondrial insult) and activation of p38-mitogen-activated protein kinase.[337] The effect of brucein D may result from interaction with death receptor Fas inducing the activation of caspase 8 via Fas-associated death domain (FADD) and the stimulation of caspase 3.[372] Another possible apoptotic pathway would involve caspase 8-induced activation of Bid mitochondrial insult and release of cytochrome c, activation of caspases 9 and 3, and cleavage of PARP.[373] The activation of p38 may explain the downregulation of c-Myc by quassinoids since p38 regulates c-Myc transcription.[374] A derivative of bruceantin named NBT-272 (CS 2.176) downregulated c-Myc in MB-derived cells and inhibited the survival of human medulloblastoma (DAOY), human neuroblastoma (WAC-2), human retinoblastoma (WERI), and Wilms' tumor (Wit-49) cells, with IC_{50} values equal to 13.6 nmol/L, 4.8 nmol/L, 4.4 nmol/L and 4.9 nmol/L, respectively, accompanied by a reduction in phosphorylated ERK and[375] hence c-Myc downregulation. Eurycomanone inhibited the growth of A549 cells (wild-type p53), with an IC_{50} value equal to 5.1 µg/mL, via reduction of p53 levels,[376] possibly on account of reduction of phosphorylated ERK.

2.3.5 *Dictamnus Dasycarpus* Turcz.

History This plant was first described by Nicolas Stepanowitsch Turczaninow in *Bulletin de la Société Impériale des Naturalistes de Moscou*, published in 1842.

Synonyms *Aquilegia fauriei* H. Lév., *Dictamnus albus* subsp. *dasycarpus* (Turcz.) Kitag., *Dictamnus albus* subsp. *dasycarpus* (Turcz.) L. Winter, *Dictamnus albus* var. *dasycarpus* (Turcz.) T.N. Liou & Y.H. Chang

Family Rutaceae Juss., 1789

Common Names Burning bush, bag sun (Korean), bai xian, bai xian pi (Chinese)

Habitat and Description This is a perennial herb which grows to 1 m tall in the grassy areas of Russia, China, Mongolia and Korea. The plant causes dermatitis. The stems are hairy. The leaves are compound. The rachis is winged and bears three to six pairs of leaflets plus a terminal one. The leaflets are elliptic, 3–10 cm × 1.5–5 cm, crenulate and dotted with oil glands. Inflorescences and flowers have follicles with opaque dark brown globose to ellipsoid sessile or stalked glands. The inflorescence is a conspicuous terminal raceme which grows to 25 cm long. The calyx comprises five minute sepals which are connate at base. The corolla consists of five petals which are pinkish with purple venations, lanceolate, 2–2.5 cm × 0.5–1 cm and arranged in a curious orchid or cesalpinaceaous-like fashion. The androecium includes 10 stamens which are exerted and recurved. The gynoecium consists of five carpels which are free. The fruits are series of membranaceous, somewhat crustaceous, brownish-beaked follicles which are 1–2 cm long. The seeds are minute, globose and glossy (Figure 2.14).

Medicinal Uses In China the plant is used to treat skin infections and itchiness, and jaundice. Likewise, Koreans use the plant to treat jaundice but also for flu and arthritis.

Phytopharmacology The plant contains the furanocoumarins psoralen, xanthotoxin[377] and scopoletin;[378] the flavonoids quercetin, isoquercitrin, rutin[378] and wogonin;[379] a series of phenolic glycosides;[380] the furan derivatives fraxinellone,[381] 6β-hydroxyfraxinellone, isofraxinellone, calodendrolide,[382] fraxinellonone,[379] 9α-hydroxyfraxinellone-9-*O*-β-D-glucoside, dictamnusine, dictamdiol, dictamdiol A, dictamdiol B and 9β-hydroxyfraxinellone;[383] the guaiane sesquiterpene dictamnol,[384]

■ **FIGURE 2.14** *Dictamnus dasycarpus* Turcz.

the sesquiterpene glycoside dictamnosides A–E,[385] F, G[386] and H–N;[387] furoquinoline alkaloids such as dictamnine,[381] haplopine[382] and skimmianine;[379] the quinoline alkaloid dasycaryne,[388,389] and triterpene limonoids such as rutaevin,[390] obacunone, 7α-acetylobacunol, 7α-acetyldihydronomilin and limonin.[382,389] The essential oil contains the terpenes syn-7-hydroxy-7-anisylnorbornene, pregeijerene and geijerene.[391]

The dermatological properties of the plant are provided by the counter-irritant properties of psoralen and xanthotoxin.[377]

Proposed Research Pharmacological study of limonin derivatives for the treatment of cancer.

Rationale Limonoid triterpenes have attracted a great deal of interest on account of their abilities to repress the growth of several malignancies.[392] Structurally, limonoids are formed by loss of four terminal carbons of the side chain in the apotirucallane skeleton and then cyclized to form the 17-β furan ring which participates in the cytotoxic activity.[265,392–394] Limonoids with C28–C19 ether bond are powerfully cytotoxic. Such limonoids are 12-hydroxyamoorastatone (CS 2.177), 12-hydroxyamoorastatin (CS 2.178) and 12-acetoxyamoorastatin (CS 2.179) isolated *Melia azedarach* L. (family Meliaceae Juss.), which abated the growth of human lung adenocarcinoma epithelial (A549) cells, with IC_{50} values equal to $9.2 \times 10^{-1}\,\mu g/mL$, $4 \times 10^{-2}\,\mu g/mL$ and $10^{-2}\,\mu g/mL$, respectively,[395] suggesting that a ketone at C11 and an epoxy at C14–C15 are favorable for activity. Indeed, the 14,15β-epoxide is required for alkylation of amine

■ **CS 2.177** 12-Hydroxyamoorastatone

■ **CS 2.178** 12-Hydroxyamoorastatin

■ **CS 2.179** 12-Acetoxyamoorastatin

■ **CS 2.180** 28-Isobutylsendanin

■ **CS 2.181** 28-Deacetylsendanin

and/or thiol sites of cellular targets.[393] In addition, the 28-isobutylsendanin (CS 2.180) and 28-deacetylsendanin (CS 2.181) from *Melia azedarach* L. (family Meliaceae Juss.) with a ketone at C11 were powerfully cytotoxic against mouse leukemia (P388) cells, with IC_{50} values equal to 0.03 μg/mL and 0.02 μg/mL, respectively,[396] with a better activity for the limonoid with a hydroxyl moiety at C28. A series of C11 keto-limonoids isolated

■ **CS 2.182** 12-Deacetyltrichilin I

■ **CS 2.183** 1-Acetyltrichilin H

■ **CS 2.184** 3-Deacetyltrichilin H

■ **CS 2.185** 1-Acetyl-3- deacetyltrichilin H

from *Melia azedarach* L (family Meliaceae Juss.) inhibited the growth of P388 cells, with IC_{50} values as follows: 12-deacetyltrichilin I (CS 2.182), 0.01 µg/mL; 1-acetyltrichilin H (CS 2.183), 0.4 µg/mL; 3-deacetyltrichilin H (CS 2.184), 0.04 µg/mL; 1-acetyl-3-deacetyltrichilin H (CS 2.185),

0.4 μg/mL; 1-acetyl-2-deacetyltrichilin H (CS 2.186), 0.6 μg/mL; trichilin H (CS 2.187), 0.1 μg/mL; trichilin D (CS 2.188), 0.05 μg/mL; 1,12-diacetyltrichilin B (CS 2.189), 0.4 μg/mL and aphanastatin (CS 2.190), 0.06 μg/mL.[397] These results showed that an ester at C28 is beneficial

■ **CS 2.186** 1-Acetyl-2- deacetyltrichilin H

■ **CS 2.187** Trichilin H

■ **CS 2.188** Trichilin D

■ **CS 2.189** 1,12-Diacetyltrichilin B

whereas acetylation of C1 hydroxyl reduces the activity and the C2 and C12 substitutions are not necessary for activity. Toosendanin (CS 2.191) isolated from *Melia toosendan* Siebold & Zucc. (family Meliaceae Juss.) inhibited the survival of the following cell lines (IC_{50} values in parentheses): human prostate cancer (PC-3; 1.2×10^{-7} M), human hepatocellular carcinoma (BEL-7404; 2.6×10^{-8} M), human neuroblastoma (SH-SY5Y; 1.5×10^{-7} M), human glioblastoma (U251; 3.3×10^{-8} M), human promyelocytic leukemia (HL-60; 6.1×10^{-9}) and human histiocytic lymphoma (U937; 5.4×10^{-9} M),[398] suggesting that limonoids may be useful against blood cell malignancies. Trichilin H and toosendanin from *Melia toosendan* Siebold & Zucc. (family Meliaceae Juss.) inhibited the growth of human nasopharyngeal carcinoma (KB) cells, with IC_{50} values equal to 0.1 μg/mL and 3.8 μg/mL, respectively,[399] confirming that an ester at C28 is beneficial.

A second interesting type of cytotoxic limonoid lacks C19–C28 esters and presents an α,β-unsaturated 3 ketone on ring A, which brings about cytotoxicity.[265,393,394] One such limonoid is azadirone (CS 2.192) from *Azadirachta indica* A. Juss (family Meliaceae Juss.), which presents a Δ_{14-15} instead of an epoxy on ring D and was shown to be lethal to the following cell lines (with IC_{50} values in parentheses): U251 (2.5 μM), multidrug-resistant breast carcinoma (MCF-7/ADR; 0.06 μM), human

■ **CS 2.190** Aphanastatin

■ **CS 2.191** Toosendanin

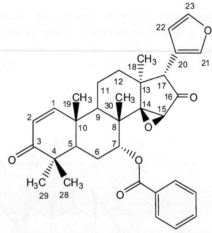

■ **CS 2.192** Azadirone

■ **CS 2.193** 7-Benzoylnimbocinol

■ **CS 2.194** Epoxyazadiradione

■ **CS 2.195** 7-Deacetyl-7-benzoylepoxyazadiradione

colon cancer (SW620; 7 μM), human lung cancer (H522; 6 μM), human melanoma (M14; 0.08 μM), human ovary adenocarcinoma (SK-OV-3; 5 μM), human prostate carcinoma (DU-145; 0.05 μM) and human renal carcinoma (A498; 2 μM),[394] confirming that the lack of C12 substitution does not reduce the cytotoxic effects against breast, skin and prostate cancers. 7-Benzoylnimbocinol (CS 2.193), epoxyazadiradione (CS 2.194), 7-deacetyl-7 benzoylepoxyazadiradione (CS 2.195) from *Azadirachta indica* A. Juss (family Meliaceae Juss.) moderately inhibited the growth of HL-60 cells, with IC$_{50}$ values equal to 5.3 μM, 9.3 μM and 3.1 μM,

■ **CS 2.196** Dysobinin

■ **CS 2.197** Azadiradione

■ **CS 2.198** Epoxyazadiradione

■ **CS 2.199** Mahonin

respectively[400] suggesting that a ketone in C16 is not favorable for activity. Dysobinin (CS 2.196), azadiradione (CS 2.197) and epoxyazadiradione (CS 2.198) isolated from a member of the genus *Chisocheton* Blume (family Meliaceae Juss.) were deleterious to the following cell lines (respective IC$_{50}$ values in parentheses): KB (3.1 µg/mL, 9.3 µg/mL, 12.8 µg/mL), human small-cell lung cancer (NCI-H187; 1.6 µg/mL, 6.4 µg/mL, 7.5 µg/mL) and human breast adenocarcinoma (MCF-7; 2.1 µg/mL, 7.1 µg/mL, 4.6 µg/mL),[401] confirming the negative effect of a ketone in C16. From the same plant, mahonin (CS 2.199) inhibited the growth of NCI-H187 and MCF-7 cells, with IC$_{50}$ values equal to 15.6 µg/mL and 18.4 µg/mL,

■ **CS 2.200** Anthothecol

■ **CS 2.201** Malleastrone A

■ **CS 2.202** Malleastrone B

respectively, and was inactive against KB cells,[401] implying that an ester in C6 lowers the cytotoxic activity. Anthothecol (CS 2.200) from *Khaya anthotheca* (Welw.) C. DC. (family Meliaceae Juss.) was toxic to P388 cells, with an IC$_{50}$ value equal to 1.2 μg/mL.[393]

Malleastrone A (CS 2.201) and B (CS 2.202) isolated from a member of the genus *Malleastrum* (Baill.) J.F. Leroy (family Meliaceae Juss.) presents

■ **CS 2.203** Proxylocarpin A

■ **CS 2.204** Proxylocarpin B

■ **CS 2.205** Proxylocarpin E

a ketone at C7 which is necessary for activity.[265,393,394] Malleatrone A and B inhibited the survival of the following cell lines (respective IC$_{50}$ values in parentheses): human ovarian cancer (A2780; 0.4 μM, 0.6 μM), human breast carcinoma (MDA-MB-435; 0.4 μM, 0.3 μM), human colon cancer (HT-29; 0.2 μM, 0.2 μM), human non-small-cell lung carcinoma (H522-T1; 0.2 μM, 0.2 μM) and human histiocytic lymphoma (U937; 0.2 μM, 0.1 μM).[402] Limonoids with a α,β-unsaturated 3 ketone in ring A but lacking of a proper furan ring at C17 have the tendency to be less potent. For instance, proxylocarpins A, B and E (CS 2.203–205) from *Xylocarpus granatum* J. Koenig (family Meliaceae Juss.) were mildly cytotoxic to the following cell lines (respective IC$_{50}$ values in parentheses): A549 (19 μM, 13.5 μM, 7.8 μM), A2780 (0.3 μM, 0.9 μM, 2.5 μM), human intestinal

■ **CS 2.206** 12-O-Methylvolkensin

■ **CS 2.207** 28-Deoxonimbolide

■ **CS 2.208** Ohchinin acetate

■ **CS 2.209** 1-O-Deacetylohchinolide B

adenocarcinoma cell (HCT-8; 2.8 μM, 4.5 μM, 3.8 μM), human liver cancer (BEL-7402; 2.7 μM, 5.5 μM, 4.4 μM) and human gastric cancer (BGC-823; 3.3 μM, 6.7 μM, 4.6 μM).[403]

Seco-C limonoids are moderately cytotoxic. This is exemplified with 12-O-methylvolkensin (CS 2.206) from *Melia toosendan* Siebold & Zucc. (family Meliaceae Juss.), which stopped the growth of KB cells, with an IC$_{50}$ value equal to 8.7 μg/mL.[398] 28-Deoxonimbolide (CS 2.207) and ohchinin acetate (CS 2.208) from *Azadirachta indica* A. Juss (family Meliaceae Juss.) prevented the proliferation of HL-60 cells, with IC$_{50}$ values equal to 2.7 μM and 9.9 μM, respectively.[400] 1-O-Deacetylohchinolide B (CS 2.209) and 1-O-deacetylohchinolide A (CS 2.210) from *Melia azedarach* L (family Meliaceae Juss.) were toxic against human epithelial

■ **CS 2.210** 1-*O*-Deacetylohchinolide A

■ **CS 2.211** Chisonimbolinin C

cancer (Hela S3) cells, with IC_{50} values equal to 0.1 μM and 2.4 μM, respectively.[404] The cytotoxic activity of seco-C limonoids is further reduced when the 17β furan ring is absent. This is the case with chisonimbolinin C and D (CS 2.211 and CS 2.212) isolated from a member of the genus *Chisocheton* Blume (family Meliaceae Juss.), which weakly compromised the growth of human epitheloid cervix carcinoma (Hela) cells, with IC_{50} values equal to 13 μM and 32 μM, respectively.[405]

Seco-D rings are mildly cytotoxic: gedunin from the bark of *Xylocarpus granatum* J. Koenig (family Meliaceae Juss.) inhibited the survival of human colon cancer (CaCo-2) cells, with an IC_{50} value equal to

■ **CS 2.212** Chisonimbolinin D

■ **CS 2.213** Gedunin

■ **CS 2.214** 7-Deacetyl-7-benzoylgedunin

■ **CS 2.215** 7-Deacetylgedunin

16.8 μM.[406] Gedunin (CS 2.213) and 7-deacetyl-7-benzoylgedunin (CS 2.214) from *Azadirachta indica* A. Juss (family Meliaceae Juss.) reduced the lifespan of HL-60 cells, with IC_{50} values equal to 5.9 μM and 2.9 μM, respectively,[400] evidencing the fact that an ester in C7, and especially a bulky one, is beneficial for cytotoxic activity.[265,393,394] This is substantiated with 7-deacetylgedunin (CS 2.215) from *Xylocarpus granatum* J. Koenig (family Meliaceae Juss.), which weakly inhibited the growth of human lung tumor (CHAGO) and human hepatocellular liver carcinoma (HepG2) cells, with IC_{50} values of 16 μM and 10.2 μM, respectively.[407]

■ **CS 2.216** 11β-Hydroxygedunin

■ **CS 2.217** 7-Deacetoxy-7α,11β-dihydroxygedunin

■ **CS 2.218** 7-Deacetoxy-7α-hydroxygedunin

■ **CS 2.219** 11-Oxogedunin

11β-Hydroxygedunin (CS 2.216), 7-deacetoxy-7α,11β-dihydroxygedunin (CS 2.217), gedunin, 7-deacetoxy-7α-hydroxygedunin (CS 2.218) and 11-oxogedunin (CS 2.219) from *Cedrela sinensis* A. Juss. (family Meliaceae Juss.) inhibited the growth of P388 cells, with IC_{50} values equal to 5.4 µg/mL, 7.8 µg/mL, 3.3 µg/mL, 4.5 µg/mL and 3 µg/mL, respectively,[408] confirming that a ketone at C11 and an ester at C7 ameliorate the cytotoxic potencies of limonoids.

■ **CS 2.220** Prieurianin

■ **CS 2.221** 14,15β-Epoxyprieurianin

■ **CS 2.222** Odontadenin A

Prieurianin (CS 2.220) and 14,15β-epoxyprieurianin (CS 2.221) with seco-A,B rings from *Guarea guidonia* (L.) Sleumer (family Meliaceae Juss.) abated the survival of P388 cells, with IC_{50} values equal to 4.4 µg/mL and 0.4 µg/mL, respectively,[409] confirming that C14-C15 epoxy is beneficial for cytotoxic activity. The seco A-D limonoid odontadenin A (CS 2.222) isolated from a plant (mis)identified as *Odontadenia macrantha* (Willd. ex Roem. & Schult.) Markgr. (family Apocynaceae Juss.) was cytotoxic against A2780 cells, with an IC_{50} value equal to 3.2 µg/mL.[410]

CS 2.223 Limonin

CS 2.224 Azadirachtin

Although massive evidence is available on the cytotoxic effect of limonoids, there is a dearth of information about their apoptotic mechanism of action, which may very well imply mitochondrial insult linked to an increase in the level of pro-apoptotic Bax on account of NF-κB inhibition. In fact, toosendanin is well known to induce the release of cytochrome c from damaged mitochondria.[411] Limonin (CS 2.223) from *Citrus reticulata* Blanco (family Rutaceae Juss.) weakly inhibited the growth of human colon adenocarcinoma (SW480) cells, with an IC_{50} value of 54.7 μM, and this was associated with an increase in the level of pro-apoptotic Bax, a decreased level of anti-apoptotic Bcl-2 protein (and hence an increase in cytoplasmic Ca^{2+} levels), a fall in mitochondrial membrane potential, mitochondrial insult, release of cytochrome c and activation of caspase 3.[412]

Note that the anti-apoptotic Bcl-2 protein is known to inhibit inositol 1,4,5-trisphosphate receptor ($InsP_3R$)-induced release of calcium in the cytoplasm from the endoplasmic reticulum.[413] Toosendanin, like limonin, increased the cytoplasmic levels of Ca^{2+} in NG108-15cells[414–416] and destroyed primary rat hepatocytes, with an IC_{50} value of 14.9 μM, via apoptosis with increased levels in ROS, a fall in mitochondrial potential, release of cytochrome c, activation of caspases 9 and 3, activation of anti-apoptotic ERK and activation of pro-apoptotic JNK.[417] Note that JNK, which is inhibited by NF-κB, induces pro-apoptotic Bax translocation to mitochondria and therefore apoptosis via release of cytochrome c and thereby activation of caspase 3.[300] In fact, azadirachtin (CS 2.224) from *Azadirachta indica* A. Juss. (family Meliaceae Juss.) at a dose of 10 μM inhibited tumor necrosis factor-α (TNF-α)-induced NF-κB activation by

■ **CS 2.225** Nimbolide

■ **CS 2.226** Obacunone

preventing TNF-α from binding to its receptor (TNFR) in U937 cells.[418] TNF-α binds to the TNFR and sequentially recruits tumor necrosis factor receptor type 1-associated death domain protein (TRADD), Fas-associated death domain (FADD) and caspase-8, leading to activation of caspases 9 and 3 and therefore the cleavage of PARP. Nimbolide (CS 2.225) from *Azadirachta indica* A. Juss (family Meliaceae Juss.) at a dose of 10 μM inhibited the TNF-α-induced NF-κB activation and hence the transcription of the anti-apoptotic proteins Bcl-2, Bcl-xL, IAP-1 and IAP-2 in human chronic myeloid leukaemia (KBM-5) cells. This was effected via blockade of IκBα phosphorylation by IκB kinase β (IKKβ) through its interaction with cysteine at position 179,[419] and consequent pro-apoptotic Bax activation and mitochondrial insult. Obacunone (CS 2.226) from *Citrus aurantium* L. (family Rutaceae Juss.) mitigated the viability of human pancreatic cancer (PANC-28) cells (mutant p53) by 42% at a dose of 100 μM, with elevated levels of Bax, release of cytochrome c, activation of caspase 3, and PARP cleavage,[420] independently of p53. In addition, obacunone inhibited the growth of SW480 cells (mutant p53), with an IC_{50} value equal to 56.2 μM, via apoptosis, increase of pro-apoptotic Bax, release of cytochrome c, caspase 9 and 3 activation, DNA fragmentation and induction of p21[421] without p53 activation.

2.3.6 *Pulsatilla Chinensis* (Bunge) Regel

History This plant was first described by Eduard August von Regel in *Tent. Fl.Ussur.*, published in 1861.

Synonyms *Anemone chinensis* Bunge, *Anemone pulsatilla* var. chinensis (Bunge) Finet & Gagnep.

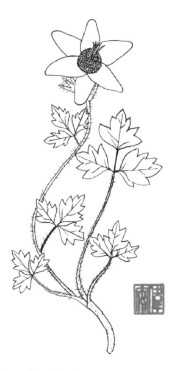

■ **FIGURE 2.15** *Pulsatilla chinensis* (Bunge) Regel

Family Ranunculaceae Juss., 1789

Common Name Bai tou weng (Chinese)

Habitat and Description This is a small perennial herb which grows wild in the grassy spots of East Russia, China and Korea. The rhizome is erect and about 1 cm across. The leaves are simple and basal. The petiole is 5–15 cm long and covered with long, soft hairs. The blade is ovate, 4.5–15 cm×6.5–20 cm, trifoliolate and hairy beneath. The flowers are solitary and showy. The perianth consists of a calyx comprising five sepals which are violet, oblong and 2.5–4.5 cm×1–2 cm. The stamens are numerous. The gynoecium consists of several free carpels. The infructescence is globose, 1 cm in diameter and made of numerous, 0.5-cm-long achenes, which are spindle-shaped and pilose with a 3.5–6.5 cm plumose beak formed by a persistent style (Figure 2.15).

Medicinal Uses In Korea and China the plant is used to treat amebiasis, malaria, gonorrhea, diarrhea and to induce urination.

Phytopharmacology The plant contains lupane triterpenes, such as pulsatillic acid, 23-hydroxy-betulinic acid,[422] betulinic acid, lupeol, betulin and 3-oxo-23-hydroxybetulinic acid,[423] a series of triterpene saponins including pulsatilloside D, E and G,[422,424–428] the lignans (+)-pinoresinol and β-peltatin,[425] the furanocoumarin 8-methoxy-psoralen[429] and ranunculin.[430]

Proposed Research Pharmacological study of lup-28-al-20(29)-en-3-one and other 3-keto lupane derivatives for the treatment of lung cancer.

Rationale Lupane triterpenes have generated a burst of interest as potential chemotherapeutic agents for the treatment of brain, blood and skin malignancies and one can ascertain with confidence that most of what is understood today about lupane chemotherapeutic potential comes from the study of betulinic acid isolated from members of the genus *Betula* L. (family Betulaceae, Gray) and lupeol which is common in edible fruits and vegetables. Indeed, an emerging body of evidence has conceived the notion that lupeol induces apoptosis preferably in skin, pancreatic and prostate malignancies by interfering with the G protein Ras and inhibiting the enzymatic activity of DNA topoisomerase and polymerase.[431]

The following compounds were shown to inhibit the growth of mouse melanoma (B16 2F2) cells via apoptosis (IC_{50} values obtained are in parentheses): Lupeol (Lup-20(29)-en-3β-ol) (CS 2.227) (38 μM), lupenone (CS 2.228) (25.4 μM), lupenyl acetate (CS 2.229) (22.7 μM), betulin (CS 2.230) (27.4 μM), betulin diacetate (CS 2.231) (36.8 μM), betulinic acid (7.9 μM), betulinic acid methylester (CS 2.232) (4.9 μM), lup-28-al-20(29)-ene-3β-ol

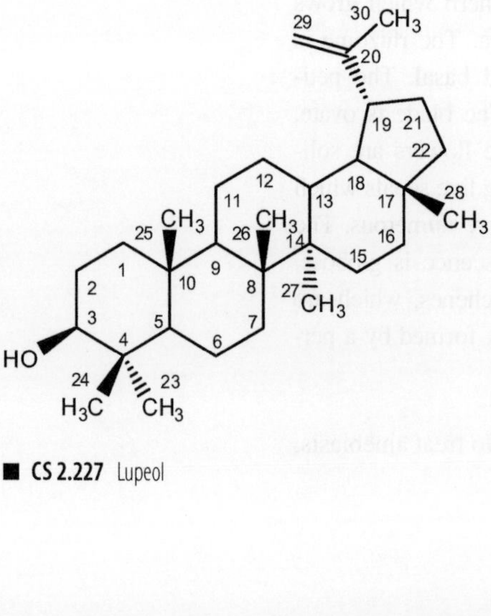

■ **CS 2.227** Lupeol

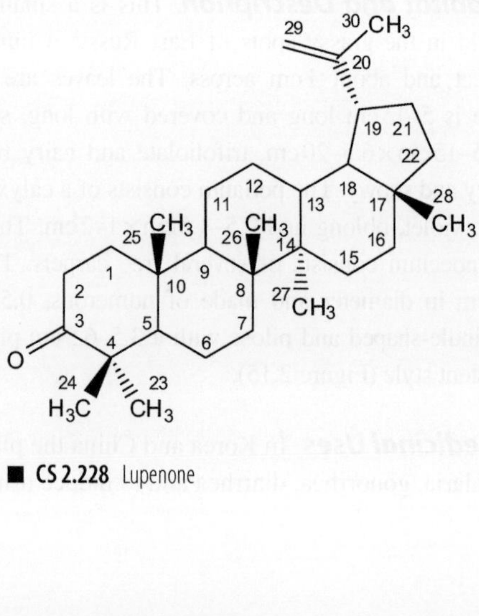

■ **CS 2.228** Lupenone

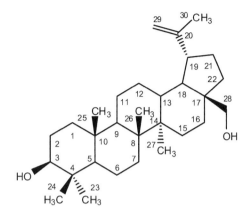

■ **CS 2.229** Lupenyl acetate

■ **CS 2.230** Betulin

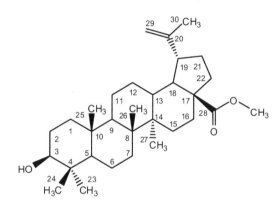

■ **CS 2.231** Betulin diacetate

■ **CS 2.232** Betulinic acid methylester

(CS 2.233) (6.4 µM) and lup-28-al-20(29)-ene-3-one (CS 2.234) (4.1 µM). These figures suggest that a carbonyl group at C28 is beneficial for apoptotic activity[432] in melanoma cells. Lupeol and C3 ester derivatives including 3β-(3R acetoxyhexadecanoyloxy)-lup-20(29)-ene (CS 2.235), 3β-(3R-acetoxyhexadecanoyloxy)-29-*nor*-lupan-20-one (CS 2.236) and 3β-(3-ketohexadecanoyloxy)-29-*nor*-lupan-20-one (CS 2.237) inhibited the enzymatic activity of DNA polymerase β, with IC$_{50}$ values equal to 6.4 µM, 3.8 µM and 5.3 µM, respectively.[433]

The roles of DNA polymerase β and topoisomerase II are to repair damaged DNA and to resolve DNA topological issues, and inhibition of these enzymes by lupeol should promote apoptosis via pro-apoptotic protein

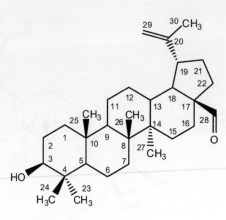

■ **CS 2.233** Lup-28-al-20(29)-ene-3β-ol

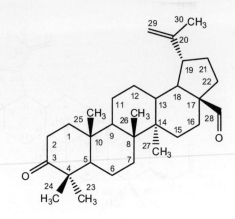

■ **CS 2.234** Lup-28-al-20(29)-ene-3—one

■ **CS 2.235** 3β-(3R-Acetoxyhexadecanoyloxy)-lup-20(29)-ene

■ **CS 2.236** 3β-(3-ketohexadecanoyloxy)-29-*nor*-lupan-20-one

■ **CS 2.237** 3β-(3R-acetoxyhexadecanoyloxy)-29-*nor*-lupan-20-one

(p53) accumulation in p53 wild-type cancer cells and/or mitogen-activated protein kinase (MAPK) p38.[434,435] In fact, Hata et al. (2003) observed that lupeol at a dose of 10 μM increased the level of phosphorylated MAPK p38 in B16 2F2 cells.[436] Saleem et al. (2005) studied the effect of lupeol against human pancreatic adenocarcinoma cells (AsPC-1) and found that apoptosis at a dose of 35 μM was induced by inhibition of G protein Ras.[437] G protein Ras promotes the survival of cancer cells via two anti-apoptotic pathways – the Ras/ERK and the Ras/PI3K/Akt pathways[438] – and its inhibition by lupeol is followed by the inhibition of protein kinase B (Akt) and therefore blockade of NF-κB activation via decrease in the phosphorylation of IκBα protein, decreased levels of phospho-p38 MAPK and, incongruously, ERK activation.[439] Note that G protein Ras decreases Fas receptor at the transcriptional level,[440] hence the possible enhancement of Fas-receptor susceptibility, and therefore of the death signaling pathway, by lupeol. In fact, lupeol induced apoptosis in androgen-sensitive human prostate adenocarcinoma (LNCaP) cells, with an IC_{50} value equal to 21 μmol/L, via the Fas receptor death signaling pathway with cleavage of PARP and activation of caspase 8.[441] It is also useful to understand that inhibition of G protein Ras by lupeol may explain the inhibition of topoisomerase II.[442]

The oxidation of the C28 methyl of lupeol (Lup-20(29)-en-3β-ol) into a carboxylic acid provides betulinic acid, which inhibits the enzymatic activity of topoisomerase I, induces apoptosis (notably via mitochondrial insults and ROS generation)[443,444] and suppresses carcinogen-induced NF-κB activation by inhibiting IκBα kinase.[445] Betulinic at first drew a great deal of interest as a potential agent for the treatment of melanoma since it selectively induced apoptosis and DNA fragmentation in human melanoma (MEL-1, MEL-2 and MEL-4) cells, with IC_{50} values equal

■ **CS 2.238** Betulinic acid

■ **CS 2.239** 23-Hydroxybetulinic acid

to 1.1 μg/mL, 2 μg/mL and 4.8 μg/mL, respectively.[446] Betulinic acid (CS 2.238), 23-hydroxybetulinic acid (CS 2.239) and 3-oxo-23-hydroxy-betulinic acid (CS 2.240) isolated from *Pulsatilla chinensis* (Bge) Regel abrogated the growth of mouse melanoma (B16) cells, with IC_{50} values equal to 76 μg/mL, 32 μg/mL and 22.5 μg/mL, respectively, by mitochondrial insult and increase of ROS.[447] Comparable mechanisms were observed in human melanoma (UISO-Mel-1) where betulinic acid at a dose of 8 μg/mL induced mitochondrial insult and generated ROS, and hence increased levels of phosphorylated p38.[448,449] Betulinic acid at a dose of 10 μg/mL induced Akt phosphorylation and boosted the levels of the anti-apoptotic induced myeloid leukemia cell differentiation protein (Mcl-1) in human melanoma cells M20,[450] probably on account of increased levels of ROS, which are known to activate Akt.[451]

Comparable mechanisms were observed in human melanoma (UISO-Mel-1) where betulinic acid at a dose of 8 μg/mL induced mitochondrial insult and generated ROS and hence increased levels of phosphorylated p38.[448,449] Betulinic acid at a dose of 10 μg/mL induced Akt phosphorylation and boosted the levels of the anti-apoptotic induced Mcl-1 protein in M20 cells.[450] This was probably caused by increased levels of ROS, which are known to activate Akt.[451] Note that the betulinic acid derivative impressic acid (CS 2.241) from *Acanthopanax koreanum* Nakai (family Araliaceae, Juss.) inhibited TNFα-induced NF-κB transcription activity in human hepatocellular liver carcinoma (HepG2) cells that had been transfected, with an IC_{50} value of 4.9 μM.[452]

■ **CS 2.240** 3-Oxo-23-Hydroxybetulinic acid

■ **CS 2.241** Impressic acid

Another captivating oncological feature of betulinic acid is its ability to command the death of several brain cancer cells[453] via mitochondrial insults.[454] For instance, human glioblastoma (SK-55), human glioblastoma (U-343) and human medulloblastoma (D283) cells exposed to betulinic acid were killed, with IC_{50} values equal to 6 µg/mL, 5 µg/mL and 3 µg/mL, respectively.[455] Neuroblastoma (SH-EP) cells exposed to 10 µg/mL of betulinic acid became apoptotic with upregulation of pro-apoptotic Bax and Bcl-xL, fall of mitochondrial membrane potential and activation of the FLICE, CPP32 (caspase 3) and therefore PARP cleavage.[456] This mitochondrial mechanism was further observed in SH-EP cells, where betulinic acid provoked direct mitochondrial insults resulting in cytochrome c and apoptosis inducing factor (AIF) release, caspases 3 and 8 activation and PARP cleavage.[457] In DU-145 prostate cancer cells betulinic acid increased the levels of ROS, DNA damage and caspase 3 activation, interacted with topoisomerase I and hampered its interaction with oxidatively damaged DNA, but the generation of ROS and subsequent DNA damage did not induce stabilization of the topoisomerase I–DNA cleavable complex.[458] In addition, betulinic acid inhibited the growth of LNCaP and RKO colon cancer cells at a dose of 15 µg/mL by an interesting cellular mechanism involving the degradation of cyclin D1 and the transcription factor specificity proteins 1 (Sp1), 3 (Sp3) and 4 (Sp4) by the proteasome, which resulted in a decrease in survivin and NF-κB.[459,460] Note that in human colorectal carcinoma (HCT-116) cells betulinic acid inhibited IκBα kinase, which normally phosphorylates IκBα and brings about NF-κB inactivation.[461]

In regards to the structure–activity relationship of lupane triterpenoids, several lines of evidence demonstrate that cytotoxic potencies and the spectrum of activities are enhanced with a carboxylic or a carbonylic group at C28, oxidation of the 3β-hydroxy into a ketone and unsaturation of the C20-C29 bond, whereas acylation of the 3β-hydroxy and esterification of C28 reduces cytotoxic activity.[462,463] In fact, a ketone at C3 increases the binding affinity of lupanes to topoisomerases I and II, while oxidation of C30 to an aldehyde boosts the inhibition of topoisomerase I.[464]

These favorable chemical features are exemplified with ochraceolides A, B and C (CS 2.242–2.244) from a member of the genus *Kokoona* Thwaites (family Celastraceae, R. Br.), which annihilated mouse leukemia (P388) cells, with IC_{50} values equal to 0.2 μg/mL, 7.8 μg/mL and 0.5 μg/mL, respectively.[465] From the same genus, ochraceolide D (CS 2.245) inhibited the growth of human fibrosarcoma (HT1080) and human glioblastoma (U373) cells, with IC_{50} values equal to 14.7 μg/mL and 12.2 μg/mL, respectively; the respective values for ochraceolide E (CS 2.246) against the same cell lines were 3.9 μg/mL and 8.6 μg/mL.[466] Ochraceolide A and dihydroochraceolide A (CS 2.247) isolated from a member of the genus *Lophopetalum* Wight ex Arn. (family Celastraceae, R. Br.) were lethal to the following cell lines (respective IC_{50} values in parentheses): human breast cancer (ZR-751; 4.5 μg/mL, 3 μg/mL) and U373 (6.7 μg/mL, 9.1 μg/mL),[467] confirming that the reduction of the C3 ketone into a hydroxyl group mitigates the cytotoxic potencies of lupanes.

An example of a potent cytotoxic lupane with carboxylic groups in C17 is 3-*epi*-betulinic acid (CS 2.248) isolated from members of the genus

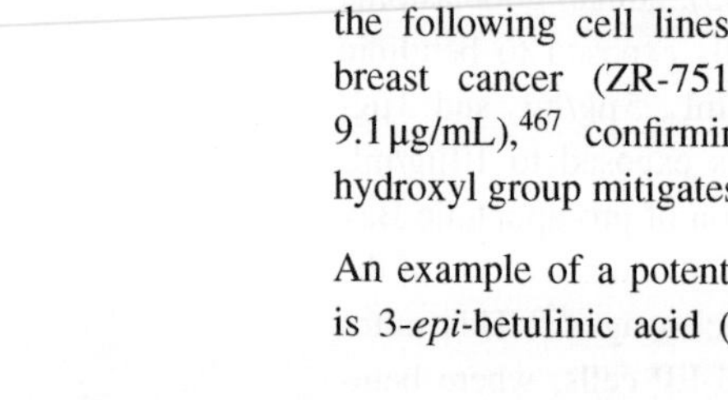

■ **CS 2.242** Ochraceolide A

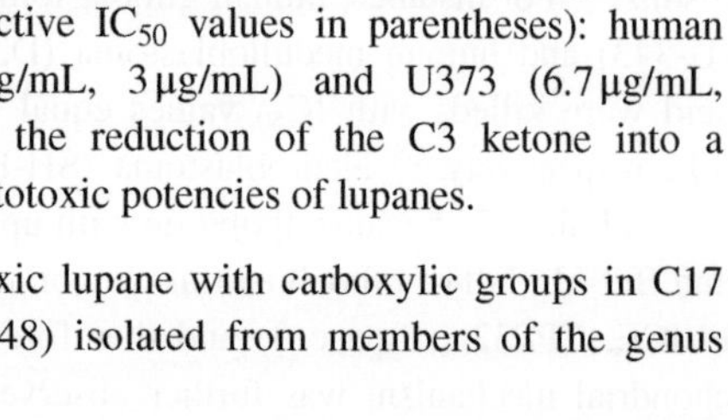

■ **CS 2.243** Ochraceolide B

■ **CS 2.244** Ochraceolide C

■ **CS 2.245** Ochraceolide D

Maytenus Molina (family Celastraceae, R. Br.). This inhibited the growth of human epitheloid cervix carcinoma (Hela) and human laryngeal carcinoma (Hep2) cells, with IC_{50} values equal to 2.1 μg/mL and 3.1 μg/mL, respectively, whereas 28,30-dihydroxy-3-oxolup-20(29)-ene (CS 2.249), isolated from the same plant and substituted by a hydroxyl group at C28, inhibited the growth of the same cells, with IC_{50} values equal to 4 μg/mL and 7.1 μg/mL, respectively.[468] The beneficial effects of a ketone at C3 is exemplified by 6β-hydroxylup-20(29)-en-3-oxo-27,28-dioic acid (CS 2.250) from *Viburnum odoratissimum* Ker. Gawl (family Adoxaceae, E. Mey.), which destroyed human stomach cancer (NUGC) cells by 80% at a dose of 10 μM.[469] Likewise, lup-20(29)-en-11α-ol-25,3β lactone (CS 2.251), 3-deoxybetulonic acid (CS 2.252) and lupenyl acetate isolated from *Coussarea paniculata* (Vahl) Standl. (family Rubiaceae, Juss.) displayed weak cytotoxic effects against human ovarian cancer (A2780) cells, with IC_{50} values equal to 18 μg/mL, 17.4 μg/mL and 16 μg/mL, respectively,[470] suggesting that the β hydroxyl at C3 should be free if it is not oxidized and that the removal of the hydroxyl group at C3 reduces the cytotoxic potentials of lupanes.

The 3-keto seco-A lupane 6-dehydroxy-20,29-dihydroviburolide (CS 2.253) synthesized from viburolide that had been obtained from a member of the genus *Viburnum* L. (family Adoxaceae, E. Mey.) induced the apoptosis of PC-3 cells, with an IC_{50} value equal to 12.3 μM, whereas betulinic acid had an IC_{50} value superior to 30 μM.[471] The benefit of a C28 carbonyl moiety and a ketone group at C3 is obvious in the case of lup-28-al-20(29)-en-3-one (CS 2.254), which abated the survival of a bewildering array of cells (IC_{50} values in parentheses): human promyelocytic leukemia (HL-60; 0.4 μM), human histiocytic lymphoma (U937; 1.5 μM), human

■ **CS 2.246** Ochraceolide E

■ **CS 2.247** Dihydroochraceolide A

■ **CS 2.248** 3-*epi*-Betulinic acid

■ **CS 2.249** 28,30-Dihydroxy-3-oxolup-20(29)-ene

erythromyeloblastoid leukemia (K562; 1.8 μM), human melanoma (G361; 9.4 μM), human skin melanoma (SK-MEL-28; 9.3 μM), human neuroblastoma (GOTO; 5.2 μM), human neuroblastoma (NB-I; 5.8 μM), normal human lung fibroblast (W138; 17.3 μM), human osteosarcoma (MG63; 17.9 μM), human osteosarcoma (Saos-2; 13.5 μM), human lung adenocarcinoma epithelial (A549; 2.3 μM), human lung giant-cell carcinoma (Lu-65; 2.4 μM), human lung large-cell carcinoma (Lu-99; 0.8 μM), human stomach cancer (SH-10-TC; 5.8 μM), human gastric cancer (MKN-45;

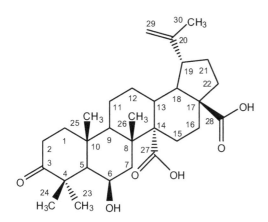

■ **CS 2.250** 6β-Hydroxylup-20(29)-en-3-oxo-27,28-dioic acid

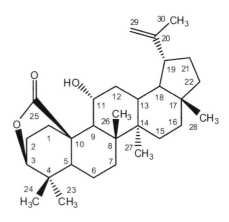

■ **CS 2.251** Lup-20(29)-en-11α-ol-25,3β lactone

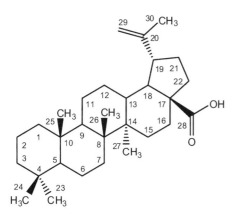

■ **CS 2.252** 3-Deoxybetulonic acid

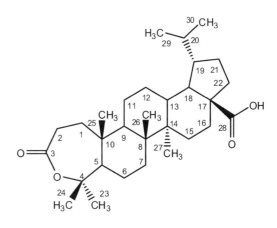

■ **CS 2.253** 6-Dehydroxy-20,29-dihydroviburolide

9.6 µM), human colorectal adenocarcinoma (HCT-15; 7.2 µM), human pancreatic carcinoma (MIA PaCa-2; 8.1 µM), human renal carcinoma (ACHN; 9.4 µM), human hepatocellular liver carcinoma (HepG2; 9.3 µM), human breast adenocarcinoma (MCF-7; 14.7 µM), human epitheloid cervix carcinoma (Hela; 2.1 µM), human epidermoid cervical carcinoma (CaSki; 6.3 µM), human ovarian adenocarcinoma (OVK-18; 10.2 µM), human bladder carcinoma (EJ-1; 16.2 µM) and human bladder cancer (T24; 12.5 µM), indicating high cytotoxic potencies for lung and leukemia cells via the induction of apoptosis and topoisomerase I inhibition.[472]

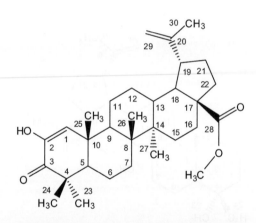

■ **CS 2.254** Lup-28-al-20(29)-en-3-one

■ **CS 2.255** Betulonic acid

■ **CS 2.256** 2-Hydroxy-3-oxolupa-1,20(29)-dien-28-oic acid

■ **CS 2.257** Methyl 2-hydroxy-3-oxolupa-1,20(29)-dien-28-oate

The importance of a ketone group in C3 was further confirmed by an experiment conducted by Urban et al. (2004) where betulonic acid (CS 2.255), 2-hydroxy-3-oxolupa-1,20(29)-dien-28-oic acid (CS 2.256), methyl 2-hydroxy-3-oxolupa-1,20(29)-dien-28-oate (CS 2.257), and 2-acetoxy-3-oxolupa-1,20(29)-dien-28-oic acid (CS 2.258) inhibited the growth of the following cell lines (respective IC$_{50}$ values in parentheses): A549 (146 μmol/L, 15 μmol/L, 10 μmol/L, 11 μmol/L, 11 μmol/L), DU-145 (196 μmol/L, 36 μmol/L, 23 μmol/L, 9 μmol/L, 18 μmol/L), MCF-7 (143 μmol/L, 29 μmol/L, 14 μmol/L, 6 μmol/L, 17 μmol/L) and human skin melanoma (SK-MEL-2; 21 μmol/L, 26 μmol/L, 23 μmol/L, 11 μmol/L,

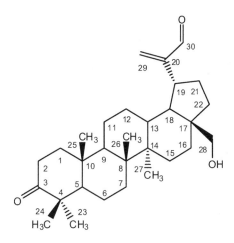

■ **CS 2.258** 2-Acetoxy-3-oxolupa-1,20(29)-dien-28-oic acid

■ **CS 2.259** 28-Hydroxy-3-oxo-lup-20-(29)-en-30-al

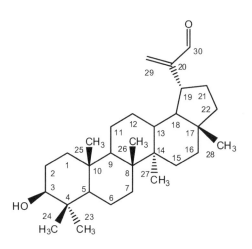

■ **CS 2.260** 3-Hydroxy-lup-20(29)-en-30-al

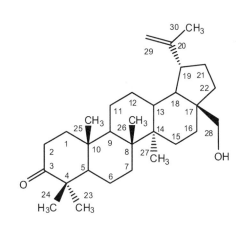

■ **CS 2.261** 28-Hydroxy-lup-20(29)-en-3-one

24 μmol/L), suggesting that a C28 carbonyl rather than a carboxyl is in fact necessary for potent cytotoxic effects.[461]

The oxidation of the C30 methyl into an aldehyde or a carboxylic acid does not abrogate the cytotoxic potencies of lupanes. For instance, 28-hydroxy-3-oxo-lup-20(29)-en-30-al (CS 2.259), 3-hydroxy-lup-20-(29)-en-30-al (CS 2.260) and 28-hydroxy-lup-20(29)-en-3-one (CS 2.261) from *Acacia mellifera* (Vahl) Benth. (family Fabaceae, Lindl.) destroyed human bronchopulmonary non-small-cell lung carcinoma (NSCLC-L6)

■ **CS 2.262** 3β-Acetoxy-lup-20(29)-en-30-al

■ **CS 2.263** 3β-Acetoxy-lup-20(29)-en-30-oic acid

■ **CS 2.264** 3β-Acetoxy-lupan-30-oic acid

■ **CS 2.265** 6-β,30-Dihydroxy-3-oxolup 20(29)-en-28-oic acid

cells, with IC$_{50}$ values equal to 15 μg/mL, 11 μg/mL and 30 μg/mL, respectively.[473,474] 28-Hydroxy-3-oxo-lup-20(29)-en-30-al isolated from a member of the family Celastraceae R. Br. inhibited the growth of a range of cell lines (respective IC$_{50}$ values in parentheses): HepG2 (12.6 μg/mL), MCF-7 (2.9 μg/mL), A549 (16.4 μg/mL), HL-60 (1.6 μg/mL), human hepatocellular carcinoma (Hep3B; 4.7 μg/mL), human breast adenocarcinoma (MDA-MB-231; 7.9 μg/mL), human gingival cancer (CA9-22; 1.4 μg/mL).[475] The positive influence of a carbonyl in C30 was further reinforced by Hata et al. (2008) where 3β-acetoxy-lup-20(29)-en-30-al (CS 2.262), 3β-acetoxy-lup-20(29)-en-30-oic acid (CS 2.263) and 3β-acetoxy-lupan-30-oic acid (CS 2.264) were selectively cytotoxic against the following

■ **CS 2.266** 20 Hydroxy-3-oxolup-28-oic acid

cell lines (respective IC_{50} values in parentheses): HL-60 (5.1 µM, 2.1 µM, 3.4 µM), U937 (7.7 µM, 4.4 µM, 4.7 µM) and Lu-99 (3 µM, 1.5 µM, 3 µM), via induction of apoptosis, suggesting as well that Δ_{28-29} favors cytotoxic activity.[476] The positive effect of Δ_{28-29} was concurrently exemplified with 6-β,30-dihydroxy-3-oxolup 20(29)-en-28-oic acid (CS 2.265) and 20 hydroxy-3-oxolup-28-oic acid (CS 2.266) isolated from the genus *Viburnum* L. family Adoxaceae, E. Mey., which negated the multiplication of P388 cells (respective IC_{50} values of 6.5 µg/mL and 35 µg/mL) and human colon cancer (HT-29) cells (9 µg/mL and 29 µg/mL).[477]

2.3.7 *Clinopodium Chinense* (Benth.) Kuntze

History This plant was first described by Carl Ernst Otto Kuntze in *Revisio Generum Plantarum,* published in 1891.

Synonyms *Calamintha chinensis* Benth., *Calamintha clinopodium* var. chinensis (Benth.) Miq., *Satureja chinensis* (Benth.) Briq.

Family Lamiaceae Martinov, 1820

Common Name Feng lun cai (Chinese)

Habitat and Description *Clinopodium chinense* is a tall perennial herb that grows in the grassy spots and forests of China, Korea, Japan and Taiwan. The main stem is 1 m tall, quadrangular and hairy. The leaves are simple, exstipulate and simple. The petiole is 0.3–0.9 cm long. The leaf blade is dentate, ovate, 2–4 cm × 1.5–2.5 cm, papery, hairy beneath, rounded at base and obtuse at apex. The inflorescence is an axillary head

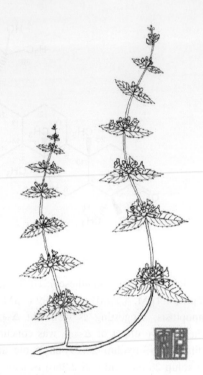

■ **FIGURE 2.16** *Clinopodium chinense* (Benth.) Kuntze

of flowers from 1–3 cm in diameter. The calyx is narrowly tubular, longitudinally veined, purplish, 0.5 cm long, hairy, bilobed, the upper lobe trifid, the lower bifid and longer. The corolla is purplish, 1 cm long, hairy and has two lips, the upper lobe straight and emarginate, the lower lobe trifid. The androecium consists of four stamens of different length. The gynoecium consists of two clefts at apex. The fruits are ovoid nutlets which are minute, fawn, and packed in the membranous and persistent calyx (Figure 2.16).

Medicinal Uses In Taiwan the plant is used to break fever, to treat venereal infection, to facilitate digestion, to check bleeding, to regulate menses and to invigorate.

Phytopharmacology The plant abounds with oleanane saponins such as clinoposaponins VI, IX–XIV, and XVII[478,479] and buddlejasaponin IV[478], and elaborates some phenylpropanoids including clinopodic acids A–I[480] and rosmarinic acid.[480]

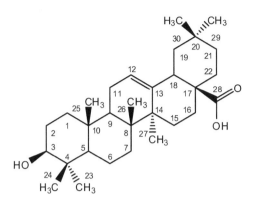

■ **CS 2.267** Oleanolic acid

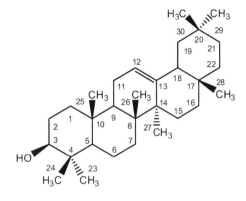

■ **CS 2.268** β-Amyrin

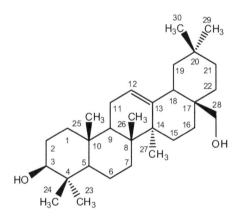

■ **CS 2.269** Erythrodiol

The medicinal properties have not been substantiated but it seems likely that the role of phenyl propanoids in the digestive system effects the involvement of oleanane saponins and rosmarinic in the anti-inflammatory/antibacterial properties.

Proposed Research Pharmacological study of tingenone and derivatives for the treatment of glioblastoma.

Rationale Oleananes, ursanes and friedelanes are pentacyclic triterpenes of compelling chemotherapeutic value. In regard to the cytotoxic properties of oleananes, oleanolic acid (CS 2.267), β-amyrin (CS 2.268) and erythrodiol (CS 2.269) inhibited the growth of mouse melanoma

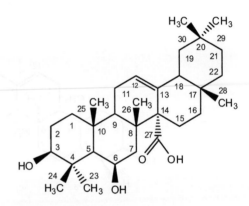

■ **CS 2.270** 3β-Hydroxy-olean-12-en-27-oic acid

■ **CS 2.271** 3β,6β-Dihydroxy-olean-12-en-27-oic acid

■ **CS 2.272** 3β-Acetoxy-olean-12-en-27-oic acid

■ **CS 2.273** 3β-Hydroxyolean-12-en-27-oic acid

(B16 2F2) cells, with IC$_{50}$ values equal to 4.8 μM, 48.8 μM and 33.4 μM, respectively,[432] indicating that a carboxylic group at C17 is beneficial for cytotoxic activity against melanoma. In addition, a carboxylic group is beneficial in C14 with a Δ_{12-13}, as seen with 3β-hydroxy-olean-12-en-27-oic acid (CS 2.270), 3β,6β-dihydroxy-olean-12-en-27-oic acid (CS 2.271) and 3β-acetoxy-olean-12-en-27-oic acid (CS 2.272) isolated from *Astilbe chinensis* (Maxim.) Franch. & Sav. (family Saxifragaceae Juss.), which inhibited the growth human epithelial ovarian cancer (HO-8910) cells, human epitheloid cervix carcinoma (Hela) and human promyelocytic leukemia (HL-60) cells.[481] The beneficial effect of C14 carboxylic acid was further observed with 3β-hydroxyolean-12-en-27-oic acid (CS 2.273) from

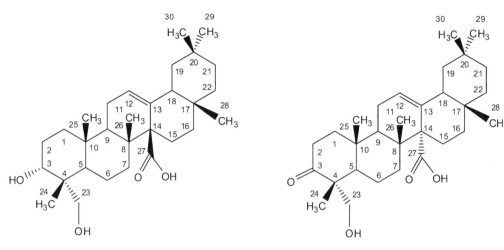

■ **CS 2.274** Aceriphyllic acid A

■ **CS 2.275** 23-Hydroxy-3-oxoolean-12-en-27-oic acid

Astilbe chinensis (Maxim.) Franch. & Sav. (family Saxifragaceae Juss.), which was cytotoxic against HO-8910, Hela, HL-60 and human colon adenocarcinoma (CoLo 205) cells, with IC_{50} values equal to 8 μg/mL, 3.9 μg/mL, 3.6 μg/mL, and 7.3 μg/mL, respectively.[482] Aceriphyllic acid A (CS 2.274) and 23-hydroxy-3-oxoolean-12-en-27-oic acid (CS 2.275) from *Aceriphyllum rossii* Engler. nullified the growth of Lewis lung carcinoma (LLC) cells, with IC_{50} values equal to 8.3 μM and 6.5 μM, respectively.[483]

The changes in location of methyl groups between oleanane, ursane and friedelane triterpenes have tremendous impacts on the cellular pathways responsible for the death of cancer cells.[484] Structure–activity relationship studies on the apoptotic potencies of oleananes have established that the A ring C3 ketone Δ_{1-2} and C ring C12 Δ_{9-11} conjugated systems are the main pharmacophores since they invite nucleophilic attack[485] and result in the synthesis of 2-cyano-3,12-dioxoolean-1,9-dien-28-oic acid (CS 2.276), 2-cyano-3,12-dioxoolean-1,9-dien-28-imidazolide ester (CS 2.277) and 2-cyano-3,12-dioxoolean-1,9-dien-28-methyl ester, which potently destroy multiple myeloma. The first two of these compounds induced apoptosis in human multiple myeloma (U266) cells, with IC_{50} values equal to 0.7 μM and 0.3 μM, respectively.[486] Indeed, the apoptotic activity of 2-cyano-3,12-dioxoolean-1,9-dien-28-oic acid and congeners is due to their ability to interact with the thiol groups of cysteine amino acids in proteins and glutathione (GSH). In addition, the oleanane framework confers a steric hindrance with the binding to specific proteins.[484] 2-Cyano-3,12-dioxoolean-1,9-dien-28-oic acid at a dose of 5 μM induced a Fas-like

■ **CS 2.276** 2-Cyano-3,12-dioxoolean-1,9-dien-28-oic acid

■ **CS 2.277** 2-Cyano-3,12-dioxoolean-1,9-dien-28-imidazolideester

■ **CS 2.278** 2-Cyano-3,12-dioxoolean-1,9-dien-28-methyl ester

apoptosis in human osteogenic sarcoma (SaOS-2) cells with caspase-8 activation, cleavage of Bid, translocation of cleaved Bid to the mitochondria (where it recruits pro-apoptotic Bax), mitochondrial insult, release of cytochrome c, caspase 3 activation and DNA fragmentation.[487]

In addition, 2-cyano-3,12-dioxoolean-1,9-dien-28-methyl ester (CS 2.278) inhibited the survival of HL-60, human myeloid (BC-1), and human promyelocytic leukemia (NB4) cells, with IC_{50} values of 0.4 μM, 0.4 μM, and 0.2 μM, respectively, brought about by induction of pro-apoptotic Bax, mitochondrial insult, caspase 3 activation and inhibition of ERK, suggesting the possibility of a parallel interaction with the Ras/Raf/ERK/MERK pathway.[488] In human histiocytic lymphoma (U937) cells, 2-cyano-3,12-dioxoolean-1,9-dien-28-oic acid decreased the cytoplasmic levels of glutathione (GSH) resulting in the accumulation of ROS, which stimulated JNK and MAPK p38, and hence activation of caspase 8, probable cleavage of Bid and mitochondrial insult, release of cytochrome c, and activation of caspase 3.[489]

The involvement of ROS was further evidenced by 2-cyano-3,12-dioxoolean-1,9-dien-28-oic acid and 2-cyano-3,12-dioxoolean-1,9-dien-28-imidazolide ester, which induced apoptosis in U266 cells by decreasing the cytoplasmic levels of GSH, resulting in an increase in ROS and hence a decrease in anti-apoptotic FLICE-like inhibitory protein (FLIP). As the latter normally blocks the activation of caspase 8, this decrease would cause its activation, followed by probable cleavage of Bid, mitochondrial insult and caspase 3 activation.[486] Note that the ROS generated by 1 µmol/L of 2-cyano-3,12-dioxoolean-1,9-dien-28-methyl ester inhibited NF-κB activation in human chronic myeloid leukemia (KBM-5) cells and hence downregulation of anti-apoptotic FLIP and Bcl-2 proteins.[490] In addition, cysteine-179 in IκB kinases is a nucleophile that reacts with the conjugated systems of 2-cyano-3,12-dioxoolean-1,9-dien-28-imidazolide ester and becomes unable to phosphorylate IκBα.[484]

Hela cells treated with 1 µM of 2-cyano-3,12-dioxoolean-1,9-dien-28-methyl ester interacted covalently with Janus-activated kinase-1 (JAK1) on cysteine 1077 and therefore inhibited STAT3 phosphorylation and activation.[484] Note that STAT3 inhibition results in pro-apoptotic protein (p53) activation followed by CD95 stimulation and hence caspase 8 activation, cleavage of Bid and mitochondrial insult in wild-type p53 cancer cells. 2-Cyano-3,12-dioxoolean-1,9-dien-28-methyl ester also formed covalent bonds with STAT3 on cysteine 259.[485]

The induction of ROS in cancer cells by 2-cyano-3,12-dioxoolean-1,9-dien-28-oic acid and congener appears to result from two main mechanisms, which involve the nucleophilic attack of thiol groups. In human Burkitt's lymphoma (Ramos) cells, 2-cyano-3,12-dioxoolean-1,9-dien-28-oic acid at a dose of 2.5 µM reacted with the thiol groups of mitochondrial proteins producing misfolding, aggregation in the mitochondrial inner membrane, opening of unregulated mitochondrial permeability transition (PT) pores, release of cytochrome c, inhibition of the respiratory chain at complex III resulting in the generation of mitochondrial ROS and apoptosis.[491] Note that ROS induce the opening of mitochondrial PT pores.[492] Likewise, 2-cyano-3,12-dioxoolean-1,9-dien-28-oic acid and derivatives react with the thiol group of GSH thereby increasing levels of ROS.[485] 2-Cyano-3,12-dioxoolean-1,9-dien-28-methyl ester at a dose of 5 µM lowered the levels of GSH. This resulted in the accumulation of ROS in human prostate adenocarcinoma (LNCaP) and human prostate cancer (PC-3) cells and hence in the inhibition of anti-apoptotic protein Akt.[493]

Oleananes lacking an A ring C3 ketone Δ_{1-2} and/or a C ring C12 ketone Δ_{9-11} conjugated system induce apoptosis: 3β-hydroxyolean-12-en-27-oic acid induced apoptosis in CoLo 205, cells with DNA fragmentation, a decrease in anti-apoptotic Bcl-2, increase in pro-apoptotic Bax, a fall in mitochondrial membrane potential and activation of caspase 3.[482] This mechanism is pro-apoptotic protein (p53)-independent. Maslinic acid (CS 2.279) (2α,3β-diydroxyolean-12-en-28-oic acid) from *Olea europaea* L. (family Oleaceae Hoffmanns. & Link) inhibited the growth of human colon cancer (HT-29) cells, with an IC_{50} value equal to 28.8 μg/mL, by inducing the expression of pro-apoptotic Bax and the repression of anti-apoptotic Bcl-2, mitochondrial insult, release of cytochrome c and activation of caspases 9 and 3.[494,495] One could reasonably frame the hypothesis that ROS and/or DNA damage could be upstream of these apoptotic mechanisms. In fact, among the first observation pertaining to the apoptotic mode of action of oleananes was the inhibition of DNA polymerase β by C17 carboxylic acid derivatives. Deng et al. (1999) recorded that oleanolic acid isolated from *Baeckea gunniana* Schauer ex Walp. (family Myrtaceae Juss.) inhibited the enzymatic activity of DNA polymerase β, with an IC_{50} value equal to 3.7 μM.[496] 3β,16β,23-Triacetoxyolean-12-en-28-oic acid (CS 2.280) from *Couepia polyandra* (Kunth) Rose (family Chrysobalanaceae R. Br.) inhibited the enzymatic activity of DNA

■ **CS 2.279** Maslinic acid

■ **CS 2.280** 3β,16β,23-Triacetoxyolean-12-en-28-oic acid

polymerase β, with IC$_{50}$ values equal to 13 μM.[497] Note that ursanes with carboxylic groups in C17 share the same activity as 3β-hydroxyurs-12,19(29)-dien-28-oic acid (CS 2.281), 3β-hydroxyurs-18,20(30)-dien-28-oic acid (CS 2.282) and ursolic acid (CS 2.283) isolated from *Baeckea gunniana* Schauer ex Walp. (family Myrtaceae Juss.), which inhibited the enzymatic activity of DNA polymerase β, with IC$_{50}$ values equal to 3.2 μM, 2.5 μM, and 4.8 μM, respectively.[496]

Ursane and oleanane triterpenes differ by the position of C29 and C30 methyl moieties. Ursolic acid from *Calluna vulgaris* (L.) Hull (family Ericaceae Juss.) inhibited the growth of mouse melanoma (B16) cells, with an IC$_{50}$ value of 7.7 μM,[498] suggesting that a C17 carboxylic acid may be of value against melanoma. The opening of the ursane framework yielded

■ **CS 2.281** 3β-Hydroxyurs-12,19(29)-dien-28-oic acid

■ **CS 2.282** 3β-Hydroxyurs-18,20(30)-dien-28-oic acid

■ **CS 2.283** Ursolic acid

seco-derivatives, such as 5β-(1-methyl-2-ethyl)-10α-(3-aminopropyl)-des-A-urs-12-en-28-oic acid (CS 2.284), which destroyed rat prostatic basal epithelial (NRP-152) cells, with an IC_{50} value equal to $0.3\,\mu M$.[499] α-Amyrin (CS 2.285), uvaol (CS 2.286) and ursolic acid inhibited the growth of B16 2F2 cells, with IC_{50} values equal to $50\,\mu M$, $46.8\,\mu M$ and $3.1\,\mu M$, respectively,[432] implying that a carboxylic group at C17 and a Δ_{12-13} are beneficial. This inference was further supported by the fact that 3β-hydroxy-urs-12-en-27-oic acid (CS 2.287) and β-peltoboykinolic acid (CS 2.288) from *Astilbe chinensis* (Maxim.) Franch. & Sav. (family Saxifragaceae Juss.) and *Aceriphyllum rossii* (Oliv.) Engler (family Saxifragaceae Juss.) were cytotoxic against HO-8910, Hela, HL-60

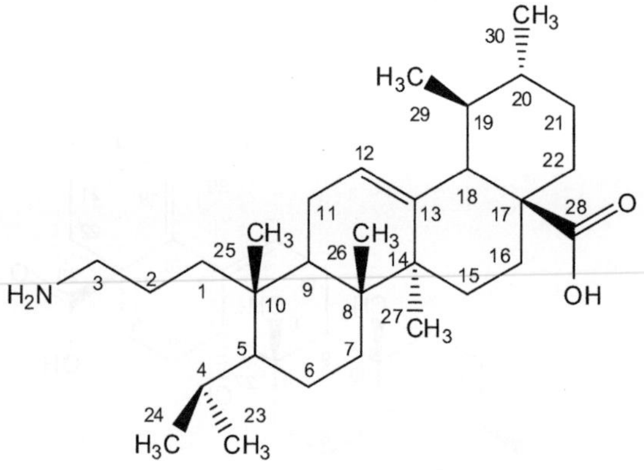

■ **CS 2.284** 5β-(1-Methyl-2-ethyl)-10α-(3-aminopropyl)-des-A-urs-12-en-28-oic acid

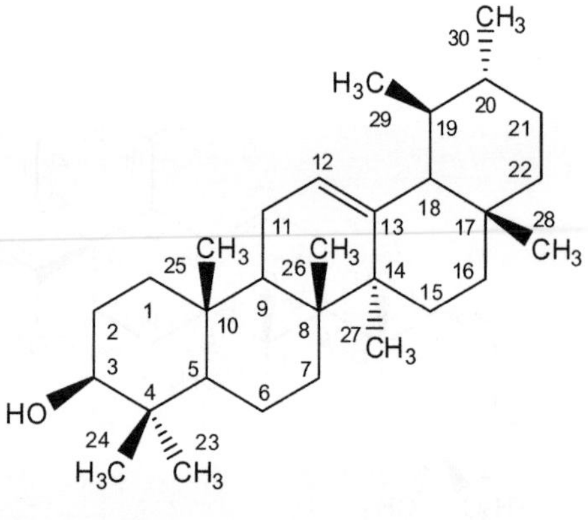

■ **CS 2.285** α-Amyrin

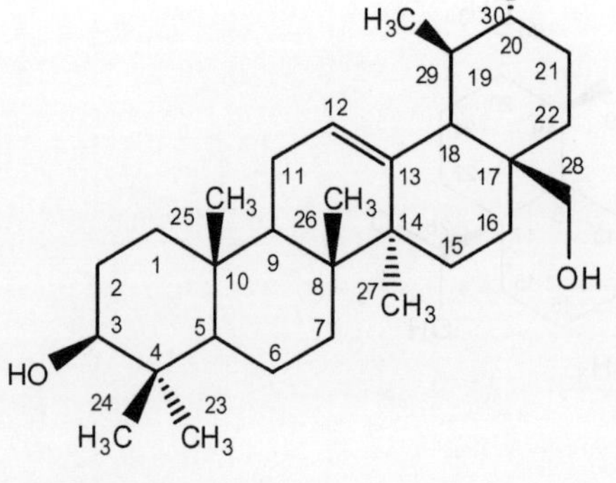

■ **CS 2.286** Uvaol

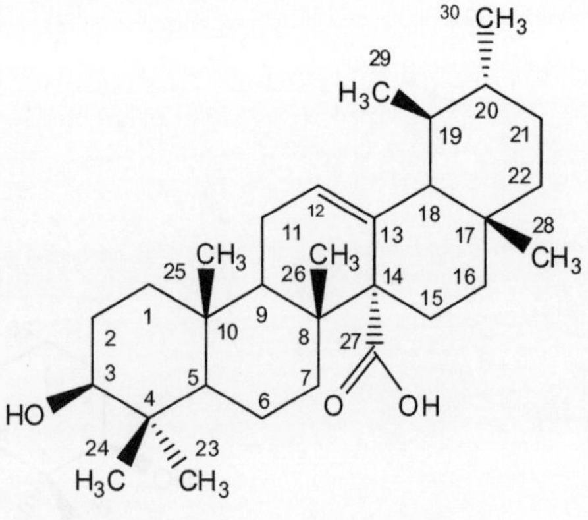

■ **CS 2.287** 3β-Hydroxy-urs-12-en-27-oic acid

Lewis lung carcinoma (LLC) cells.[481,483] 3β,20α-Dihydroxyurs-21-en-28-oic acid (CS 2.289), ursolic acid and 3β,27-dihydroxy-12-ursen-28-oic acid (CS 2.290) from *Nerium oleander* L. (family Apocynaceae Juss.) inhibited the growth of human normal lung fibroblast (WI-38) cells, with IC_{50} values equal to 3.4 μM, 1.8 μM and 7.2 μM, respectively,[500] suggesting that a Δ_{20-21} with a carboxylic group at C17 enhance the toxicity of ursanes against non-malignant cells. A ketone at C12 is a positive chemical feature, as exemplified with microfokienoxane C (CS 2.291) and

■ **CS 2.288** β-Peltoboykinolic acid

■ **CS 2.289** 3β,20α-Dihydroxyurs-21-en-28-oic acid

■ **CS 2.290** 3β,27-Dihydroxy-12-ursen-28-oic acid

■ **CS 2.291** Microfokienoxane C

■ **CS 2.292**　3β,28-Dihydroxy-11α-methoxyurs-12-ene

■ **CS 2.293**　Urs-12-ene-3β,22α-diol

■ **CS 2.294**　3β,11α,19α,24,30-Pentahydroxy-
20β,28-epoxy-28β-methoxy-ursane

■ **CS 2.295**　1α,3β-Dihydroxy-olean-18-ene

3β,28-dihydroxy-11α-methoxyurs-12-ene (CS 2.292) isolated from *Microtropis fokienensis* Dunn (family Celastraceae R. Br.), which destroyed human hepatocellular liver carcinoma (HepG2) cells, with IC_{50} values equal to 3.8 μg/mL and 4.6 μg/mL, respectively.[501] The cytotoxicity of ursanes is further enhanced by unsaturations which impose planarity to the framework: the $\Delta_{12\text{-}13}$ urs-12-ene-3β,22α-diol (CS 2.293) isolated from *Nardophyllum bryoides* (Lam.) Cabrera (Asteraceae Bercht. & J. Presl) destroyed murine lung adenocarcinoma (LM3) cells, with an IC_{50} value equal to 0.8 μM,[502] whereas the $\Delta_{18\text{-}19}$ 3β,11α,19α,24,30-pentahydroxy-20β, 28-epoxy-28β-methoxy-ursane (CS 2.294) and 1α,3β-dihydroxy-olean-18-ene (CS 2.295) isolated from a member of the genus *Juglans* L.

■ **CS 2.296** 3,4-Secofriedelan-4-oxo-3-oic acid

■ **CS 2.297** 2α,3β-Dihydroxy-D:A-friedoolean-28-oic acid

(family Juglandaceae DC. ex Perleb) inhibited the growth of two cell lines (respective IC_{50} values in parentheses): murine melanoma (B16F10; 1.5 μM, 0.3 μM), human laryngeal carcinoma (Hep-2; 5.5 μM, 5.4 μM), human breast adenocarcinoma (MCF-7; 1.2 μM, 3.8 μM) and human glioblastoma (U87; 3.2 μM, 1.3 μM).[503]

Friedelanes are closely related to oleananes and ursanes from which they differ by the positions of methyl groups in C5, C9 and C13. The presence of a carboxylic at C17 seems to be again a positive cytotoxic feature for friedelane as the synthetic friedelane derivative 3,4-secofriedelan-4-oxo-3-oic acid (CS 2.296) weakly inhibited the growth of MCF-7, human non-small-cell lung cancer (H460) and human glioblastoma (SF268) cells, with IC_{50} values equal to 26.9 μM, 32.8 μM and 24.6 μM, respectively.[504] 2α,3β-Dihydroxy-D:A-friedoolean-28-oic acid (pluricostatic acid) (CS 2.297) and 3-oxo-friedelan-28 oic acid (CS 2.298) isolated from *Marila pluricostata* Standl. & L.O. Williams (family Calophyllaceae J. Agardh) inhibited the growth of human breast adenocarcinoma (MCF-7),[505] suggesting that hydroxyl or ketone moieties in ring A and C17 carboxylic acid enhance the cytotoxic properties of friedelanes. These beneficial effects are obvious with 3β,23-epoxy-friedelan-28-oic acid (CS 2.299), canophyllic acid (CS 2.300), and 3-oxo-friedelan-28-oic acid from *Calophyllum inophyllum* L. (family Calophyllaceae J. Agardh), which inhibited HL-60 cells, with IC_{50} values equal to 10.6 μM, 4.6 μM and 2.6 μM, respectively.[506]

Members of the family Celastraceae R. Br. elaborate a series of aromatic and polyoxygenated friedelanes known as celastroids or quinone methides which elicit potent cytotoxic effects. One such compound is isoiguesterin

■ **CS 2.298** 3-Oxo-friedelan-28 oic acid

■ **CS 2.299** 3β,23-Epoxy-friedelan-28-oic acid

■ **CS 2.300** Canophyllic acid

■ **CS 2.301** Isoiguesterin

(CS 2.301) isolated from *Salacia madagascariensis* (Lam.) DC. (family Celastraceae, R. Br.), which nullified the growth of mouse leukemia (P388) cells, with an IC_{50} value equal to 0.2 μg/mL.[507] Other examples of cytocidal celastroids are 17-(methoxycarbonyl)-28-*nor*-isoiguesterin (CS 2.302) and 28-hydroxyisoiguesterin (CS 2.303) from *Salacia kraussii* (Harv.) Harv. (family Celastraceae R. Br.), which inhibited the growth of HT-29 cells, with IC_{50} values equal to 2.3 μg/mL and 6 μg/mL, respectively.[508] Tingenone (CS 2.304), 20-hydroxy-20-*epi*-tingenone (CS 2.305) and celastrol (CS 2.306) isolated from a member of the genus *Kokoona* Thwaites (family Celastraceae R. Br.) prevented the proliferation of a broad array of cancer cells (respective IC_{50} values in parentheses): human

■ **CS 2.302** 17-(Methoxycarbonyl)-28-*nor*-isoiguesterin

■ **CS 2.303** 28-Hydroxyisoiguesterin

■ **CS 2.304** Tingenone

■ **CS 2.305** 20-Hydroxy-20-*epi*-tingenone

■ **CS 2.306** Celastrol

lymphoma (BC-1; 0.3 μg/mL, 0.9 μg/mL, 0.3 μg/mL), human fibrosarcoma (HT1080; 0.7 μg/mL, 2.1 μg/mL, 0.5 μg/mL), human lung adenocarcinoma (Lu-1; 0.6 μg/mL, 2.2 μg/mL, 1.3 μg/mL), human melanoma (MEL-2; 0.2 μg/mL, 1.5 μg/mL, 0.6 μg/mL), human colon carcinoma (Col-2; 0.9 μg/mL, 5.8 μg/mL, 2.3 μg/mL), human nasopharyngeal carcinoma (KB; 0.5 μg/mL, 1.7 μg/mL, 1.2 μg/mL), mouse leukemia (P388; 0.1 μg/mL, 2 μg/mL, 3.5 μg/mL), human epithelial carcinoma (A-431; 0.7 μg/mL, 1.1 μg/mL, 0.6 μg/mL), androgen sensitive human prostate cancer (LNCaP; 0.2 μg/mL, 1 μg/mL, 0.3 μg/mL), human breast cancer (ZR-751; 0.6 μg/mL, 1 μg/mL, 0.4 μg/mL) and human glioblastoma (U373; 0.2 μg/mL, 0.8 μg/mL, 0.4 μg/mL).[509] These results suggest that a free C3 hydroxy and a C2 ketone conjugated with Δ_{1-10} account for the cytotoxicity of celastroids. 6-Oxotingenol, 6-oxopristimerol, pristimerin and tingenone isolated from members of the genus *Maytenus* Molina (Celastraceae R. Br.) abated the survival of the following cell lines (respective IC_{50} values in parentheses): mouse lymphocytic leukemia (L1210; 6 μg/mL, 2.8 μg/mL, 0.3 μg/mL, 0.1 μg/mL), mouse leukemia (P388; 2.6 μg/mL, 1.5 μg/mL, 0.1 μg/mL, 0.04 μg/mL) and human nasopharyngeal carcinoma (KB; 30 μg/mL, 2.8 μg/mL, 0.5 μg/mL, 0.2 μg/mL.[510] Other celastroids of oncohaematological value are 22β-hydroxytingenone and celastrol, which, along with tingenone and pristimerin killed human multiple myeloma (RPMI8226) cells, with IC_{50} values equal to 18.3 μM, 3 μM, 7 μM and 16.6 μM, respectively.[511] 23-*nor*-22-Hydroxy-6-oxo-tingenol (CS 2.307), 3-methoxy-22-β-hydroxy-6-oxo-tingenol (CS 2.308) and 22β-hydroxy-7,8-dihydro-6-oxo-tingenol (CS 2.309) isolated from *Maytenus amazonica* Mart. (family Celastraceae R. Br.) killed P388, A549, HT-29 and human

■ **CS 2.307** *23-nor-22-Hydroxy-6-oxo-tingenol*

■ **CS 2.308** *3-Methoxy-22-β-hydroxy-6-oxo-tingenol*

skin melanoma (SK-MEL-28) cells,[512] showing some selectivity of C6 keto tingenols against colon cancer and a beneficial effect of C22 beta hydroxylation. From the same plant, amazoquinone (CS 2.310), 7-hydroxy-7,8-dihydro-tingenone (CS 2.311), 7,8-dihydro-6-oxo-tingenol (CS 2.312) and 23-oxo-isotingenone (CS 2.313) exhibited cytotoxic potencies against the following cell lines (respective IC_{50} values in parentheses): P388 (2.5 μg/mL, 5 μg/mL, 5 μg/mL, 5 μg/mL), A549 (5 μg/mL, 5 μg/mL, 5 μg/mL, 5 μg/mL), HT-29 (5 μg/mL, 5 μg/mL, 10 μg/mL, 10 μg/mL) and human melanoma (MEL-28; 10 μg/mL, 10 μg/mL, 10 μg/mL, 10 μg/mL.[513] The main pharmacophore of 6-oxotingenol (CS 2.314) and 6-oxopristimerol

■ **CS 2.309** 22β-Hydroxy-7,8-dihydro-6-oxo-tingenol

■ **CS 2.310** Amazoquinone

■ **CS 2.311** 7-Hydroxy-7,8-dihydro-tingenone

■ **CS 2.312** 7,8-Dihydro-6-oxo-tingenol

■ **CS 2.313** 23-Oxo-isotingenone

■ **CS 2.314** 6-Oxotingenol

■ **CS 2.315** 6-Oxopristimerol

■ **CS 2.316** Pristimerin

(CS 2.315) is obviously the conjugated C6 ketone Δ_{7-8} system, which invites nucleophilic attack.[514] In fact, 6-oxo celastroids in general are attacked by nucleophiles at C6.[515] Pristimerin (CS 2.316) itself is subject to nucleophilic attacks at C2 in the A ring and C6 in the B ring,[516] and other quinone methides bind to GSH and other cellular nucleophiles including proteins and DNA.[517] In addition, the pseudoplanar framework of celastroids such as netzahualcoyonol (CS 2.317) allow some levels of DNA

■ **CS 2.317** Netzahualcoyonol

■ **CS 2.318** 3-Methyl-6-oxotingenol

■ **CS 2.319** 8-*epi*-6-Deoxoblepharodol

■ **CS 2.320** Pristimerol

intercalation and reaction with the DNA nucleophilic group.[518] Note that 3-methyl-6-oxotingenol (CS 2.318) was inactive against P388 and mouse lymphocytic leukemia (L1210) cells but was toxic to KB cells, with an IC_{50} value equal to $11\,\mu g/mL$,[510] suggesting that a free hydroxyl at C3 is a structural requirement for the haematological cytotoxicity of celastroids. Double bonds in the celastroid framework favor planarity and are beneficial for cytotoxicity. This is exemplified with 8-*epi*-6-deoxoblepharodol (CS 2.319) and pristimerol (CS 2.320), which elicited cytotoxic effects against Hela, Hep-2 and African green monkey kidney (Vero) cells, with

■ **CS 2.321** Scutione

IC$_{50}$ values equal to 1.6 μg/mL, 5.9 μg/mL, 2.2 μg/mL and 0.5 μg/mL, 2.8 μg/mL, 1.4 μg/mL, respectively.[519] Scutione (CS 2.321) isolated from *Maytenus scutioides* (Griseb.) Lourteig & O'Donell (family Celastraceae R. Br.) abated the growth of Hela, Hep-2 and Vero cells, with respective IC$_{50}$ values equal to 4.9 μg/mL, 5.6 μg/mL and 7.2 μg/mL.[520]

Some evidence suggests that celastroids interfere with NF-κB-induced transcription of anti-apoptotic proteins via manifold routes. Celastrol at a dose of 2.5 μM in human chronic myeloid leukemia (KBM-5) cells inhibited the phosphorylation of IκBα and hence NF-κB inactivation and reduction in anti-apoptotic IAP1, IAP2, Bcl-2, Bcl-xL, c-FLIP, and survivin protein expression.[521] Cysteine-179 in IκB kinases is a nucleophile which may very well react with the conjugated system of celastrol. Also, Cdc37 and Hsp90 aggregation with IKK complex is necessary for the phosphorylation of IκBα.[522] Indeed, celastrol induced apoptosis in human pancreatic carcinoma (PANC-1) cells, with an IC$_{50}$ value equal to 3 μmol/L, by hampering the binding of heat shock protein 90 (Hsp90) with its cochaperone Cdc37.[523] Celastrol induced apoptosis in the following cell lines (IC$_{50}$ values in parentheses): human keratinocytes (HaCaT; 1.1 μM), human embryonic kidney (HEK293; 2.9 μM), human foreskin fibroblast (HS68; 6.8 μM), human hepatic (WRL-68; 6.3 μM) and rat embryonic heart (H9C2; 3.3 μM).[524] In HaCaT cells celastrol increased the levels of pro-apoptotic Bax protein, induced a fall of mitochondrial membrane potential and activation of caspases 8, 9 and 3, and inhibited NF-κB activation by negating the phosphorylation of IκBα and hence the downregulation of anti-apoptotic Bcl-2 and Bcl-xL proteins.[524] Additionally, pristimerin at a dose of 5 μmol/L induced apoptosis in PC-3 cells via direct nucleophilic attack at C6 by the hydroxyl group of the N-terminal threonine of the

■ **CS 2.322** Acetyl-11-keto-β-boswellic acid

proteasome β5 subunit.[526] In KBM-5 cells 200 nM of pristimerin inhibited transforming growth factor beta activated kinase-1 (TAK1) and therefore the phosphorylation of IκBα and the transcription of anti-apoptotic Bcl-2, Bcl-xL, survivin, anti-apoptotic induced myeloid leukemia cell differentiation protein (Mcl-1), cyclin D1 and c-Myc.[525] In addition, pristimerin at a dose of 3 μmol/L abated the viability of human breast adenocarcinoma (MDA-MB-231) cells by 80%, with caspase 3 activation, PARP cleavage and release of cytochrome c, followed by a fall in mitochondrial membrane potential.[526] Tingenone, 22β-hydroxytingenone, pristimerin and celastrol inhibited the polymerization of tubulin, with IC_{50} values equal to 8 μM, 12 μM, 8 μM and 2 μM, respectively.[511]

With respect to the mechanism of action of friedelanes, some evidence suggests the involvement of Ca^{2+} from the endoplasmic reticulum and possible inhibition of STAT3. Acetyl-11-keto-β-boswellic acid (CS 2.322) from *Boswellia serrata* Roxb. ex Colebr. (family Burseraceae Kunth) mitigated the survival of HL-60 and human T cell lymphoblast-like cell (CCRF-CEM) cells, via apoptosis downstream of the ceramide dependent (CD95) receptor (Fas), and repressed the expression of topoisomerases I and II.[527] Note that the levels of free cytoplasmic Ca^{2+} in human polymorphonuclear leukocytes (PMNL) cells were boosted by acetyl-11-keto-β-boswellic acid at a dose of 30 μM, probably from endoplasmic reticulum leakage and hence MAPK p38 activation, which results in C/EBP-homologous protein (CHOP) induction, superoxide production[528] and therefore increases the levels of reactive oxygen species (ROS). Acetyl-11-keto-β-boswellic acid repressed the growth of LNCaP and PC-3 cells by 80% at a dose of 10 μg/mL, with activation of pro-apoptotic JNK via a probable increase in free cytoplasmic Ca^{2+} and hence activation of CHOP inducing the transcription of death receptor (DR5) protein, resulting in the activation of caspases 8 and 3 and cleavage of PARP.[529] It would be interesting to measure the effects of acetyl-11-keto-β-boswellic acid against inositol 1,4,5-trisphosphate ($InsP_3$) receptor ($InsP_3Rs$)-induced release of calcium.[413] Human multiple myeloma (U266) cells exposed to 50 μmol/L of acetyl-11-keto-β-boswellic acid became apoptotic due to the inhibition of JAK2, the consequent inactivation of STAT3 and the resulting reduced transcription of anti-apoptotic Bcl-2, Bcl-xL, Mcl-1 and pro-proliferative cyclin D1,[530] possibly brought about by the C11 keto Δ_{12-13} moiety of acetyl-11-keto-β-boswellic reacting with JAK2 thiol groups.[484]

REFERENCES

[183] Murata T, Shinohara M, Hirata T, Kamiya K, Nishikawa M, Miyamoto M. New triterpenes of *Alisma plantago-aquatica* L. var. Orientale Samuels. Tetrahedron Lett 1968;9(1):103–8.

[184] Pei-Wu G, Fukuyama Y, Yamada T, Rei W, Jinxian B, Nakagawa K. Triterpenoids from the rhizome of *Alisma plantago-aquatica*. Phytochem 1988;27(4):1161–4.

[185] Yoshikawa M, Hatakeyama S, Tanaka N, Fukuda Y, Yamahara J, Murakami N. Crude drugs from aquatic plants. I. On the constituents of *Alismatis rhizoma*. (1). Absolute stereostructures of alisols E 23-acetate, F, and G, three new protostane-type triterpenes from Chinese *Alismatis rhizoma*. Chem Pharm Bull 1993;41(11):1948–54.

[186] Yoshikawa M, Hatakeyama S, Tanaka N, Matsuoka T, Yamahara J, Murakami N. Crude drugs from aquatic plants. II. On the constituents of the rhizome of *Alisma orientale* Juzep. originating from Japan, Taiwan, and China. Absolute stereostructures of 11-deoxyalisols B and B 23-acetate. Chem Pharm Bull 1993;41(12):2109–12.

[187] Nakajima Y, Satoh Y, Katsumata M, Tsujiyama K, Ida Y, Shoji J. Terpenoids of *Alisma orientale* rhizome and the crude drug Alismatis rhizoma. Phytochem 1994;36(1):119–27.

[188] Yoshikawa M, Murakami T, Ikebata A, Ishikado A, Murakami N, Yamahara J, et al. Absolute stereostructures of alismalactone 23-acetate and alis-maketone-A 23-acetate, new seco-protostane and protostane-type triterpenes with vasorelaxant effects from Chinese *Alismatis rhizoma*. Chem Pharm Bull 1997;45(4):756–8.

[189] Yoshikawa M, Tomohiro N, Murakami T, Ikebata A, Matsuda H, Matsuda H, et al. Studies on *Alismatis rhizoma*. III. Stereostructures of new protostane-type triterpenes, alisols H, I, J-23-acetate, K-23-acetate, L-23-acetate, M-23-acetate, and N-23-acetate, from the dried rhizome of *Alisma orientale*. Chem Pharm Bull 1999;47(4):524–8.

[190] Zhao M, Xu LJ, Che CT. Alisolide, alisols O and P from the rhizome of *Alisma orientale*. Phytochem 2008;69(2):527–32.

[191] Hu XY, Guo YQ, Gao WY, Chen HX, Zhang TJ. A new triterpenoid from *Alisma orientalis*. Chinese Chem Lett 2008;19(4):438–40.

[192] Hu XY, Guo YQ, Gao WY, Zhang TJ, Chen HX. Two new triterpenes from the rhizomes of *Alisma orientalis*. J Asian Nat Prod Res 2008;10(5–6):481–4.

[193] Oshima Y, Iwakawa T, Hikino H. Alismol and alismoxide, sesquiterpenoids of *Alisma rhizomes*. Phytochem 1983;22(1):183–5.

[194] Yoshikawa M, Hatakeyama S, Tanaka N, Fukuda Y, Murakami N, Yamahara J. Orientalols A, B, and C, sesquiterpene constituents from Chinese *Alismatis rhizoma*, and revised structures of alismol and alismoxide. Chem Pharm Bull 1992;40(9):2582–4.

[195] Peng GP, Tian G, Huang XF, Lou FC. Guaiane-type sesquiterpenoids from *Alisma orientalis*. Phytochem 2003;63(8):877–81.

[196] Hikino H, Iwakawa T, Oshima Y, Nishikawa K, Murata T. Diuretic principles of *Alisma-plantago-aquatica* var. orientale rhizomes. Shoyakugaku Zasshi 1982;36:150–3.

[197] Jiang ZY, Zhang XM, Zhou J, Zhang FX, Chen JJ, Lü Y, et al. Two new sesquiterpenes from *Alisma orientalis*. Chem Pharm Bull 2007;55(6):905–7.

[198] Lee S, Kho Y, Min B, Kim J, Na M, Kang S, et al. Cytotoxic triterpenoides from Alismatis Rhizoma. Arch Pharm Res 2001;24(6):524–6.

[199] Law BYK, Wang M, Ma DL, Al-Mousa F, Michelangeli F, Cheng SH, et al. Alisol B, a novel inhibitor of the sarcoplasmic/endoplasmic reticulum Ca2+ ATPase pump, induces autophagy, endoplasmic reticulum stress, and apoptosis. Mol Cancer Ther 2010;9(3):718–30.

[200] Chen HW, Hsu MJ, Chien CT, Huang HC. Effect of alisol B acetate, a plant triterpene, on apoptosis in vascular smooth muscle cells and lymphocytes. Eur J Pharmacol 2001;419(2–3):127–38.

[201] Soucie EL, Annis MG, Sedivy J, Filmus J, Leber B, Andrews DW, et al. Myc potentiates apoptosis by stimulating bax activity at the mitochondria. Mol Cell Biol 2001;21(14):4725–36.

[202] Huang YT, Huang DM, Chueh SC, Teng CM, Guh JH. Alisol B acetate, a triterpene from *Alismatis rhizoma*, induces Bax nuclear translocation and apoptosis in human hormone-resistant prostate cancer PC-3 cells. Cancer Lett 2006;231(2):270–8.

[203] Toyokuni S, Okamoto K, Yodoi J, Hiai H. Persistent oxidative stress in cancer. FEBS Lett 1995;358:1–3.

[204] Butts BD, Kwei KA, Bowden GT, Briehl MM. Elevated basal reactive oxygen species and phospho-Akt in murine keratinocytes resistant to ultraviolet B-induced apoptosis. Mol Carcinogen 2003;37(3):149–57.

[205] Xu YH, Zhao LJ, Li Y. Alisol B acetate induces apoptosis of SGC7901 cells via mitochondrial and phosphatidylinositol 3-kinases/Akt signaling pathways. World J Gastroenterol 2009;15(23):2870–7.

[206] Rohn JL, Hueber AO, McCarthy NJ, Lyon D, Navarro P, Burgering BMTh, et al. The opposing roles of the Akt and c-Myc signalling pathways in survival from CD95-mediated apoptosis. Oncogene 1998;17(22):2811–8.

[207] Wang C, Zhang JX, Shen XL, Wan CK, Tse AKW, Fong WF. Reversal of P-glycoprotein-mediated multidrug resistance by Alisol B 23-acetate. Biochem Pharmacol 2004;68(5):843–55.

[208] Grotemeier A, Alers S, Pfisterer SG, Paasch F, Daubrawa M, Dieterle A, et al. AMPK-independent induction of autophagy by cytosolic Ca2+ increase. Cell Signal 2010;22(6):914–25.

[209] Kim J, Kundu M, Viollet B, Guan KL. AMPK and mTOR regulate autophagy through direct phosphorylation of Ulk1. Nat Cell Biol 2011;13(2):132–41.

[210] Lee S, Min B, Bae K. Chemical modification of alisol B 23-acetate and their cytotoxic activity. Arch Pharm Res 2002;25(5):608–12.

[211] Lee SM, Kwon BM, Min BS. FPTase inhibition effect of protostanes from *Alismatis rhizoma* and derivatives from alisol B 23-acetate. Korean J Pharmacog 2011;42(3):218–22.

[212] Appels NMGM, Beijnen JH, Schellens JHM. Development of farnesyl transferase inhibitors: a review. Oncologist 2005;10(8):565–78.

[213] Liu JS, Huang MF, Ayer WA, Nakashima TT. Structure of enshicine from *Schisandra henryi*. Phytochem 1984;23(5):1143–5.

[214] Li L, Xue H. Henricine, a new tetrahydrofuran lignan from *Schisandra henryi*. Planta Med 1986;6:493–4.

[215] Wang YH, Gao JP, Chen DF. Determination of lignans of Schisandra medicinal plants by HPLC. Zhongguo Zhongyao Zazhi 2003;28(12):1159–60.

[216] Chen YG, Wu ZC, Lv YP, Gui SH, Wen J, Liao XR, et al. Triterpenoids from *Schisandra henryi* with cytotoxic effect on leukemia and Hela cells in vitro. Arch Pharm Res 2003;26(11):912–6.

[217] Chen YG, Wu ZC, Gui SH, Lv YP, Liao XR, Halaweish F. Lignans from *Schisandra hernyi* with DNA cleaving activity and cytotoxic effect on leukemia and Hela cells in vitro. Fitoterapia 2005;76(3–4):370–3.

[218] Liu HT, Xu LJ, Peng Y, Yang XW, Xiao PG. Two new lignans from *Schisandra henryi*. Chem Pharm Bull 2009;57(4):405–7.

[219] Li R, Shen Y, Xiang W, Sun H. Four novel nortriterpenoids isolated from *Schisandra henryi* var. yunnanensis. Eur J Org Chem 2004;4:807–11.

[220] Chen YG, Zhang Y, Liu Y, Xue YM, Wang JH. A new triterpenoid acid from *Schisandra henryi*. Chem Nat Comp 2010;46(4):569–71.

[221] Xue YB, Yang JH, Li XN, Du X, Pu JX, Xiao WL, et al. Henrischinins A–C: three new triterpenoids from *Schisandra henryi*. Org Lett 2011;13(6):1564–7.

[222] Yoshikawa K, Kouso K, Takahashi J, Matsuda A, Okazoe M, Umeyama A, et al. Cytotoxic constituents of the fruit body of *Daedalea dickisii*. J Nat Prod 2005;68(6):911–4.

[223] Kang HM, Lee SK, Shin DS, Lee MY, Han DC, Baek NI, et al. Dehydrotrametenolic acid selectively inhibits the growth of H-ras transformed rat2 cells and induces apoptosis through caspase-3 pathway. Life Sci 2006;78(6):607–13.

[224] Ling H, Zhou L, Jia X, Gapter LA, Agarwal R, Ng KY. Polyporenic acid C induces caspase-8-mediated apoptosis in human lung cancer A549 cells. Mol Carcinogenesis 2009;48(6):498–507.

[225] Kikuchi T, Uchiyama E, Ukiya M, Tabata K, Kimura Y, Suzuki T, et al. Cytotoxic and apoptosis-inducing activities of triterpene acids from *Poria cocos*. J Nat Prod 2011;74(2):137–44.

[226] Tang W, Liu JW, Zhao WM, Wei DZ, Zhong JJ. Ganoderic acid T from *Ganoderma lucidum* mycelia induces mitochondria mediated apoptosis in lung cancer cells. Life Sci 2006;80(3):205–11.

[227] Chen NH, Zhong JJ. Ganoderic acid Me induces G1 arrest in wild-type p53 human tumor cells while G1/S transition arrest in p53-null cells. Process Biochem 2009;44(8):928–33.

[228] Zhou L, Shi P, Chen NH, Zhong JJ. Ganoderic acid Me induces apoptosis through mitochondria dysfunctions in human colon carcinoma cells. Process Biochem 2011;46(1):219–25.

[229] Min BS, Gao JJ, Nakamura N, Hattori M. Triterpenes from the spores of *Ganoderma lucidum* and their cytotoxicity against Meth-A and LLC tumor cells. Chem Pharm Bull 2000;48(7):1026–33.

[230] Gao JJ, Min BS, Ahn EM, Nakamura N, Lee HK, Hattori M. New triterpene aldehydes, lucialdehydes A – C, from *Ganoderma lucidum* and their cytotoxicity against murine and human tumor cells. Chem Pharm Bull 2002;50(6):837–40.

[231] Nomura M, Takahashi T, Uesugi A, Tanaka R, Kobayashi S. Inotodiol, a lanostane triterpenoid, from *Inonotus obliquus* inhibits cell proliferation through caspase-3-dependent apoptosis. Anticancer Res 2008;28(5A):2691–6.

[232] Li CH, Chen PY, Chang UM, Kan LS, Fang WH, Tsai KS, et al. Ganoderic acid X, a lanostanoid triterpene, inhibits topoisomerases and induces apoptosis of cancer cells. Life Sci 2005;77(3):252–65.

[233] Ries S, Biederer C, Woods D, Shifman O, Shirasawa S, Sasazuki T, et al. Opposing effects of RAS on p53: transcriptional activation of mdm2 and induction of p19(ARF). Cell 2000;103(2):321–30.

[234] Banskota AH, Tezuka Y, Phung LK, Tran KQ, Saiki I, Miwa Y, et al. Cytotoxic cycloartane-type triterpenes from *Combretum quadrangulare*. Bioorg Med Chem Lett 1998;8(24):3519–24.

[235] Smith-Kielland I, Dornish JM, Malterud KE, Hvistendahl G, Rømming Chr, Bøckman OC, et al. Cytotoxic triterpenoids from the leaves of *Euphorbia pulcherrima*. Planta Med 1996;62(4):322–5.

[236] Madureira AM, Spengler G, Molnár A, Varga A, Molnár J, Abreu PM, et al. Effect of cycloartanes on reversal of multidrug resistance and apoptosis induction on mouse lymphoma cells. Anticancer Res 2004;24(2 B):859–64.

[237] Duarte N, Ramalhete C, Varga A, Molnár J, Ferreira MJU. Multidrug resistance modulation and apoptosis induction of cancer cells by terpenic compounds isolated from *Euphorbia* species. Anticancer Res 2009;29(11):4467–72.

[238] Nam NH, Van Kiem P, Ban NK, Thao NP, Nhiem NX, Cuong NX, et al. Chemical constituents of *Mallotus macrostachyus* growing in Vietnam and cytotoxic activity of some cycloartane derivatives. Phytochem Lett 2011;4(3):348–52.

[239] Khan MTH, Khan SB, Ather A. Tyrosinase inhibitory cycloartane type triterpenoids from the methanol extract of the whole plant of *Amberboa ramosa* Jafri and their structure–activity relationship. Bioorg Med Chem 2006;14(4):938–43.

[240] Chang TS. An updated review of tyrosinase inhibitors. Int J Mol Sci 2009;10(6):2440–75.

[241] Kim JH, Baek SH, Kim DH, Choi TY, Yoon TJ, Hwang JS, et al. Downregulation of melanin synthesis by haginin A and its application to in vivo lightening model. J Invest Dermatol 2008;128(5):1227–35.

[242] Chu W, Pak BJ, Bani MR, Kapoor M, Lu SL, Tamir A, et al. Tyrosinase-related protein 2 as a mediator of melanoma specific resistance to *cis*-diamminedichloroplatinum (II): therapeutic implications. Oncogene 2000;19(3):395–402.

[243] Grougnet R, Magiatis P, Mitaku S, Loizou S, Moutsatsou P, Terzis A, et al. Seco-cycloartane triterpenes from *Gardenia aubryi*. J Nat Prod 2006;69(12):1711–4.

[244] Nuanyai T, Sappapan R, Teerawatananond T, Muangsin N, Pudhom K. Cytotoxic 3,4-seco-cycloartane triterpenes from *Gardenia sootepensis*. J Nat Prod 2009;72(6):1161–4.

[245] Nuanyai T, Sappapan R, Vilaivan T, Pudhom K. Gardenoins E–H, cycloartane triterpenes from the apical buds of *Gardenia obtusifolia*. Chem Pharm Bull 2011;59(3):385–7.

[246] Shen T, Wan W, Yuan H, Kong F, Guo H, Fan P, et al. Secondary metabolites from *Commiphora opobalsamum* and their antiproliferative effect on human prostate cancer cells. Phytochem 2007;68(9):1331–7.

[247] Shen T, Yuan HQ, Wan WZ, Wang XL, Wang XN, Ji M, et al. Cycloartane-type triterpenoids from the resinous exudates of *Commiphora opobalsamum*. J Nat Prod 2008;71(1):81–6.

[248] Sashidhara KV, Singh SP, Kant R, Maulik PR, Sarkar J, Kanojiya S, et al. Cytotoxic cycloartane triterpene and rare isomeric bisclerodane diterpenes from the leaves of *Polyalthia longifolia* var. pendula. Bioorg Med Chem Lett 2010;20(19):5767–71.

[249] Omobuwajo OR, Martin MT, Perromat G, Sevenet T, Awang K, Païs M. Cytotoxic cycloartanes from *Aglaia argentea*. Phytochem 1996;41(5):1325–8.

[250] Mohamad K, Martin MT, Leroy E, Tempête C, Sévenet T, Awang K, et al. Argenteanones C–E and argenteanols B–E, cytotoxic cycloartanes from *Aglaia argentea*. J Nat Prod 1997;60(2):81–5.

[251] Parra-Delgado H, Ramírez-Apan T, Martínez-Vázquez M. Synthesis of argentatin A derivatives as growth inhibitors of human cancer cell lines in vitro. Bioorg Med Chem Lett 2005;15(4):1005–8.

[252] Tian Z, Xu L, Chen S, Zhou L, Yang M, Chen S, et al. Cytotoxic activity of schisandrolic and isoschisandrolic acids involves induction of apoptosis. Chemother 2007;53(4):257–62.

[253] Itokawa H, Oshida Y, Ikuta A, Inatomi H, Ikegami S. Flavonol glycosides from the flowers of *Cucurbita pepo*. Phytochem 1981;20(10):2421–2.

[254] Krauze-Baranowska M, Cisowski W. Flavonols from *Cucurbita pepo* L. herb. Acta Poloniae Pharm-Drug Res 1996;53(1):53–6.

[255] Sicilia T, Niemeyer HB, Honig DM, Metzler M. Identification and stereochemical characterization of lignans in flaxseed and pumpkin seeds. J Agr Food Chem 2003;51(5):1181–8.

[256] Li W, Koike K, Tatsuzaki M, Koide A, Nikaido T. Cucurbitosides F–M, acylated phenolic glycosides from the seeds of *Cucurbita pepo*. J Nat Prod 2005;68(12):1754–7.

[257] Rauwald HW, Sauter M, Schilcher H. A 24β-ethyl-Δ7-steryl glucopyranoside from *Cucurbita pepo* seeds. Phytochem 1985;24(11):2746–8.

[258] Garg VK, Nes WR. Occurrence of Δ5-sterols in plants producing predominantly Δ7-sterols: studies on the sterol compositions of six cucurbitaceae seeds. Phytochem 1986;25(11):2591–7.

[259] Appendino G, Jakupovic J, Belloro E, Marchesini A. Multiflorane triterpenoid esters from pumpkin. An unexpected extrafolic source of PABA. Phytochem 1999;51(8):1021–6.

[260] Appendino G, Jakupovic J, Belloro E, Marchesini A. Triterpenoid p-aminobenzoates from the seeds of zucchini. Fitoterapia 2000;71(3):258–63.

[261] Ding YL, Deng XM, Cai H, Wang FS, Wang XL, Zhang YM, et al. Studies on the chemical constituents of *Cucurbita pepo* cv dayangua. Chinese Pharm J 2002;37(9):659–61.

[262] Feng XS, Wang DC, Cai H, Deng XM, Liu YR. Determination of the cucurbitacins from *Cucurbita pepo* cv dayangua by HPLC. J Chinese Med Materials 2007;30(4):418–20.

[263] Wang DC, Pan HY, Deng XM, Xiang H, Gao HY, Cai H, et al. Cucurbitane and hexanorcucurbitane glycosides from the fruits of *Cucurbita pepo* cv dayangua. J Asian Nat Prod Res 2007;9(6):525–9.

[264] Wang DC, Xiang H, Li D, Gao HY, Cai H, Wu LJ, et al. Purine-containing cucurbitane triterpenoids from *Cucurbita pepo* cv dayangua. Phytochem 2008;69(6):1434–8.

[265] Jian CC, Ming HC, Rui LN, Cordell GA, Quiz SX. Cucurbitacins and cucurbitane glycosides: structures and biological activities. Nat Prod Rep 2005;22(3):386–99.

[266] Morris Kupchan S, Gray AH, Grove MD. Tumor inhibitors. XXIII. The cytotoxic principles of merah oreganos H. J Med Chem 1967;10(3):337–40.

[267] Kupchan SM, Smith RM, Aynehchi Y, Maruyama M. Tumor inhibitors. LVI. Cucurbitacins O, P, and Q, the cytotoxic principles of *Brandegea bigelovii*. J Org Chem 1970;35(9):2891–4.

[268] Jayaprakasam B, Seeram NP, Nair MG. Anticancer and antiinflammatory activities of cucurbitacins from *Cucurbita andreana*. Cancer Lett 2003;189(1):11–16.

[269] Rodriguez N, Vasquez Y, Hussein AA, Coley PD, Solis PN, Gupta MP. Cytotoxic cucurbitacin constituents from *Sloanea zuliaensis*. J Nat Prod 2003;66(11):1515–6.

[270] Fang X, Phoebe Jr. CH, Pezzuto JM, Fong HHS, Farnsworth NR, Yellin B, et al. Plant anticancer agents, XXXIV. Cucurbitacins from *Elaeocarpus dolichostylus*. J Nat Prod 1984;47(6):988–93.

[271] Van Dang G, Rode BM, Stuppner H. Quantitative electronic structure–activity relationship (QESAR) of natural cytotoxic compounds: maytansinoids, quassinoids and cucurbitacins. Eur J Pharm Sci 1994;2(5–6):331–50.

[272] Fujita E, Nagao Y. Tumor inhibitors having potential for interaction with mercapto enzymes and/or coenzymes. A review. Bioorg Chem 1977;6(3):287–309.

[273] Duncan KLK, Duncan MD, Alley MC, Sausville EA. Cucurbitacin E-induced disruption of the actin and vimentin cytoskeleton in prostate carcinoma cells. Biochem Pharmacol 1996;52(10):1553–60.

[274] Witkowski A, Woynarowska B, Konopa J. Inhibition of the biosynthesis of deoxyribonucleic acid, ribonucleic acid and protein in HeLa S3 cells by cucurbitacins, glucocorticoid-like cytotoxic triterpenes. Biochem Pharmacol 1984;33(7):995–1004.

[275] Blaskovich MA, Sun J, Cantor A, Turkson J, Jove R, Sebti SM. Discovery of JSI-124 (cucurbitacin I), a selective Janus kinase/signal transducer and activator of transcription 3 signaling pathway inhibitor with potent antitumor activity against human and murine cancer cells in mice. Cancer Res 2003;63(6):1270–9.

[276] Buettner R, Mora LB, Jove R. Activated STAT signaling in human tumors provides novel molecular targets for therapeutic intervention. Clin Cancer Res 2002;8(4):945–54.

[277] Lin L, Deangelis S, Foust E, Fuchs J, Li C, Li PK, et al. A novel small molecule inhibits STAT3 phosphorylation and DNA binding activity and exhibits potent growth suppressive activity in human cancer cells. Mol Cancer 2010;9:217.

[278] Sun J, Blaskovich MA, Jove R, Livingston SK, Coppola D, Sebti SM. Cucurbitacin Q: a selective STAT3 activation inhibitor with potent antitumor activity. Oncogene 2005;24(20):3236–45.

[279] Sun C, Zhang M, Shan X, Zhou X, Yang J, Wang Y, et al. Inhibitory effect of cucurbitacin e on pancreatic cancer cells growth via STAT3 signaling. J Cancer Res Clin Oncol 2010;136(4):603–10.

[280] Niu G, Wright KL, Ma Y, Wright GM, Huang M, Irby R, et al. Role of Stat3 in regulating p53 expression and function. Mol Cell Biol 2005;25:7432–40.

[281] Huang WW, Yang JS, Lin MW, Chen PY, Chiou SM, Chueh FS, et al. Cucurbitacin E induces G_2/M phase arrest through STAT3/p53/p21 signaling and provokes apoptosis via Fas/CD95 and mitochondria-dependent pathways in human bladder cancer T24 cells. Ev Based Compl Alt Med 2012 Article ID 952762

[282] Siqueira JM, Gazola AC, Farias MR, Volkov L, Rivard N, De Brum-Fernandes AJ, et al. Evaluation of the antitumoral effect of dihydrocucurbitacin-B in both in vitro and in vivo models. Cancer Chemother Pharmacol 2009;64(3):529–38.

[283] Ishdorj G, Johnston JB, Gibson SB. Inhibition of constitutive activation of STAT3 by curcurbitacin-I (JSI-124) sensitized human B-Leukemia cells to apoptosis. Mol Cancer Ther 2010;9(12):3302–14.

[284] Zhang M, Zhang H, Sun C, Shan X, Yang X, Li-Ling J, et al. Targeted constitutive activation of signal transducer and activator of transcription 3 in human hepatocellular carcinoma cells by cucurbitacin B. Cancer Chemother Pharmacol 2009;63(4):635–42.

[285] Zhang M, Sun C, Shan X, Yang X, Li-Ling J, Deng Y. Inhibition of pancreatic cancer cell growth by cucurbitacin b through modulation of signal transducer and activator of transcription 3 signaling. Pancreas 2010;39(6):923–9.

[286] Morleya SJ, Jeffrey I, Bushell M, Paina VM, Clemens MJ. Differential requirements for caspase-8 activity in the mechanism of phosphorylation of eIF2K, cleavage of eIF4GI and signaling events associated with the inhibition of protein synthesis in apoptotic Jurkat T cells. FEBS Letters 2000;477:229–36.

[287] Li Y, Wang R, Ma E, Deng Y, Wang X, Xiao J, et al. The induction of G2/M cell-cycle arrest and apoptosis by cucurbitacin e is associated with increased phosphorylation of eIF2α in leukemia cells. Anti-Cancer Drugs 2010;21(4):389–400.

[288] Dong Y, Lu B, Zhang X, Zhang J, Lai L, Li D, et al. Cucurbitacin E, a tetracyclic triterpenes compound from Chinese medicine, inhibits tumor angiogenesis through VEGFR2-mediated Jak2–STAT3 signaling pathway. Carcinogenesis 2010;31(12):2097–104.

[289] Chan KT, Li K, Liu SL, Chu KH, Toh M, Xie WD. Cucurbitacin B inhibits STAT3 and the Raf/MEK/ERK pathway in leukemia cell line K562. Cancer Lett 2010;289(1):46–52.

[290] Chung JY, Uchida E, Grammer TC, Blenis J. STAT3 serine phosphorylation by ERK-dependent and -independent pathways negatively modulates its tyrosine phosphorylation. Mol Cell Biol 1997;17(11):6508–16.

[291] Duangmano S, Dakeng S, Jiratchariyakul W, Suksamrarn A, Smith DR, Patmasiriwat P. Antiproliferative effects of cucurbitacin B in breast cancer cells:

down-regulation of the c-Myc/hTERT/telomerase pathway and obstruction of the cell cycle. Int J Mol Sci 2010;11(12):5323–38.

[292] Bowman T, Broome MA, Sinibaldi D, Wharton W, Pledger WJ, Sedivy JM, et al. Stat3-mediated Myc expression is required for Src transformation and PDGF-induced mitogenesis. Proc Nat Acad Sci USA 2001;98(13):7319–24.

[293] Nakashima S, Matsuda H, Kurume A, Oda Y, Nakamura S, Yamashita M, et al. Cucurbitacin E as a new inhibitor of cofilin phosphorylation in human leukemia U937 cells. Bioorg Med Chem Lett 2010;20(9):2994–7.

[294] Yang L, Wu S, Zhang Q, Liu F, Wu P. 23,24-Dihydrocucurbitacin B induces G2/M cell-cycle arrest and mitochondria-dependent apoptosis in human breast cancer cells (Bcap37). Cancer Lett 2007;256(2):267–78.

[295] Jin HR, Jin X, Dat NT, Lee JJ. Curbitacin B suppresses the transactivation activity of RelA/p65. J Cell Biochem 2011;112(6):1643–50.

[296] Ding N, Yamashita U, Matsuoka H, Sugiura T, Tsukada J, Noguchi J, et al. Apoptosis induction through proteasome inhibitory activity of cucurbitacin D in human T-cell leukemia. Cancer 2011;117(12):2735–46.

[297] Lam TL, Wright G, Davis RE, Lenz G, Farinha P, Dang L, et al. Cooperative signaling through the signal transducer and activator of transcription 3 and nuclear factor-κB pathways in subtypes of diffuse large B cell lymphoma. Blood 2007;111(7):3701–13.

[298] Park CS, Lim H, Han KJ, Baek SH, Sohn HO, Lee DW, et al. Inhibition of nitric oxide generation by 23,24-dihydrocucurbitacin D in mouse peritoneal macrophages. J Pharmacol Exp Ther 2004;309(2):705–10.

[299] Takahashi N, Yoshida Y, Sugiura T, Matsuno K, Fujino A, Yamashita U. Cucurbitacin D isolated from Trichosanthes kirilowii induces apoptosis in human hepatocellular carcinoma cells in vitro. Int Immunopharmacol 2009;9(4):508–13.

[300] Tsuruta F, Sunayama J, Mori Y, Hattori S, Shimizu S, Tsujimoto Y, et al. JNK promotes Bax translocation to mitochondria through phosphorylation of 14-3-3 proteins. EMBO J 2004;23(8):1889–99.

[301] Chan KT, Meng FY, Li Q, Ho CY, Lam TS, To Y, et al. Cucurbitacin B induces apoptosis and S phase cell cycle arrest in BEL-7402 human hepatocellular carcinoma cells and is effective via oral administration. Cancer Lett 2010;294(1):118–24.

[302] Suzuki H, Fujita H, Mullauer L, Kuzumaki N, Konaka S, Togashi Y, et al. Increased expression of c-jungene during spontaneous hepatocarcinogenesis in LEC rats. Cancer Lett 1990;53(2–3):205–12.

[303] Yasuda S, Yogosawa S, Izutani Y, Nakamura Y, Watanabe H, Sakai T. Cucurbitacin B induces G2 arrest and apoptosis via a reactive oxygen species-dependent mechanism in human colon adenocarcinoma SW480 cells. Mol Nutr Food Res 2010;54(4):559–65.

[304] Zhang Y, Ouyang D, Xu L, Ji Y, Zha Q, Cai J, et al. Cucurbitacin B induces rapid depletion of the G-actin pool through reactive oxygen species-dependent actin aggregation in melanoma cells. Acta Biochim Biophys Sinica 2011;43(7):556–67.

[305] Duncan KLK, Duncan MD, Alley MC, Sausville EA. Cucurbitacin E-induced disruption of the actin and vimentin cytoskeleton in prostate carcinoma cells. Biochem Pharmacol 1996;52:1553–60.

[306] Boykin C, Zhang G, Chen YH, Zhang RW, Fan XE, Yang WM, et al. Cucurbitacin IIa: a novel class of anti-cancer drug inducing non-reversible actin aggregation and inhibiting survivin independent of JAK2/STAT3 phosphorylation. Br J Cancer 2011;104(5):781–9.

[307] Dakeng S, Duangmano S, Jiratchariyakul W, U-Pratya Y, Bögler O, Patmasiriwat P. Inhibition of Wnt signaling by cucurbitacin B in breast cancer cells: reduction of Wnt-associated proteins and reduced translocation of galectin-3-mediated β-catenin to the nucleus. J Cell Biochem 2012;113(1):49–60.

[308] Oh H, Mun YJ, Im SJ, Lee SY, Song HJ, Lee HS, et al. Cucurbitacins from *Trichosanthes kirilowii* as the inhibitory components on tyrosinase activity and melanin synthesis of B16/F10 melanoma cells. Planta Med 2002;68(9):832–3.

[309] Ramalhete C, Molnár J, Mulhovo S, Rosário VE, Ferreira MJU. New potent P-glycoprotein modulators with the cucurbitane scaffold and their synergistic interaction with doxorubicin on resistant cancer cells. Bioorg Med Chem 2009;17(19):6942–51.

[310] Lee KH, Imakura Y, Huang HC. Bruceoside-A, a novel antileukaemic quassinoid glycoside from *Brucea javanica*. J Chem Soc 1977;2:69–70.

[311] Lee KH, Imakura Y, Sumida Y, Wu RY, Hall IH, Huang HC. Tumor agents. 33. Isolation and structural elucidation of bruceoside-A and -B, novel antileukemic quassinoid glycosides, and brucein-D and -E from *Brucea javanica*. J Org Chem 1979;44(13):2180–5.

[312] Zhao M, Lau ST, Zhang XQ, Ye WC, Leung PS, Che CT, et al. Bruceines K and L from the ripe fruits of *Brucea javanica*. Helvetica Chimica Acta 2011;94(11):2099–105.

[313] Fukamiya N, Okano M, Miyamoto M, Tagahara K, Lee KH. Antitumor agents, 127. Bruceoside C, a new cytotoxic quassinoid glucoside, and related compounds from *Brucea javanica*. J Nat Prod 1992;55(4):468–75.

[314] Ohnishi S, Fukamiya N, Okano M, Tagahara K, Lee KH. Bruceosides D, E, and F, three new cytotoxic quassinoid glucosides from Brucea javanica. J Nat Prod 1995;58(7):1032–8.

[315] Lee KH, Hayashi N, Okano M. Antitumor agents, 65. Brusatol and cleomiscosin-A, antileukemic principles from *Brucea javanica*. J Nat Prod 1984;47(3):550–1.

[316] Sakaki T, Yoshimura S, Ishibashi M. New quassinoid glycosides, yadanziosides A – H, from *Brucea javanica*. Chem Pharm Bull 1984;32(11):4702–5.

[317] Sakaki T, Yoshimura S, Tsuyuki T. Two new quassinoid glycosides, yadanziosides N and O isolated from seeds of *Brucea javanica* (L.) MERR. Tetrahedron Lett 1986;27(5):593–6.

[318] Yoshimura S, Sakaki T, Ishibashi M. Constituents of seeds of *Brucea javanica*. Structures of new bitter principles, yadanziolides A, B, C, yadanziosides F, I, J, and L. Bull Chem Soc Jap 1985;58(9):2673–9.

[319] Sakaki T, Yoshimura S, Tsuyuki T. Structures of yadanziosides K, M, N, and O, new quassinoid glycosides from *Brucea javanica* (L.) MERR. Bull Chem Soc Jap 1986;59(11):3541–6.

[320] Sakaki T, Yoshimura S, Tsuyuki T. Yadanzioside P, a new antileukemic quassinoid glycoside from *Brucea javanica* (L.) MERR with the 3-O-(β-D glucopyranosyl) bruceantin structure. Chem Pharm Bull 1986;34(10):4447–50.

[321] Luyengi L, Suh N, Fong HHS, Pezzuto JM, Kinghorn AD. A lignan and four terpenoids from *Brucea javanica* that induce differentiation with cultured HL-60 promyelocytic leukemia cells. Phytochem 1996;43(2):409–12.

[322] Sakaki T, Yoshimura S, Ishibashi M. Structures of new quassinoid glycosides, yadanziosides A, B, C, D, E, G, H, and new quassinoids, dehydrobrusatol and dehydrobruceantinol from *Brucea javanica* (L.) Merr. Bull Chem Soc Jap 1985;58(9):2680–6.

[323] Yoshimura S, Ogawa K, Tsuyuki T, Takahashi T, Honda T. Yandanziolide D, a new C19-quassinoid isolated from *Brucea javanica* (L.) Merr. Chem Pharm Bull 1988;36(2):841–4.

[324] Lin LZ, Cordell GA, Ni CZ, Clardy J. A quassinoid from *Brucea javanica*. Phytochem 1990;29(8):2720–2.

[325] Anderson MM, O'Neill MJ, Phillipson JD, Warhurst DC. In vitro cytotoxicity of a series of quassinoids from *Brucea javanica* fruits against KB cells. Planta Med 1991;57(1):62–4.

[326] Rahman S, Fukamiya N, Tokuda H, Nishino H, Tagahara K, Lee KH, et al. Three new quassinoid derivatives and related compounds as antitumor promoters from *Brucea javanica*. Bull Chem Soc Jap 1999;72(4):751–6.

[327] Kim IH, Suzuki R, Hitotsuyanagi Y, Takeya K. Three novel quassinoids, javanicolides A and B, and javanicoside A, from seeds of *Brucea javanica*. Tetrahedron 2003;59(50):9985–9.

[328] Kitagawa I, Mahmud T, Simanjuntak P, Hori K, Uji T, Shibuya H. Indonesian medicinal plants. VIII. Chemical structures of three new triterpenoids, bruceajavanin A, dihydrobruceajavanin A, and bruceajavanin B, and a new alkaloidal glycoside, bruceacanthinoside, from the stems of *Brucea javanica* (Simaroubaceae). Chem Pharm Bull 1994;42(7):1416–21.

[329] Kim IH, Hitotsuyanagi Y, Takeya K. Quassinoid xylosides, javanicosides G and H, from seeds of *Brucea javanica*. Heterocycles 2004;63(3):691–7.

[330] Kim IH, Takashima S, Hitotsuyanagi Y, Hasuda T, Takeya K. New quassinoids, javanicolides C and D and javanicosides B–F, from seeds of *Brucea javanica*. J Nat Prod 2004;67(5):863–8.

[331] Kamperdick C, Van Sung T, Thi TT, Van TM, Adam G. (20R)-O-(3)-alpha-L-arabinopyranosyl-pregn-5-en-3beta,20-diol from *Brucea javanica*. Phytochem 1995;38(3):699–701.

[332] Chen YY, Pan QD, Li DP, Liu JL, Wen YX, Huang YL, et al. New pregnane glycosides from *Brucea javanica* and their antifeedant activity. Chem Biodiver 2011;8(3):460–6.

[333] Liu KCS, Yang SL, Roberts MF, Phillipson JD. Production of 11-hydroxycanthin-6-one and canthin-6-one by cell suspension cultures of *Brucea javanica*. J Pharm Pharmacol 1989;41:70.

[334] Ouyang MA, Wein YS, Zhang ZK, Kuo YH. Inhibitory activity against tobacco mosaic virus (TMV) replication of pinoresinol and syringaresinol lignans and their glycosides from the root of *Rhus javanica* var. roxburghiana. J Agr and Food Chem 2007;55(16):6460–5.

[335] Lang CC, Watt RA, Phillipson JD, Kirby GC, Warhurst DC. In vitro studies on the mode of action of quassinoid analogues against chloroquine resistant *Plasmodium falciparum*. J Pharm Pharmacol 1994;46:1062.

[336] O'Neill MJ, Bray DH, Boardman P, Chan KL, Phillipson JD, Warhurst DC, et al. Plants as sources of antimalarial drugs, Part 4: activity of *Brucea javanica* fruits against chloroquine-resistant *Plasmodium falciparum* in vitro and against *Plasmodium berghei* in vivo. J Nat Prod 1987;50:41–8.

[337] Lau ST, Lin ZX, Liao Y, Zhao M, Cheng CHK, Leung PS. Brucein D induces apoptosis in pancreatic adenocarcinoma cell line PANC-1 through the activation of p38-mitogen activated protein kinase. Cancer Lett 2009;281(1):42–52.

[338] Kim JA, Lau EK, Pan L, Carcache De Blanco EJ. NF-κB inhibitors from *Brucea javanica* exhibiting intracellular effects on reactive oxygen species. Anticancer Res 2010;30(9):3295–300.

[339] Cuendet M, Pezzuto JM. Antitumor activity of bruceantin: an old drug with new promise. J Nat Prod 2004;67:269–72.

[340] Arsenau JC, Wolter JM, Kuperminc M, Ruckdeschel JC. A Phase II study of bruceantin (NSC 165563) in advanced malignant melanoma. Invest New Drugs 1983;1:239–42.

[341] Hitotsuyanagi Y, Kim IH, Hasuda T, Yamauchi Y, Takeya K. A structure–activity relationship study of brusatol, an antitumor quassinoid. Tetrahedron 2006;62(17):4262–71.

[342] Okano M, Fukamiya N, Aratani T, Juichi M, Lee KH. Antitumor agents, 74. Bruceanol-A and -B, two new antileukemic quassinoids from *Brucea antidysenterica*. J Nat Prod 1985;48(6):972–5.

[343] Fukamiya N, Okano M, Tagahara K, Aratani T, Lee KH. Antitumor agents, 93. Bruceanol C, a new cytotoxic quassinoid from *Brucea antidysenterica*. J Nat Prod 1988;51(2):349–52.

[344] Imamura K, Fukamiya N, Okano M, Tagahara K, Lee KH. Bruceines D, E, and F. Three new cytotoxic quassinoids from *Brucea antidysenterica*. J Nat Prod 1993;56(12):2091–7.

[345] Imamura K, Fukamiya N, Nakamura M, Okano M, Tagahara K, Lee KH. Bruceines G and H, cytotoxic quassinoids from *Brucea antidysenterica*. J Nat Prod 1995;58(12):1915–9.

[346] Moretti C, Bhatnagar S, Beloeil JC, Polonsky J. Two new quassinoids from *Simaba multiflora* fruits. J Nat Prod 1986;49(3):440–4.

[347] Arisawa M, Kinghorn AD, Cordell GA, Farnsworth NR. Plant anticancer agents. XXIII. 6α-Senecioyloxychaparrin, a new antileukemic quassinoid from *Simaba multiflora*. J Nat Prod 1983;46(2):218–21.

[348] Polonsky J, Vakon Z, MoKetti C, Pettit GR, Herald CL, Rideout JA, et al. The antineoplastic quassinoids of *Simaba cuspidata* Spruce and *Ailanthus grandis* Prain. J Nat Prod 1980;4(4):503–9.

[349] Ozeki A, Hitotsuyanagi Y, Hashimoto E, Itokawa H, Takeya K, De Mello Alves S. Toxic quassinoids from *Simaba cedron*. J Nat Prod 1988;61(6):776–80.

[350] Kardono LBS, Angerhofer CK, Tsauri S, Padmawinata K, Pezzuto JM, Kinghorn AD. Cytotoxic and antimalarial constituents of the roots of *Eurycoma longifolia*. J Nat Prod 1991;54(5):1360–7.

[351] Itokawa H, Kishi E, Morita H, Takeya K. Cytotoxic quassinoids and tirucallane-type triterpenes from the woods of *Eurycoma longifolia*. Chem Pharm Bull 1992;40(4):1053–5.

[352] Morita H, Kishi E, Takeya K, Itokawa H, Iitaka Y. Highly oxygenated quassinoids from *Eurycoma longifolia*. Phytochem 1993;33(3):691–6.

[353] Polonsky J, Varon Z, Jacquemin H, Pettit GR. The isolation and structure of 13,18-dehydroglaucarubinone, a new antineoplastic quassinoid from *Simarouba amara*. Experientia 1978;34(9):1122–3.

[354] Tischler M, Cardellina II JH, Boyd MR, Cragg GM. Cytotoxic quassinoids from *Cedronia granatensis*. J Nat Prod 1992;55(5):667–71.

[355] Usami Y, Nakagawa-Goto K, Lang JY, Kim Y, Lai CY, Goto M, et al. Antitumor agents. 282. 2′-(R)-O-acetylglaucarubinone, a quassinoid from *Odyendyea gabonensis* as a potential anti-breast and anti-ovarian cancer agent. J Nat Prod 2010;73(9):1553–8.

[356] Polonsky J, Bhatnagar S, Moretti C. 15-Deacetylsergeolide, a potent antileukemic quassinoid from *Picrolemma pseudocoffea*. J Nat Prod 1984;47(6):994–6.

[357] Van Tri M, Polonsky J, Merienne C, Sevenet T. Soularubinone, a new antileukemic quassinoid from *Soulamea tomentosa*. J Nat Prod 1981;44(3):279–84.

[358] Lumonadio L, Atassi G, Vanhaelen M, Vanhaelen-Fastre R. Antitumor activity of quassinoids from *Hannoa klaineana*. J Ethnopharmacol 1991;31(1):59–65.

[359] Cassady JM, Suffness M. Terpenoids antitumor agents. In: Cassady JM, Dourous JD, editors. Anticancer agents based on natural products models. New York, NY: Academic Press; 1980.

[360] Kupchan SM, Lacadie JA. Dehydroailanthinone, a new antileukemic quassinoid from *Pierreodendron kerstingii*. J Org Chem 1975;40:654–6.

[361] Liao LL, Kupchan SM, Horwitz SB. Mode of action of the antitumor compound bruceantin, an inhibitor of protein synthesis. Mol Pharmacol 1976;12:167–76.

[362] Wall ME, Wani MC. Plant antitumor agents. 17. Structural requirements for antineoplastic activity in quassinoids [2]. J Med Chem 1978;21(12):1186–8.

[363] Valeriote FA, Corbett TH, Grieco PA, Moher ED, Collins JL, Fleck TJ. Anticancer activity of glaucarubinone analogues. Oncology Res 1998;10(4):201–8.

[364] Fresno M, Gonzales A, Vazquez D, Jimenez A. Bruceantin, a novel inhibitor of peptide bond formation. Biochim Biophys Acta 1978;518:104–12.

[365] Considine RT, Willingham Jr. W, Chaney SG. Structure–activity relationships for binding and inactivation of rabbit reticulocyte ribosomes by quassinoid antineoplastic agents. Eur J Biochem 1983;132(1):157–63.

[366] Kupchan SM, Britton RW, Lacadie JA, Ziegler MF, Sigel CW. The isolation and structural elucidation of bruceantin and bruceantinol, new potent antileukemic quassinoids from *Brucea antidysenterica*. J Org Chem 1975;40:648–54.

[367] Kupchan SM, Britton RW, Ziegler MF, Sigel CW. Bruceantin, a new potent antileukemic simaroubolide from *Brucea antidysenterica*. J Org Chem 1973;38:178–9.

[368] Morré DJ, Grieco PA. Glaucarubolone and simalikalactone D, respectively, preferentially inhibit auxin-induced and constitutive components of plant cell enlargement and the plasma membrane NADH oxidase. International Journal of Plant Sci 1999;160(2):291–7.

[369] Mata-Greenwood E, Cuendet M, Sher D, Gustin D, Stock W, Pezzuto JM. Brusatol-mediated induction of leukemic cell differentiation and G1 arrest is associated with down-regulation of c-myc. Leukemia 2002;16(11):2275–84.

[370] Donjerković D, Mueller CM, Scott D,W. Steroid- and retinoid-mediated growth arrest and apoptosis in WEHI-231 cells: role of NF-κB, c-Myc and CKI p27(Kip1). Eur J Immunol 2000;30(4):1154–61.

[371] Cuendet M, Gills JJ, Pezzuto JM. Brusatol-induced HL-60 cell differentiation involves NF-κB activation. Cancer Lett 2004;206(1):43–50.

[372] Jin Z, El-Deiry WS. Overview of cell death signaling pathways. Cancer Biol Ther 2005;4:139–63.

[373] Saelens X, Festjens N, Vande Walle L, van Gurp M, van Loo G, Vandenabeele P. Toxic proteins released from mitochondria in cell death. Oncogene 2004;23:2861–74.

[374] Kim MJ, Park BJ, Kang YS, Kim HJ, Park JH, Kang JW, et al. Downregulation of FUSE-binding protein and c-myc by tRNA synthetase cofactor p38 is required for lung cell differentiation. Nat Gen 2003;34(3):330–6.

[375] Castelletti D, Fiaschetti G, Dato VD, Ziegler U, Kumps C, Preter KD, et al. The quassinoid derivative NBT-272 targets both the AKT and ERK signaling pathways in embryonal tumors. Mol Cancer Ther 2010;9(12):3145–57.

[376] Wong PF, Cheong WF, Shu MH, Teh CH, Chan KL, Abubakar S. Eurycomanone suppresses expression of lung cancer cell tumor markers, prohibitin, annexin 1 and endoplasmic reticulum protein 28. Phytomed 2012;19(2):138–44.

[377] Komissarenko NF. Furocoumarins of *Dictamnus dasycarpus*. Chem Nat Comp 1968;4(6):319.

[378] Komissarenko NF, Levashova IG, Nadezhina TP. Flavonoids and coumarins of *Dictamnus dasycarpus*. Chem Nat Comp 1983;19(4):502.

[379] Du CF, Yang XX, Tu PF. Studies on chemical constituents in bark of Dictamnus dasycarpus. Zhongguo Zhongyao Zazhi 2005;30(21):1663–6.

[380] Chang J, Xuan LJ, Xu YM, Zhang JS. Cytotoxic terpenoid and immunosuppressive phenolic glycosides from the root bark of *Dictamnus dasycarpus*. Planta Med 2002;68(5):425–9.

[381] Yu SM, Ko FK, Su MJ, Wu TS, Wang ML, Huang TF, et al. Vasorelaxing effect in rat thoracic aorta caused by fraxinellone and dictamine isolated from the Chinese herb *Dictamnus dasycarpus* Turcz: comparison with cromakalim and Ca2+ channel blockers. Naunyn-Schmiedeberg's Arch Pharmacol 1992;345(3):349–55.

[382] Zhao W, Wolfender JL, Hostettmann K, Xu R, Qin G. Antifungal alkaloids and limonoid derivatives from *Dictamnus dasycarpus*. Phytochem 1998;47(1):7–11.

[383] Jeong SY, Sang HS, Young CK. Neuroprotective limonoids of root bark of *Dictamnus dasycarpus*. J Nat Prod 2008;71(2):208–11.

[384] Takeuchi N, Fujita T, Goto K, Morisaki N, Osone N, Tobinaga S. Dictamnol, a new trinor-guaiane type sesquiterpene, from the roots of *Dictamnus dasycarpus* Turcz. Chem Pharm Bull 1993;41(5):923–5.

[385] Zhao W, Wolfender JL, Hostettmann K, Li HY, Stoeckli-Evans H, Xu R, et al. Sesquiterpene glycosides from *Dictamnus dasycarpus*. Phytochem 1998;47(1):63–8.

[386] Zhao WM, Wang SC, Hostettmann K, Qin GW, Xu RS. Two novel sesquiterpene diglycosides from *Dictamnus dasycarpus*. Chinese Chem Lett 1999;10(7):563–6.

[387] Chang J, Xuan LJ, Xu YM, Zhang JS. Seven new sesquiterpene glycosides from the root bark of *Dictamnus dasycarpus*. J Nat Prod 2001;64(7):935–8.

[388] Chen J, Tang JS, Tian J, Wang YP, Wu FE. Dasycarine, a new quinoline alkaloid from *Dictamnus dasycarpus*. Chinese Chem Lett 2000;11(8):707–8.

[389] Yang JL, Liu LL, Shi YP. Limonoids and quinoline alkaloids from *Dictamnus dasycarpus*. Planta Med 2011;77(3):271–6.

[390] Wang Z, Xu F, An S. Chemical constituents from the root bark of *Dictamnus dasycarpus* Turcz. China J Chinese Materia Medica 1992;17(9):551–2.

[391] Lei J, Yu J, Yu H, Liao Z. Composition, cytotoxicity and antimicrobial activity of essential oil from *Dictamnus dasycarpus*. Food Chem 2008;107(3):1205–9.

[392] Tan QG, Luo XD. Meliaceous limonoids: chemistry and biological activities. Chemical Reviews 2011;111(11):7437–522.

[393] Pettit GR, Barton DHR, Herald CL, Polonsky J, Schmidt JM, Connolly JD. Evaluation of limonoids against the murine P388 lymphocytic leukemia cell line. J Nat Prod 1983;46(3):379–90.

[394] Nanduri S, Thunuguntla SSR, Nyavanandi VK, Kasu S, Kumar PM, Ram PS, et al. Biological investigation and structure–activity relationship studies on azadirone from *Azadirachta indica* A. Juss. Bioorg Med Chem Lett 2003;13(22):4111–5.

[395] Ahn J-W, Choi S-U, Lee C-O. Cytotoxic limonoids from *Melia azedarach* var. japonica. Phytochem 1994;36(6):1493–6.

[396] Itokawa H, Qiao ZS, Hirobe C, Takeya K. Cytotoxic limonoids and tetranortriterpenoids from *Melia azedarach*. Chem Pharm Bull 1995;43(7):1171–5.

[397] Takeya K, Qiao ZS, Hirobe C, Itokawa H. Cytotoxic trichilin-type limonoids from *Melia azedarach*. Bioorg Med Chem 1996;4(8):1355–9.

[398] Zhang B, Wang ZF, Tang MZ, Shi YL. Growth inhibition and apoptosis-induced effect on human cancer cells of toosendanin, a triterpenoid derivative from Chinese traditional medicine. Invest New Drugs 2005;23(6):547–53.

[399] Tada K, Takido M, Kitanaka S. Limonoids from fruit of *Melia toosendan* and their cytotoxic activity. Phytochem 1999;51(6):787–91.

[400] Kikuchi T, Ishii K, Noto T, Takahashi A, Tabata K, Suzuki T, et al. Cytotoxic and apoptosis-inducing activities of limonoids from the seeds of *Azadirachta indica* (Neem). J Nat Prod 2011;74(4):866–70.

[401] Maneerat W, Laphookhieo S, Koysomboon S, Chantrapromma K. Antimalarial, antimycobacterial and cytotoxic limonoids from *Chisocheton siamensis*. Phytomed 2008;15(12):1130–4.

[402] Murphy BT, Brodie P, Slebodnick C, Miller JS, Birkinshaw C, Randrianjanaka LM, et al. Antiproliferative limonoids of a *Malleastrum* sp. from the Madagascar rainforest. J Nat Prod 2008;71(3):325–9.

[403] Cui J, Deng Z, Xu M, Proksch P, Li Q, Lin W. Protolimonoids and limonoids from the Chinese mangrove plant *Xylocarpus granatum*. Helvetica Chimica Acta 2009;92(1):139–50.

[404] Zhou H, Hamazaki A, Fontana JD, Takahashi H, Esumi T, Wandscheer CB, et al. New ring C-seco limonoids from Brazilian *Melia azedarach* and their cytotoxic activity. J Nat Prod 2004;67(9):1544–7.

[405] Yang MH, Wang JS, Luo JG, Wang XB, Kong LY. Tetranortriterpenoids from *Chisocheton paniculatus*. J Nat Prod 2009;72(11):2014–8.

[406] Uddin SJ, Nahar L, Shilpi JA, Shoeb M, Borkowski T, Gibbons S, et al. Gedunin, a limonoid from *Xylocarpus granatum*, inhibits the growth of CaCo-2 colon cancer cell line in vitro. Phytother Res 2007;21(8):757–61.

[407] Pudhom K, Sommit D, Nuclear P, Ngamrojanavanich N, Petsom A. Protoxylocarpins F – H, protolimonoids from seed kernels of *Xylocarpus granatum*. J Nat Prod 2009;72(12):2188–91.

[408] Mitsui K, Saito H, Yamamura R, Fukaya H, Hitotsuyanagi Y, Takeya K. Hydroxylated gedunin derivatives from *Cedrela sinensis*. J Nat Prod 2006;69(9):1310–4.

[409] Lukačova V, Polonsky J, Moretti C, Pettit GR, Schmidt JM. Isolation and structure of 14,15β-epoxyprieurianin from the South American tree *Guarea guidona*. J Nat Prod 1982;45(3):288–94.

[410] Chaturvedula VSP, Schilling JK, Wisse JH, Miller JS, Evans R, Kingston DGI. A new cytotoxic limonoid from *Odontadenia macrantha* from the Suriname rainforest. Magnetic Resonance Chem 2003;41(2):139–42.

[411] Tang MZ, Wang ZF, Shi YL. Involvement of cytochrome c release and caspase activation in toosendanin-induced PC12 cell apoptosis. Toxicology 2004;201:31–8.

[412] Chidambara Murthy KN, Jayaprakasha GK, Kumar V, Rathore KS, Patil BS. Citrus limonin and its glucoside inhibit colon adenocarcinoma cell proliferation through apoptosis. J Agr Food Chem 2011;59(6):2314–23.

[413] Chen R, Valencia I, Zhong F, McColl KS, Roderick HL, Bootman MD, et al. Bcl-2 functionally interacts with inositol 1,4,5-trisphosphate receptors to regulate calcium release from the ER in response to inositol 1,4,5-trisphosphate. J Cell Biol 2004;166(2):193–203.

[414] Shi YL, Fuyura K, Wang WP, Terakawa S, Xu K, Yamagishi S. Calcium conductance increase by toosendanin in neuroblastoma×glioma hybrid cells. Chinese Sci Bull 1993;38:825–8.

[415] Li MF, Wu Y, Wang ZF, Shi YL. Toosendanin, a triterpenoid derivative, increases Ca2+ current in NG108-15 cells via L-type channels. Neurosci Res 2004;49:197–203.

[416] Li MF, Shi YL. The long-term effect of toosendanin on current through nifedipine-sensitive Ca2+ channels in NG108-15 cells. Toxicon 2004;42:53–60.

[417] Zhang Y, Qi X, Gong L, Li Y, Liu L, Xue X, et al. Roles of reactive oxygen species and MAP kinases in the primary rat hepatocytes death induced by toosendanin. Toxicol 2008;249(1):62–8.

[418] Thoh M, Kumar P, Nagarajaram HA, Manna SK. Azadirachtin interacts with the tumor necrosis factor (TNF) binding domain of its receptors and inhibits TNF-induced biological responses. J Biol Chem 2010;285(8):5888–95.

[419] Gupta SC, Prasad S, Reuter S, Kannappan R, Yadav VR, Ravindran J, et al. Modification of cysteine 179 of IκBα kinase by nimbolide leads to down-regulation of NF-κB-regulated cell survival and proliferative proteins and sensitization of tumor cells to chemotherapeutic agents. J Biol Chem 2010;285(46):35406–35417.

[420] Chidambara Murthy KN, Jayaprakasha GK, Patil BS. Apoptosis mediated cytotoxicity of citrus obacunone in human pancreatic cancer cells. Toxicol in Vitro 2011;25(4):859–67.

[421] Chidambara Murthy KN, Jayaprakasha GK, Patil BS. Obacunone and obacunone glucoside inhibit human colon adenocarcinoma (SW480) cells by the induction of apoptosis. Food Chem Toxicol 2001;49(7):1616–25.

[422] Ye WC, Ji NN, Zhao SX, Liu JH, Ye T, McKervey MA, et al. Triterpenoids from *Pulsatilla chinensis*. Phytochem 1996;42(3):799–802.

[423] Liu WK, Ho JCK, Cheung FWK, Liu BPL, Ye WC, Che CT. Apoptotic activity of betulinic acid derivatives on murine melanoma B16 cell line. Eur J Pharmacol 2004;498(1–3):71–8.

[424] Ye W, He A, Zhao S, Che CT. Pulsatilloside C from the roots of *Pulsatilla chinensis*. J Nat Prod 1998;61(5):658–9.

[425] Mimaki Y, Kuroda M, Asano T, Sashida Y. Triterpene saponins and lignans from the roots of *Pulsatilla chinensis* and their cytotoxic activity against HL-60 cells. J Nat Prod 1999;62(9):1279–83.

[426] Mimaki Y, Yokosuka A, Kuroda M, Hamanaka M, Sakuma C, Sashida Y. New bisdesmosidic triterpene saponins from the roots of *Pulsatilla chinensis.* J Nat Prod 2001;64(9):1226–9.

[427] Ye W, Zhang Q, Hsiao WWL, Zhao S, Che CT. New lupane glycosides from *Pulsatilla chinensis.* Planta Med 2002;68(2):183–6.

[428] Shi BJ, Li Q, Zhang XQ, Wang Y, Ye WC, Yao XS. Triterpene glycosides from the aerial parts of *Pulsatilla chinensis.* Yaoxue Xuebao 2007;42(8):862–6.

[429] Quan GH, Oh SR, Song HH, Lee HK, Chin YW. Preparative isolation of 8-methoxypsoralen from the rhizomes of *Pulsatilla chinensis* using high-speed counter-current chromatography. J Korean Soc Appl Biol Chem 2011;54(4):623–7.

[430] Zhang XQ, Liu AR, Xu LX. Determination of ranunculin in *Pulsatilla chinensis* and synthetic ranunculin by reversed phase HPLC. Acta Pharm Sinica 1990;25(12):932–5.

[431] Chaturvedi PK, Bhui K, Shukla Y. Lupeol: connotations for chemoprevention. Cancer Lett 2008;263(1):1–13.

[432] Hata K, Hori K, Takahashi S. Differentiation- and apoptosis-inducing activities by pentacyclic triterpenes on a mouse melanoma cell line. J Nat Prod 2002;65(5):645–8.

[433] Chaturvedula VSP, Zhou BN, Gao Z, Thomas SJ, Hecht SM, Kingston DGI. New lupane triterpenoids from *Solidago canadensis* that inhibit the lyase activity of DNA polymerase β. Bioorg Med Chem 2004;12:6271–5.

[434] Wada SI, Iida A, Tanaka R. Screening of triterpenoids isolated from *Phyllanthus flexuosus* for DNA topoisomerase inhibitory activity. J Nat Prod 2001;64(12):1545–7.

[435] Thornton TM, Rincon M. Non-classical p38 map kinase functions: cell cycle checkpoints and survival. Int J Biol Sci 2009;5(1):44–52.

[436] Hata K, Hori K, Takahashi S. Role of p38 MAPK in Lupeol-induced B16 2F2 mouse melanoma cell differentiation. J Biochem 2003;134(3):441–5.

[437] Saleem M, Kaur S, Kweon M, Adhami VM, Afaq F, Mukhtar H. Lupeol, a fruit and vegetable based triterpene, induces apoptotic death of human pancreatic adenocarcinoma cells via inhibition of Ras signaling pathway. Carcinogenesis 2005;26(11):1956–64.

[438] Xue L, Murray JH, Tolkovsky AM. The Ras/phosphatidylinositol 3-kinase and Ras/ERK pathways function as independent survival modules each of which inhibits a distinct apoptotic signaling pathway in sympathetic neurons. J Biol Chem 2000;275(12):8817–24.

[439] Chen G, Hitomi M, Han J, Stacey DW. The p38 pathway provides negative feedback for ras proliferative signaling. J Biol Chem 2000;275(50):38973–38980.

[440] Urquhart JL, Meech SJ, Marr DG, Shellman YG, Duke RC, Norris DA. Regulation of Fas-mediated apoptosis by N-ras in melanoma. J Invest Dermatoly 2002;119(3):556–61.

[441] Saleem M, Kweon M, Yun J, Adhami VM, Khan N, Syed DN, et al. A novel dietary triterpene lupeol induces Fas-mediated apoptotic death of

androgen-sensitive prostate cancer cells and inhibits tumor growth in a xenograft model. Cancer Res 2005;65:11203–11213.

[442] Chen G, Templeton D, Suttle DP, Stacey DW. Ras stimulates DNA topoisomerase II alpha through MEK: a link between oncogenic signaling and a therapeutic target. Oncogene 1999;18(50):7149–60.

[443] Shimizu S, Takada M, Umezawa K, Imoto M. Requirement of caspase-3(-like) protease-mediated hydrogen peroxide production for apoptosis induced by various anticancer drugs. J Biol Chem 1998;273:26900–269007.

[444] Chowdhury AR, Mandal S, Mittra B, Sharma S, Mukhopadhyay S, Majumder HK. Betulinic acid, a potent inhibitor of eukaryotic topoisomerase I: identification of the inhibitory step, the major functional group responsible and development of more potent derivatives. Med Sci Monit 2002;8:254–6.

[445] Takada Y, Aggarwal BB. Betulinic acid suppresses carcinogen-induced NF-kappa B activation through inhibition of I kappa B alpha kinase and p65 phosphorylation: abrogation of cyclooxygenase-2 and matrix metalloprotease-9. J Immunol 2003;171(6):3278–86.

[446] Pisha E, Chai H, Lee IS, Chagwedera TE, Farnsworth NR, Cordell GA, et al. Discovery of betulinic acid as a selective inhibitor of human melanoma that functions by induction of apoptosis. Nature Med 1995;1(10):1046–51.

[447] Liu WK, Ho JCK, Cheung FWK, Liu BPL, Ye WC, Che CT. Apoptotic activity of betulinic acid derivatives on murine melanoma B16 cell line. Eur J Pharm 2004;498(1–3):71–8.

[448] Tan Y, Yu R, Pezzuto JM. Betulinic acid-induced programmed cell death in human melanoma cells involves mitogen-activated protein kinase activation. Clin Cancer Res 2003;9(7):2866–75.

[449] Eiznhamer DA, Xu ZQ. Betulinic acid: a promising anticancer candidate. Drugs 2004;7(4):359–73.

[450] Selzer E, Thallinger C, Hoeller C, Oberkleiner P, Wacheck V, Pehamberger H, et al. Betulinic acid-induced Mcl-1 expression in human melanoma – Mode of action and functional significance. Molecular Med 2002;8(12):877–84.

[451] Butts BD, Kwei KA, Bowden GT, Briehl MM. Elevated basal reactive oxygen species and phospho-Akt in murine keratinocytes resistant to ultraviolet B-induced apoptosis. Mol Carcinogenesis 2003;37(3):149–57.

[452] Kim JA, Yang SY, Song SB, Kim YH. Effects of impressic acid from acanthopanax koreanum on NF-κB and PPARγ activities. Arch Pharm Res 2011;34(8):1347–51.

[453] Schmidt ML, Kuzmanoff KL, Ling-Indeck L, Pezzuto JM. Betulinic acid induces apoptosis in human neuroblastoma cell lines. Eur J Cancer 1997;33(12):2007–10.

[454] Fulda S, Debatin KM. Betulinic acid induces apoptosis through a direct effect on mitochondria in neuroectodermal tumors. Med Pediatric Oncol 2000;35(6):616–8.

[455] Tolstikova TG, Sorokina IV, Tolstikov GA, Tolstikov AG, Flekhter OB. Biological activity and pharmacological prospects of lupane terpenoids: I. Natural lupane derivatives. Russian J Bioorg Chem 2006;32(1):37–49.

[456] Fulda S, Friesen C, Los M, Scaffidi C, Mier W, Benedict M, et al. Betulinic acid triggers CD95 (APO-1/Fas)-and p53-independent apoptosis via activation of caspases in neuroectodermal tumors. Cancer Res 1997;57(21):4956–64.

[457]   Fulda S, Scaffidi G, Susin SA, Krammer PH, Kroemer G, Peter ME, et al. Activation of mitochondria and release of mitochondrial apoptogenic factors by betulinic acid. J Biol Chem 1998;273(51):33942–33948.

[458]   Ganguly A, Das B, Roy A, Sen N, Dasgupta SB, Mukhopadhayay S, et al. Betulinic acid, a catalytic inhibitor of topoisomerase I, inhibits reactive oxygen species-mediated apoptotic topoisomerase I-DNA cleavable complex formation in prostate cancer cells but does not affect the process of cell death. Cancer Res 2007;67(24):11848–11858.

[459]   Chintharlapalli S, Papineni S, Ramaiah SK, Safe S. Betulinic acid inhibits prostate cancer growth through inhibition of specificity protein transcription factors. Cancer Res 2007;67(6):2816–23.

[460]   Chintharlapalli S, Papineni S, Lei P, Pathi S, Safe S. Betulinic acid inhibits colon cancer cell and tumor growth and induces proteasome-dependent and -independent downregulation of specificity proteins (Sp) transcription factors. BMC Cancer 2011;11:371.

[461]   Takada Y, Aggarwal BB. Betulinic acid suppresses carcinogen-induced NF-κB activation through inhibition of IκBα kinase and p65 phosphorylation: abrogation of cyclooxygenase-2 and matrix metalloprotease-9. J Immunol 2003;171(6):3278–86.

[462]   Kommera H, Kaluderović GN, Kalbitz J, Paschke R. Lupane triterpenoids-betulin and betulinic acid derivatives induce apoptosis in tumor cells. Invest New Drugs 2011;29(2):266–72.

[463]   Urban M, Sarek J, Klinot J, Korinkova G, Hajduch M. Synthesis of A-seco derivatives of betulinic acid with cytotoxic activity. J Nat Prod 2004;67(7):1100–5.

[464]   Abdel Bar FM, Khanfar MA, Elnagar AY, Liu H, Zaghloul AM, Badria FA, et al. Rational design and semisynthesis of betulinic acid analogues as potent topoisomerase inhibitors. J Nat Prod 2009;72(9):1643–50.

[465]   Ngassapa OD, Soejarto DD, Che CT, Pezzuto JM, Farnsworth NR. New cytotoxic lupane lactones from *Kokoona ochracea*. J Nat Prod 1991;54(5):1353–9.

[466]   Ngassapa O, Soejarto DD, Pezzuto JM, Farnsworth NR, Che CT. Further lupane lactones from *Kokoona ochracea*. J Nat Prod 1993;56(10):1676–81.

[467]   Sturm S, Gil RR, Chai HB, Ngassapa OD, Santisuk T, Reutrakul V, et al. Lupane derivatives from *Lophopetalum wallichii* with farnesyl protein transferase inhibitory activity. J Nat Prod 1996;59(7):658–63.

[468]   Núñez MJ, Reyes CP, Jiménez IA, Moujir L, Bazzocchi IL. Lupane triterpenoids from *Maytenus* species. J Nat Prod 2005;68(7):1018–21.

[469]   Shen YC, Prakash CVS, Wang LT, Chien CT, Hung MC. New vibsane diterpenes and lupane triterpenes from *Viburnum odoratissimum*. J Nat Prod 2002;65(7):1052–5.

[470]   Prakash Chaturvedula VS, Schilling JK, Johnson RK, Kingston DGI. New cytotoxic lupane triterpenoids from the twigs of *Coussarea paniculata*. J Nat Prod 2003;66(3):419–22.

[471]   Ku YL, Rao GV, Chen CH, Wu C, Guh JH, Lee SS. A novel secobetulinic acid 3,4-lactone from *Viburnum aboricolum*. Helvetica Chimica Acta 2003;86(3):697–702.

[472]   Hata K, Hori K, Ogasawara H, Takahashi S. Anti-leukemia activities of lup-28-al-20(29)-en-3-one, a lupane triterpene. Toxicol Lett 2003;143(1):1–7.

[473] Mutai C, Abatis D, Vagias C, Moreau D, Roussakis C, Roussis V. Cytotoxic lupane-type triterpenoids from *Acacia mellifera*. Phytochem 2004;65(8):1159–64.

[474] Mutai C, Abatis D, Vagias C, Moreau D, Roussakis C, Roussis V. Lupane triterpenoids from *Acacia mellifera* with cytotoxic activity. Molecules 2007;12(5):1035–44.

[475] Chen IH, Du YC, Lu MC, Lin AS, Hsieh PW, Wu CC, et al. Lupane-type triterpenoids from *Microtropis fokienensis* and *Perrottetia arisanensis* and the apoptotic effect of 28-hydroxy-3-oxo-lup-20(29)-en-30-al. J Nat Prod 2008;71(8):1352–7.

[476] Hata K, Ogawa S, Makino M, Mukaiyama T, Hori K, Iida T, et al. Lupane triterpenes with a carbonyl group at C-20 induce cancer cell apoptosis. J Nat Med 2008;62(3):332–5.

[477] El-Gamal AA. Cytotoxic lupane-, secolupane-, and oleanane-type triterpenes from *Viburnum awabuki*. Nat Prod Res 2008;22(3):191–7.

[478] Mori F, Miyase T, Ueno A. Oleanane-triterpene saponins from *Clinopodium chinense* Var. parviflorum. Phytochem 1994;36(6):1485–8.

[479] Miyase T, Matsushima Y. Saikosaponin homologues from *Clinopodium* spp. The structures of clinoposaponins XII–XX. Chem Pharm Bull 1997;45(9):1493–7.

[480] Murata T, Sasaki K, Sato K, Yoshizaki F, Yamada H, Mutoh H, et al. Matrix metalloproteinase-2 inhibitors from *Clinopodium chinense* var. parviflorum. J Nat Prod 2009;72(8):1379–84.

[481] Sun HX, Ye YP, Pan YJ. Cytotoxic oleanane triterpenoids from the rhizomes of *Astilbe chinensis* (Maxim.) Franch. et Savat. J Ethnopharmacol 2004;90(2–3):261–5.

[482] Tu J, Sun HX, Ye YP. 3β-hydroxyolean-12-en-27-oic acid: a cytotoxic, apoptosis-inducing natural drug against COLO-205 cancer cells. Chem Biodivers 2006;3(1):69–78.

[483] Van LTK, Tran MH, Phuong TT, Tran MN, Jin CK, Jang HS, et al. Oleanane-type triterpenoids from *Aceriphyllum rossii* and their cytotoxic activity. J Nat Prod 2009;72(8):1419–23.

[484] Sporn MB, Liby K, Yore MM, Suh N, Albini A, Honda T, et al. Platforms and networks in triterpenoid pharmacology. Drug Dev Res 2007;68(4):174–82.

[485] Ahmad R, Raina D, Meyer C, Kufe D. Triterpenoid CDDO-methyl ester inhibits the Janus-activated kinase-1 (JAK1)→signal transducer and activator of transcription-3 (STAT3) pathway by direct inhibition of JAK1 and STAT3. Cancer Res 2008;68(8):2920–6.

[486] Ikeda T, Nakata Y, Kimura F, Sato K, Anderson K, Motoyoshi K, et al. Induction of redox imbalance and apoptosis in multiple myeloma cells by the novel triterpenoid 2-cyano-3, 12-dioxoolean-1,9-dien-28-oic acid. Mol Cancer Ther 2004;3(1):39–45.

[487] Ito Y, Pandey P, Sporn MB, Datta R, Kharbanda S, Kufe D. The novel triterpenoid CDDO induces apoptosis and differentiation of human osteosarcoma cells by a caspase-8 dependent mechanism. Mol Pharmacol 2001;59(5):1094–9.

[488] Konopleva M, Tsao T, Ruvolo P, Stiouf I, Estrov Z, Leysath CE, et al. Novel triterpenoid CDDO-Me is a potent inducer of apoptosis and differentiation in acute myelogenous leukemia. Blood 2002;99(1):326–35.

[489] Ikeda T, Sporn M, Honda T, Gribble GW, Kufe D. The novel triterpenoid CDDO and its derivatives induce apoptosis by disruption of intracellular redox balance. Cancer Res 2003;63(17):5551–8.

[490]   Shishodia S, Sethi G, Konopleva M, Andreeff M, Aggarwal BB. A synthetic triterpenoid, CDDO-Me, inhibits IκBα kinase and enhances apoptosis induced by TNF and chemotherapeutic agents through down-regulation of expression of nuclear factor κB-regulated gene products in human leukemic cells. Clinical Cancer Res 2006;12(6):1828–38.

[491]   Brookes PS, Morse K, Ray D, Tompkins A, Young SM, Hilchey S, et al. The triterpenoid 2–cyano-3,12-dioxooleana-1,9-dien-28-oic acid and its derivatives elicit human lymphoid cell apoptosis through a novel pathway involving the unregulated mitochondrial permeability transition pore. Cancer Res 2007;67(4):1793–802.

[492]   Verses AE, Kowaltowski AJ, Grijalba MT, Meinicke AR, Castilho RF. The role of reactive oxygen species in mitochondrial permeability transition. Biosci Rep 1997;17:43–52.

[493]   Deeb D, Gao X, Jiang H, Janic B, Arbab AS, Rojanasakul Y, et al. Oleanane triterpenoid CDDO-Me inhibits growth and induces apoptosis in prostate cancer cells through a ROS-dependent mechanism. Biochem Pharmacol 2010;79(3):350–60.

[494]   Reyes-Zurita FJ, Rufino-Palomares EE, Lupiáñez JA, Cascante M. Maslinic acid, a natural triterpene from *Olea europaea* L., induces apoptosis in HT29 human colon-cancer cells via the mitochondrial apoptotic pathway. Cancer Lett 2009;273(1):44–54.

[495]   Reyes-Zurita FJ, Pachón-Peña G, Lizárraga D, Rufino-Palomares EE, Cascante M, Lupiáñez JA. The natural triterpene maslinic acid induces apoptosis in HT29 colon cancer cells by a JNK-p53-dependent mechanism. BMC Cancer 2011;11:154.

[496]   Deng JZ, Starck SR, Hecht SM. DNA polymerase β inhibitors from *Baeckea gunniana*. J Nat Prod 1999;62(12):1624–6.

[497]   Prakash Chaturvedula VS, Gao Z, Hecht SM, Jones SH, Kingston DGI. A new acylated oleanane triterpenoid from *Couepia polyandra* that inhibits the lyase activity of DNA polymerase β. J Nat Prod 2003;66(11):1463–5.

[498]   Es-saady D, Simon A, Ollier M, Maurizis JC, Chulia AJ, Delage C. Inhibitory effect of ursolic acid on B16 proliferation through cell cycle arrest. Cancer Lett 1996;106(2):193–7.

[499]   Finlay HJ, Honda T, Gribble GW, Danielpour D, Benoit NE, Suh N, et al. Novel A-ring cleaved analogs of oleanolic and ursolic acids which affect growth regulation in NRP.152 prostate cells. Bioorg Med Chem Lett 1997;7(13):1769–72.

[500]   Fu L, Zhang S, Li N, Wang J, Zhao M, Sakai J, et al. Three new triterpenes from *Nerium oleander* and biological activity of the isolated compounds. J Nat Prod 2005;68(2):198–206.

[501]   Chen IH, Chang FR, Wu CC, Chen SL, Hsieh PW, Yen HF, et al. Cytotoxic triterpenoids from the leaves of *Microtropis fokienensis*. J Nat Prod 2006;69(11):1543–6.

[502]   Sánchez M, Mazzuca M, Veloso MJ, Fernández LR, Siless G, Puricelli L, et al. Cytotoxic terpenoids from *Nardophyllum bryoides*. Phytochem 2010;71(11–12):1395–9.

[503]   Yang H, Cho HJ, Sim SH, Chung YK, Kim DD, Sung SH, et al. Cytotoxic terpenoids from *Juglans sinensis* leaves and twigs. Bioorg Med Chem Lett. 2012;22(5):2079–83.

[504] Moiteiro C, Manta C, Justino F, Tavares R, Curto MJM, Pedro M, et al. Hemisynthetic secofriedelane triterpenes with inhibitory activity against the growth of human tumor cell lines in vitro. J Nat Prod 2004;67(7):1193–6.

[505] Olmedo DA, López-Pérez JL, del Olmo E, Vásquez Y, San Feliciano A, Gupta MP. A new cytotoxic friedelane acid – pluricostatic acid – and other compounds from the leaves of *Marila pluricostata*. Molecules 2008;13(11):2915–24.

[506] Li YZ, Li ZL, Yin SL, Shi G, Liu MS, Jing YK, et al. Triterpenoids from *Calophyllum inophyllum* and their growth inhibitory effects on human leukemia HL-60 cells. Fitoterapia 2010;81(6):586–9.

[507] Sneden AT. Isoiguesterin, a new antileukemic bisnortriterpene from *Salacia madagascariensis*. J Nat Prod 1981;44(4):503–7.

[508] Figueiredo JN, Räz B, Séquin U. Novel quinone methides from *Salacia kraussii* with in vitro antimalarial activity. J Nat Prod 1998;61(6):718–23.

[509] Ngassapa O, Soejarto DD, Pezzuto JM, Farnsworth NR. Quinone-methide triterpenes and salaspermic acid from *Kokoona ochracea*. J Nat Prod 1994;57(1):1–8.

[510] Shirota O, Morita H, Takeya K, Itokawa H, Iitaka Y. Cytotoxic aromatic triterpenes from *Maytenus ilicifolia* and *Maytenus chuchuhuasca*. J Nat Prod 1994;57(12):1675–81.

[511] Morita H, Hirasawa Y, Muto A, Yoshida T, Sekita S, Shirota O. Antimitotic quinoid triterpenes from *Maytenus chuchuhuasca*. Bioorg Med Chem Lett 2008;18(3):1050–2.

[512] Chávez H, Valdivia E, Estévez-Braun A, Ravelo AG. Structure of new bioactive triterpenes related to 22-β-hydroxyl-tingenone. Tetrahedron 1998;54(44):13579–13590.

[513] Chávez H, Estévez-Braun A, Ravelo AG, González AG. New phenolic and quinine-methide triterpenes from *Maytenus amazonica*. J Nat Prod 1999;62(3):434–6.

[514] González AG, Alvarenga NL, Estévez-Braun A, Ravelo AG, Estévez-Reyes R. Oxidation of natural targets by dimethyl dioxirane: regio and stereospecific reactions on enol double bond of bioactive nor quinone methide triterpenes. Tetrahedron 1996;52(32):10667–10672.

[515] Setzer WN. A theoretical investigation of cytotoxic activity of celastroid triterpenoids. J Mol Modeling 2009;15(2):197–201.

[516] Yang H, Landis-Piwowar KR, Lu D, Yuan P, Li L, Prem-Veer Reddy G, et al. Pristimerin induces apoptosis by targeting the proteasome in prostate cancer cells. J Cell Biochem 2008;103(1):234–44.

[517] Bolton JL, Turnipseed SB, Thompson JA. Influence of quinone methide reactivity on the alkylation of thiol and amino groups in proteins: studies utilizing amino acid and peptide models. Chem Biol Interact 1997;107:185–200.

[518] Setzer WN, Holland MT, Bozeman CA, Rozmus GF, Setzer MC, Moriarity DM, et al. Isolation and frontier molecular orbital investigation of bioactive quinine-methide triterpenoids from the bark of *Salacia petenensis*. Planta Med 2001;67(1):65–9.

[519] Rodríguez FM, López MR, Jiménez IA, Moujir L, Ravelo AG, Bazzocchi IL. New phenolic triterpenes from *Maytenus blepharodes*. Semisynthesis of 6-deoxoblepharodol from pristimerin. Tetrahedron 2005;61(9):2513–9.

[520] González AG, Alvarenga NL, Ravelo AG, Bazzocchi IL, Ferro EA, Navarro AG, et al. Scutione, a new bioactive norquinonemethide triterpene from *Maytenus scutioides* (Celastraceae). Bioorg Med Chem 1996;4(6):815–20.

[521] Sethi G, Kwang SA, Pandey MK, Aggarwal BB. Celastrol, a novel triterpene, potentiates TNF-induced apoptosis and suppresses invasion of tumor cells by inhibiting NF-κB-regulated gene products and TAK1-mediated NF-κB activation. Blood 2007;109(7):2727–35.

[522] Chen G, Cao P, Goeddel DV. TNF-induced recruitment and activation of the IKK complex require Cdc37 and Hsp90. Mol Cell 2002;9(2):401–10.

[523] Zhang T, Hamza A, Cao X, Wang B, Yu S, Zhan CG, et al. A novel Hsp90 inhibitor to disrupt Hsp90/Cdc37 complex against pancreatic cancer cells. Mol Cancer Ther 2008;7(1):162–70.

[524] Zhou LL, Lin ZX, Fung KP, Cheng CHK, Che CT, Zhao M, et al. Celastrol-induced apoptosis in human HaCaT keratinocytes involves the inhibition of NF-κB activity. Eur J Pharmacol 2011;670(2–3):399–408.

[525] Lu Z, Jin Y, Chen C, Li J, Cao Q, Pan J. Pristimerin induces apoptosis in imatinib-resistant chronic myelogenous leukemia cells harboring T315I mutation by blocking NF-κB signaling and depleting Bcr-Abl. Mol Cancer 2010;9:112.

[526] Wu CC, Chan ML, Chen WY, Tsai CY, Chang FR, Wu YC. Pristimerin induces caspase-dependent apoptosis in MDA-MB-231 cells via direct effects on mitochondria. Mol Cancer Ther 2005;4(8):1277–85.

[527] Hoernlein RF, Orlikowsky Th, Zehrer C, Niethammer D, Sailer ER, Simmet Th, et al. Acetyl-11-keto-β-boswellic acid induces apoptosis in HL-60 and CCRF-CEM cells and inhibits topoisomerase I. J Pharmacol Exp Ther 1999;288(2):613–9.

[528] Altmann A, Fischer L, Schubert-Zsilavecz M, Steinhilber D, Werz O. Boswellic acids activate p42(MAPK) and p38 MAPK and stimulate Ca(2+) mobilization. Biochem Biophys Res Commun 2002;290(1):185–90.

[529] Lu M, Xia L, Hua H, Jing Y. Acetyl-keto-beta-boswellic acid induces apoptosis through a death receptor 5-mediated pathway in prostate cancer cells. Cancer Res 2008;68(4):1180–6.

[530] Kunnumakkara AB, Nair AS, Sung B, Pandey MK, Aggarwal BB. Boswellic acid blocks signal transducers and activators of transcription 3 signaling, proliferation, and survival of multiple myeloma via the protein tyrosine phosphatase SHP-1. Mol Cancer Res 2009;7(1):118–28.

Chapter 3

Phenolics

■ INTRODUCTION

Reactive oxygen species (ROS) are produced by the mitochondrial metabolism of oxygen and their deleterious effects are quickly nullified by glutathione. However, some phenolics have a tendency to trigger the generation of massive amounts of ROS, resulting in apoptosis and/or cell cycle arrest. In 'wild-type' p53 cancer cells, ROS cause severe DNA damage, which in turn stimulates the pro-apoptotic protein (p53) and hence apoptosis and the arrest of mitosis. Once the p53 protein has been activated by ROS, pro-apoptotic Bcl-2-associated X protein (Bax) and p21 are activated. The translocation of pro-apoptotic Bax provokes irremediable mitochondrial insult, leakage of cytochrome c, activation of caspases 9 and 3, and apoptosis; blockage of cell division results from p21 activation. This type of cytotoxic mechanism was observed with: acacetin from members of the genus *Populus* L. (3.1.1), shikonin from *Lithospermum erythrorhizon* Siebold & Zucc. (3.2.1) and both rhein and emodin from *Rheum officinale* Baill. (3.2.2). ROS compel the phosphorylation of the following mitogen-activated protein (MAP) kinases: p38 mitogen-activated protein kinase (p38 MAPK), c-Jun N-terminal kinase (JNK) and extracellular signal-regulated kinase (ERK1/2). Flavokawain B from *Alpinia pricei* Hayata (3.1.1) boosts the cytoplasmic levels of ROS and hence activation of p38 MAPK, translocation of pro-apoptotic proteins Bax and Bim, a decrease in Bcl-2, mitochondrial insult, release of cytochrome c, activation of caspases 9 and 3, and apoptosis. Activation of p38 MAPK by ROS also results in the inhibition of protein kinase B (Akt), the activation of the pro-apoptotic p53 and caspase 8, and hence caspase 3 activation and apoptosis. Luteolin from members of the genus *Capsicum* L. (3.1.1), tricetin from *Eucalyptus globulus* Labill. (3.1.1), protoapigenone from *Thelypteris torresiana* (Gaudich.) Alston (3.1.1) and shikonin from *Lithospermum erythrorhizon* Siebold & Zucc. (3.2.1) precipitate apoptosis via a ROS-induced activation of JNK followed by the cleavage of Bim, translocation of pro-apoptotic Bax and therefore mitochondrial insult. In parallel, activation of JNK is followed by an increase in death receptor (DR5) which then activates caspase 8.

Christophe Wiart: Lead Compounds from Medicinal Plants for the Treatment of Cancer
DOI: http://dx.doi.org/10.1016/B978-0-12-398371-8.00003-9

An interesting feature of phenolics is their tendency to activate or inhibit ERK1/2, which is either anti-apoptotic or pro-apoptotic depending on the type of malignancies concerned. The activation of ERK1/2 favors the survival of cancer cells and its inhibition results in apoptosis, as observed with quercetin from *Allium sativum* L. (3.1.1) and α-mangostin from *Garcinia mangostana* L. (3.1.3). However, some malignant cells become apoptotic upon ERK1/2 activation, as is found with fisetin from members of the genus *Fragaria* L. (3.1.1) and β-hydroxyisovalerylshikonin from a member of the genus *Lithospermum* L. (3.2.1).

In 'wild-type' p53 cancer cells, phenolics such as tricetin from *Eucalyptus globulus* Labill. (3.1.1), oroxylin A from members of the genus *Scutellaria* L. (3.1.1), eupatilin from *Artemisia asiatica* Nakai ex Pamp. (3.1.1), arcommunol A from *Artocarpus communis* J.R. Forst. & G. Forst. (3.1.1), sappanchalcone from *Caesalpinia sappan* L. (3.1.1), β-lapachone from *Tabebuia avellanedae* Lorentz ex Griseb. (3.2.1), aloe-emodin from *Aloe vera* (L.) Burm.f. (3.2.2) and isochaihulactone from *Bupleurum scorzonerifolium* Willd. (3.3.1) stimulate pro-apoptotic p53, inducing apoptosis via the 'intrinsic pathway'. This involves the activation of p53, which stimulates Bax-induced mitochondrial insult and therefore leakage of cytochrome c and activation of caspase 3. The pro-apoptotic p53 is controlled by Akt, which has an immense influence on cell survival. Interestingly, some phenolics, such as wogonin from *Scutellaria baicalensis* Georgi (3.1.1), fisetin from members of the genus *Fragaria* L. (3.1.1), apigenin from *Petroselinum crispum* (Mill.) Nyman ex A.W. Hill. (3.1.1), quercetin from *Allium sativum* L. (3.1.1), myricetin from *Brassica oleracea* L. (3.1.1) and gambogenic acid from *Garcinia hanburyi* Hook.f. (3.1.3), inhibit Akt. Other cellular targets through which phenolics induce apoptosis include nuclear factor kappa-light-chain-enhancer of activated B cells (NF-κB), cyclins, topoisomerase II, and Ca^{2+} homeostasis and microtubules. Note that fisetin from members of the genus *Fragaria* L. (3.1.1), xanthohumol from *Humulus sativus* L. (3.1.1), scoparone from *Artemisia capillaris* Thunb. (3.1.2), plumbagin from *Plumbago zeylanica* L. (3.2.1), manassantin A from *Saururus chinensis* (Lour.) Baill. (3.3.1), gomisin N from a member of the genus *Schisandra* Michx (3.3.1) and honokiol from *Houpoea officinalis* (Rehder & E.H. Wilson) N.H. Xia & C.Y. Wu (3.3.1) abate the transcriptional activity of NF-κB. In addition, phenolics such as 1-hydroxy-3,7,8-trimethoxyxanthone from *Gentianopsis paludosa* (Munro ex Hook. f.) Ma (3.1.3), psorospermin from *Psorospermum febrifugum* Spach (3.1.3), gambogic acid from *Garcinia hanburyi* Hook.f. (3.1.3), plumbagin from *Plumbago zeylanica* L. (3.2.1), shikonin from *Lithospermum erythrorhizon* Siebold & Zucc. (3.2.1), β-lapachone from

Tabebuia avellanedae Lorentz ex Griseb. (3.2.1), emodin from *Rheum officinale* Baill. (3.2.2), and alizarin 2-methyl ether from *Rubia cordifolia* L. (3.2.2) inhibit the enzymatic activity of topoisomerase II, leading to DNA damage. The ROS generated by some phenolics induce endoplasmic reticulum (ER) stress and therefore leakage and accumulation of free cytoplasmic Ca^{2+}, which induces mitochondrial insult and activation of the Ca^{2+}-dependent protease calpain, CCAAT/enhancer binding protein homologous protein (CHOP), DR5 and caspase 8, as exemplified with wogonin from *Scutellaria baicalensis* Georgi (3.1.1), genistein from *Glycine max* (L.) Merr. (3.1.1), morin from *Chlorophora tinctoria* (L.) Gaudich. ex Benth. (3.1.1), α-mangostin from *Garcinia mangostana* L. (3.1.3), β-lapachone from *Tabebuia avellanedae* Lorentz ex Griseb. (3.2.1), aloe-emodin from *Aloe vera* (L.) Burm.f. (3.2.2) and rhein from *Rheum officinale* Baill. (3.2.2). Of oncological interest is the fact that phenolics like casticin from *Achillea millefolium* L (3.1.1), ferulenol from *Ferula communis* L. (3.1.2), and isochaihulactone from *Bupleurum scorzonerifolium* Willd. (3.3.1) interfere with tubulin homeostasis and therefore abrogate the division of cancer cells. In this chapter morusin, hedyotiscone C, gaudichaudione H, 5-hydroxy-2-methoxy-1,4-naphthoquinone, damnacanthal and justicidin A and their derivatives are identified as lead phenolics in the treatment of cancer.

Topic # 3.1

Benzopyrones

3.1.1 *Morus Australis* **Poir.**

History This plant was first described by Jean Louis Marie Poiret, in *Encyclopédie Méthodique, Botanique*, published in 1796.

Synonyms *Morus acidosa* Griff., *Morus alba* var. *indica* Bureau, *Morus alba* var. *nigriformis* Bureau, *Morus alba* var. *stylosa* Bureau, *Morus australis* var. *hastifolia* (F.T. Wang & T. Tang ex Z.Y. Cao) Z.Y. Cao, *Morus australis* var. *incisa* C.Y. Wu, *Morus australis* var. *inusitata* (H. Lév.)

C.Y. Wu, *Morus australis* var. *linearipartita* Z.Y. Cao, *Morus australis* var. *oblongifolia* Z.Y. Cao, *Morus bombycis* Koidz., *Morus bombycis* var. *angustifolia* Koidz., *Morus bombycis* var. *bifida* Koidz., *Morus bombycis* var. *longistyla* Koidz., *Morus bombycis* var. *tiliifolia* Koidz., *Morus cavaleriei* H. Lév., *Morus formosensis* Hotta, *Morus hastifolia* F.T. Wang & T. Tang ex Z.Y. Cao, *Morus inusitata* H. Lév., *Morus longistylus* Diels, *Morus nigriformis* (Bureau) Koidz., *Morus stylosa* var. *ovalifolia* Ser.

Family Moraceae Gaudich, 1835

Common Name Ji sang (Chinese)

Habitat and Description *Morus australis* is a shrub that grows in the forests of China, Korea, Japan, Taiwan, Burma and India. The bark is fibrous and used to make paper. The stems are articulate, tereste and exude a pure white latex. The leaves are simple, alternate and stipulate. The stipules are linear. The petiole is 1–1.5 cm long and covered with a few hairs. The blade is cordate at base, acuminate at apex, serrate, hairy beneath, 5–15 cm × 1–10 cm and presents three to five pairs of secondary nerves. The male inflorescence is an axillary catkin which is 1–1.5 cm long. The female inflorescence is globose, 1 cm across and hairy. The male flowers are minute and include four imbricate, ovate and green sepals around four stamens with yellow anthers. The females' flowers are minute and comprise four imbricate, dark green, oblong sepals which enclose a pyriform ovary which develops a bifid stigma. The fruit is a red syncarp which is 1 cm in diameter and edible (Figure 3.1).

Medicinal Uses In China, Korea and Japan the plant is used to treat sexual impotence, coughs and diabetes, and to promote the discharge of urine.

Phytopharmacology The plant produces: a number of triterpenes, including 3β-[(*m*-methoxybenzoyl)oxy]urs-12-en-28-oic acid,[1] β-amyrin, ursolic acid and betulinic acid;[1] the benzofuran derivatives mulberrofuran D,[2] sanggenofuran A,[3] mongolisin C and mulberrofurans E,G,F and Q;[4] moracins C and M;[5] a series of flavonoids including quercetin,[1] australones A[1] and B,[6] morusin,[1,3,5] australisines A–C,[3,4] cyclomorusin,[3] kuwanons G and H,[3,4] sanggenols N and O,[3] chalcomoracin,[4] 5,7,2',4'-tetrahydroxy-3-methoxyflavone, kuwanone C,[5] and morachalcone A;[5] the stilbenes oxyresveratrol and 4'-(2-methyl-2-buten-4-yl)oxyresveratrol;[5] and austrafurans A–C.[7]

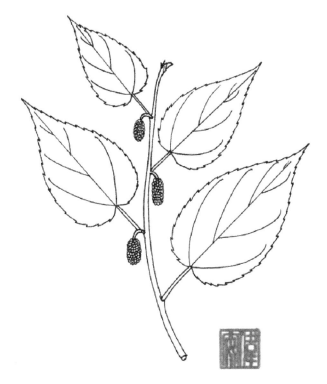

■ **FIGURE 3.1** *Morus australis* Poir.

The medicinal properties have not yet been substantiated.

Proposed Research Pharmacological study of morusin and derivatives for the treatment of colon cancer.

Rationale Flavonoids impose the death of cancer cells by activating p53. However, the development of these γ-benzopyrones as chemotherapeutic and/or chemoprotective agents is quite questionable. As a matter of fact, flavonoids are cytotoxic and apoptotic at high doses in vitro but are, in vivo, quickly degraded by an impressive arsenal of entero-hepatic enzymes acquired throughout the evolutionary process as well as colonic bacterial flora. These limitations pose a striking challenge: to enhance the cytotoxic activity of flavonoids via hemisynthetic modifications and to ensure that these flavonoids enter cancer cells with intact structures by the development of pro-drugs or 'pro-flavonoids' that would undoubtly lead to the discovery of potent chemotherapeutic agents.

■ **CS 3.1** Chrysin

■ **CS 3.2** Pinocembrin

■ **CS 3.3** Galangin

■ **CS 3.4** Baicalein

The flavonoid chrysin (CS 3.1), which occurs in members of the genus *Passiflora* L. (family Passifloraceae Juss. ex Roussell), at a dose of 10 μM, destroyed 60% of human histiocytic lymphoma (U937) cells via inhibition of protein kinase B (Akt) and hence downregulation of nuclear factor kappa-light-chain-enhancer of activated B cells (NF-κB) and X-linked inhibitor of apoptosis protein (XIAP), probable translocation of pro-apoptotic Bcl-2-associated X protein (Bax) and mitochondrial insult, followed by the release of cytochrome c, caspase 3 activation and cleavage of the pro-oncogenic phospholipase C gamma 1 (PLC-γ1).[8] Chrysin at a dose of 40 μM boosted tumor necrosis factor α (TNFα)-mediated apoptosis in human colorectal carcinoma (HCT-116), human hepatocellular liver carcinoma (HepG2) and human nasopharyngeal carcinoma (NCE) cells to 40%, 25% and 30%, respectively, with caspases 8 and 3 activation, inhibition of both IκBα degradation and subsequent p65 nuclear translocation and downregulation of anti-apoptotic FLICE inhibitory protein (c-FLIP),[9] the latter probably being caused by protein kinase B (Akt) inactivation. The reduction of chrysin at Δ_{2-3} produced pinocembrin (CS 3.2), which occurs naturally in *Alpinia galanga* (L.) Willd. (family Zingiberaceae Martinov). This repressed the growth of a range of cell lines by varying amounts (values in parentheses): human normal fibroblast (LF-1; 30%), human umbilical vein endothelial (HUVEC; 50%), human cervical carcinoma (SiHa; 40%), human epitheloid cervix carcinoma (Hela; 5%), human breast adenocarcinoma (MCF-7; 40%), human colorectal carcinoma (HCT-116; 10%) and human colon adenocarcinoma (SW480; 20%) at a dose of 200 μM.[10] In HCT-116, pinocembrin induced apoptosis via translocation of pro-apoptotic Bax to the mitochondria, mitochondrial insult, loss of mitochondrial membrane potential MMP, release of cytochrome c and activation of caspases 9 and 3,[10] suggesting that the unsaturation between C2 and C3 favors the apoptotic activity of flavonoids. Hydroxylation of chrysin at C3 produces galangin (CS 3.3), which is found in *Alpinia officinarum* Hance (family Zingiberaceae Martinov). Galangin at a dose of 100 μM nullified the growth of human promyelocytic leukemia (HL-60) cells (p53 null), with caspase 3 activation and DNA fragmentation.[11] The apoptotic activity of galangin might involve mitochondrial insult since a fall in mitochondrial membrane potential would result in an increase inBax, the release of cytochrome c and apoptosis inducing factor (AIF), followed by activation of caspases 3 and 9, as observed in HepG2, human hepatocellular carcinoma (Hep3B) and human hepatoma (PLC/PRF/5) cells, where IC$_{50}$ values equal to 134 μmol/L, 87.3 μmol/L and 79.8 μmol/L, respectively, were measured.[12] Hydroxylation of chrysin at C6 forms baicalein (CS 3.4), which occurs in members of the genus *Scutellaria* L. (family Lamiaceae Martinov). Baicalein destroyed human lung carcinoma

■ **CS 3.5** Acacetin

■ **CS 3.6** Wogonin

(CH27) cells at a dose of 80 µM, with a decrease in cyclins B and D and cdk4, a decrease in Bcl-2 expression and activation of caspase 3.[13] In fact, 50 µM of baicalein completely prevented the phosphorylation of Akt in human multiple myeloma (U266) cells,[14] thereby effecting apoptosis.[15] Introduction of a methoxy group at C4' into the chrysin framework creates the flavonoid acacetin, which occurs in members of the genus *Populus* L. (family Salicaceae Mirb.). Acacetin (CS 3.5) abated the growth of human gastric carcinoma (AGS) cells (wild-type p53), with an IC_{50} value of 40 µM. This was brought about by the generation of reactive oxygen species (ROS), activation of pro-apoptotic protein (p53), and therefore entry of pro-apoptotic Bax into the mitochondria, a resulting fall in mitochondrial membrane potential, release of cytochrome c and activation of caspase.[16]

The addition of a methoxy group at C8 of chrysin produces wogonin (CS 3.6), which is found in *Scutellaria baicalensis* Georgi (family Lamiaceae Marinov). Wogonin abated the growth of HL-60 cells by 90%, with a decrease in anti-apoptotic induced myeloid leukemia cell differentiation protein (Mcl-1), translocation of Bax to the mitochondria, and mitochondrial insult followed by activation of caspase 3 and poly (ADP-ribose) polymerase (PARP) cleavage.[17] A similar mechanism was observed in human hepatocellular carcinoma (BEL-7402) cells, where wogonin induced a boost of free cytoplasmic Ca^{2+}, a fall in mitochondrial membrane potential, the release of cytochrome c, activation of caspase 9 and therefore apoptosis, with an IC_{50} value of 80 µM.[18] Wogonin nullified the growth of human ductal breast epithelial carcinoma (T47D), human breast adenocarcinoma (SKBR-3) and human breast adenocarcinoma (MDA-MB-231) cells at a dose of 200 µM via inhibition of Akt and, hence, p27 and glycogen synthase kinase-3 β (GSK-3β) activation, downregulation of cyclin D1, pro-apoptotic Bax increase, caspase 3 activation and PARP cleavage.[19] One can reasonably make the inference that activated

■ **CS 3.7** Nor-wogonin

■ **CS 3.8** Fisetin

GSK-3β phosphorylates the anti-apoptotic-induced Mcl-1 and therefore compels its ubiquitylation and degradation by proteasomes. Wogonin and nor-wogonin (CS 3.7) destroyed HL-60 cells, with IC_{50} values equal to 67.5 μM and 21.7 μM, via apoptosis involving an increase in Bax, caspase 3 activation, PARP cleavage, a fall in mitochondrial membrane potential and the release of cytochrome c.[20] The inactivation of Akt by wogonin was further observed in MCF-7 cells, together with an increase in Bax, activation of caspases 3 and 9 and apoptosis, with an IC_{50} value equal to 71.3 μM.[21]

The removal of C5 and C8 hydroxyls and the addition of C3, C4′ and C5′ hydroxyls creates fisetin (CS 3.8), which is plentiful in members of the genus *Fragaria* L. (family Rosaceae Juss.). In human non-small-cell lung carcinoma (H1299) cells, 50 μM of fisetin imposed apoptosis via inhibition of IκB kinase (IKK) and hence blockage of IκBα phosphorylation and degradation and the resultant suppression of p65 translocation into the nucleus.[22] Fisetin inhibited the proliferation of human prostate adenocarcinoma (LNCaP), human prostate cancer (CWR22rv1) and human prostate cancer (PC-3) cells by 62%, 55% and 46%, respectively.[23] In LNCaP cells, fisetin inhibited the enzymatic activity of Akt, resulting in p21 and p27 activation, inhibition of XIAP, an increase in Bax, a decrease in Bcl-2 and Bcl-xL, mitochondrial insult, the release of cytochrome c, activation of caspases 9, 3 and 8, and apoptosis.[23] Further studies revealed that the growth of human pancreas adenocarcinoma (AsPC-1) cells was inhibited by fisetin, with an IC_{50} value of of 38 μM, via inhibition of IκB kinase (IKK), IκBα dephosphorylation and subsequent inactivation of NF-κB transcriptional activity.[24] The inhibition of IKK may very well have resulted from the inhibition of Akt. Fisetin at a dose of 120 μM reduced the growth of PC-3 and human prostate carcinoma (DU-145) cells to 65% and 45%, respectively. In PC-3 cells fisetin induced autophagy via inhibition of mammalian target of rapamycin (mTOR), which was due to inactivated Akt.[25] Fisetin induced apoptosis in human colon cancer (HT-29) cells, with an IC_{50} value equal to 43.4 µM, via apoptosis brought about by inhibition of Akt and hence decreased levels of Bcl-2 and Bcl-xL, increased Bak, GSK-3β activation, phosphorylation and degradation of β-catenin and therefore decreased transcription factor 4 (TCF4)-β-catenin transcriptional activity.[26] In addition, fisetin abated the viability of Hela cells ('wild-type' p53) by sustained activation of extracellular signal-regulated kinase (ERK1/2) and hence Fas-associated death domain-(FADD)-independent activation of caspase 8 and cell death, with an IC_{50} value of 52 µM.[27] Note that inhibiting Akt probably enhances Raf, ERK1/2 and mitogen-activated protein kinase (MEK), thereby enabling caspase 8 activation.

Hydroxylation of chrysin in C4′ forms apigenin, which occurs principally in *Petroselinum crispum* (Mill.) Nyman ex A.W. Hill. (family Apiaceae Lindl.). In HL-60 cells, apigenin abated the mitochondrial membrane potential, resulting in the generation of mitochondrial ROS, release of cytochrome c, activation of caspases 9 and 3, and apoptosis.[28] Apigenin induced apoptosis in human neuroblastoma (NUB-7) cells by activation of p53 Bax-induced mitochondrial insult, caspase 3 activation and PARP cleavage.[29] Inhibition of Akt inhibits the phosphorylation of Bcl-2-associated death promoter (Bad), which binds to Bcl-xL allowing pro-apoptotic Bax and Bak to aggregate and permeate the mitochondrial membrane.[30] In addition, apigenin repressed the proliferation of Hela cells (wild-type p53) at a dose of 74 µM by stimulating p53, and therefore the Fas receptor (CD95), activating caspase 8 and hence causing apoptosis, an increase in p21 and therefore cell arrest[31] brought about by decreases in CDK1 and cyclin B. Human breast carcinoma (MDA-MB-435) cells exposed to 40 µM of apigenin underwent apoptosis with release of cytochrome c, caspase 3 activation, DNA cleavage and apoptosis plus a decrease in p27 and therefore inhibition of the cyclin D1/CDK4 complex and cell arrest.[32] Apigenin suppressed the proliferation of HepG2, Hep3B and PLC/PRF/5 cells, with IC_{50} values equal to 8 µg/mL, 22.1 µg/mL and 22.7 µg/mL, respectively, implying that 'wild-type' p53 cells are more susceptible.[33] Apigenin (CS 3.9) was cytotoxic against a number of cell lines (IC_{50} values in parentheses): U937 (39.7 µM), HL-60 (29.2 µM),

■ **CS 3.9** Apigenin

■ **CS 3.10** Naringenin

■ **CS 3.11** KUF-7

■ **CS 3.12** Kaempferol

human acute monocytic leukemia (THP-1; 27.8 μM), Jurkat human T cell lymphoblast-like (29.6 μM) and human erythromyeloblastoid leukemia (K562; 54.4 μM).[34] In THP-1 cells, apigenin boosted the cytoplasmic levels of ROS, thereby activating p38 mitogen-activated protein kinase (p38 MAPK), ERK1/2 and the pro-apoptotic protein kinase C δ (PKCδ).[34] The pro-oxidant activity of apigenin was indeed confirmed in HepG2 cells, where there was activation of NADPH oxidase and apoptosis, with an IC_{50} value equal to 10 μM.[35] This was probably aided by PKC inhibition, since apigenin hampers the phosphorylation of Akt at serine 473 and threonine 308.[30] The saturation of apigenin at Δ_{2-3} provides naringenin (CS 3.10);[36] KUF-7 (CS 3.11) increased the levels of ROS, resulting in mitochondrial insult and hence caspase 3 activation and PARP cleavage.[37] Kaempferol (CS 3.12) is formed by the hydroxylation of apigenin at C3 and is found in *Camellia chinensis* (L.) Kuntze (family Theaceae Mirb.). At a dose of 50 μM it reduced the viability of human glioblastoma (LN229), human glioblastoma-astrocytoma, epithelial-like (U-MG87) and human glioblastoma (T98G 7) cells by 50%[38] and at a dose of 100 μM abated the growth of mouse mammary epithelial (HC11), MDA-MB-231, T47D and

■ **CS 3.13** Luteolin

MCF-7 cells by 10%, 20%, 60% and 80%, respectively.[39] In U-MG87 cells, kaempferol induced caspase 3 activation and PARP cleavage, a decrease in Bcl-2, inhibition of superoxide dismutase-1 (SOD-1) and thioreduxin-1 (TRX-1) and, hence, an increase in ROS, a fall in mitochondrial membrane potential, and mitochondrial insult leading to apoptosis.[38] The pro-oxidant activity of kaempferol was further exemplified in MCF-7 cells where ROS induced the sustained activation of ERK1/2 and therefore apoptosis.[39] At a dose of 160 μM, kaempferol reduced the percentage of viable human ovarian carcinoma (OVCAR-3), human ovarian carcinoma (A2780/CP70) and human ovarian carcinoma (A2780) cells to 30%, 40% and 50%, respectively, implying that the presence of wild-type p53 enhances the apoptotic potencies.[40] In A2780 cells, wild-type p53 inhibition of Akt, and hence activation of the pro-apoptotic proteins (p53) and Bax, inhibition of Bcl-xL, mitochondrial insult, release of cytochrome c, activation of caspases 9 and 3 and apoptosis, may account for the apoptosis imposed by kaempferol.[40]

Hydroxylation of apigenin at C5′ forms luteolin (CS 3.13), which is found in members of the genus *Capsicum* L. (family Solanaceae Juss.). TNFα-induced NF-κB activation in human colon adenocarcinoma (CoLo 205) cells was abated by 40 μM of luteolin, which compromised the interaction of p65 and cAMP response element-binding protein (CBP), resulting in the augmentation and prolongation of c-Jun N-terminal kinase (JNK) activation and hence caspase 8 activation and apoptosis.[41] The activation of JNK may account for the upregulation of death receptor (DR5) and therefore for the activation of caspase 8, Bid cleavage, mitochondrial insult and the activation of caspases 9 and 3, which was observed in Hela cells exposed to 20 μM of luteolin.[42] In addition, 10 μmol/L of luteolin inhibited proliferation of human hepatocellular carcinoma (HLF) cells by 40% with a decrease in phosphorylated STAT3.[43] Luteolin destroyed mouse neuroblastoma (Neuro2a) cells at a dose of 10 μM via the generation of

■ **CS 3.14** Genistein

■ **CS 3.15** Quercetin

ROS, phosphorylation of JNK, p38 MAPK and ERK1/2, and hence p53 phosphorylation, mitochondrial translocation of pro-apoptotic Bax, mitochondrial insult with endoplasmic reticulum stress, and caspase-12 and CCAAT/enhancer binding protein homologous protein (CHOP) activation.[44] Note that the isoflavonoid genistein (CS 3.14) from *Glycine max* (L.) Merr. (family Fabaceae Lindl.) boosted the cytoplasmic levels of ROS, leading to endoplasmic reticulum stress, Ca^{2+} leakage, activation of calpain, CHOP, DR5 and caspase 8, and apoptosis, in Hep3B cells, with an IC_{50} value of 21.4 μM.[45]

Quercetin from *Allium sativum* L. (family Amaryllidaceae J. St.Hil.) differs from luteolin by a hydroxyl group at C3. Quercetin (CS 3.15) repressed the multiplication of human lung adenocarcinoma epithelial (A549) cells (wild-type p53) and human non-small-cell lung carcinoma (H1299) cells (p53 null), at a dose of 60 μM, by 30% and 70%, respectively,[46] suggesting that wild-type p53 protein is targeted. However, quercetin also destroyed human squamous carcinoma of the tongue (SCC-9) cells (mutant p53), with an IC_{50} value of 50 μM, via caspase 3 activation.[47] In HepG2, quercetin abated the level of phosphorylated Akt and ERK1/2 and induced apoptosis, with an IC_{50} value equal to 87 μmol/L.[48] Quercetin augmented the susceptibility of DU-145, U-MG87, non-Hodgkin's B-lymphoma and human B-cell lymphoma (SUDHL4) cells to tumor necrosis factor-related apoptosis-inducing ligand (TRAIL) by blocking the phosphorylation of Akt.[49,50] Note that quercetin only allowed human brain glioblastoma (A-172) cells to grow by 64%, at a dose of 50 μM, by inhibiting Akt.[51] Morin (CS 3.16) differs from quercetin by its hydroxylation pattern at C2′ and C5′ and is found in *Chlorophora tinctoria* (L.) Gaudich. ex Benth. (family Moraceae Gaudich.) and *Maclura pomifera* (Raf.) C.K. Schneid. (family Moraceae Gaudich.). In HL-60 cells (p53 null), morin prompted an increase in free cytoplasmic Ca^{2+} and ROS, translocation of pro-Bax,

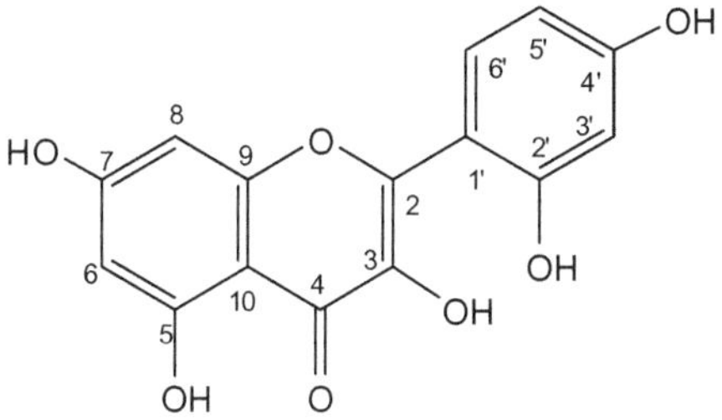

■ **CS 3.16** Morin

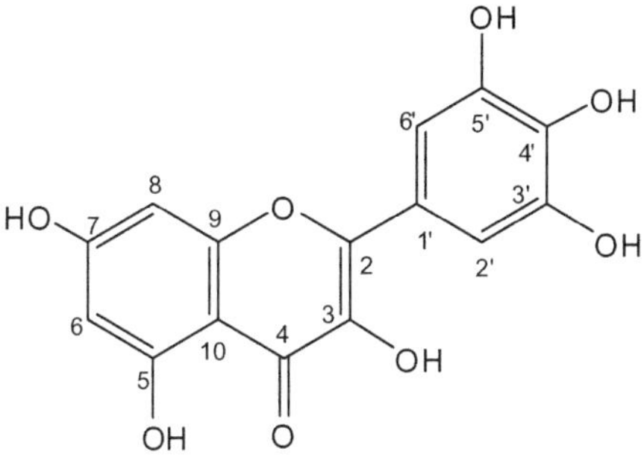

■ **CS 3.17** Myricetin

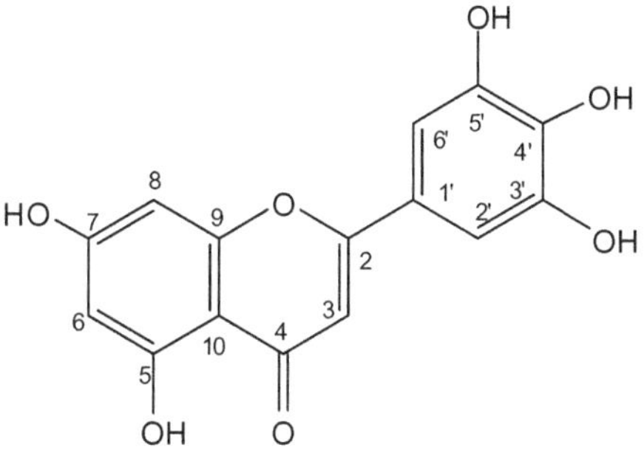

■ **CS 3.18** Tricetin

mitochondrial insult and activation of caspases 9 and 3, resulting in apoptosis with an IC_{50} of 400 μM.[52] Morin also arrested cell division via the induction of p21 and therefore the inhibition of cyclin A.[52]

The hydroxylation of quercetin at C3′ produces myricetin (CS 3.17), which occurs naturally in *Brassica oleracea* L. (family Brassicaceae Burnett). In HL-60 cells myricetin induced Bax, abated the mitochondrial membrane potential, compelled the release of cytochrome c, activated caspases 9 and 3 and resulted in apoptosis with an IC_{50} value of 42.2 μM.[53] Note that myricetin has the interesting ability to bind directly to the MEK–ATP binding site[54] and therefore blocks the multiplication of cells. In addition, myricetin attaches itself to the Akt–ATP binding site via chemical interactions between the C3 hydroxyl and asparagine 292, the C7 hydroxyl and alanine 230, and the C3′ hydroxyl group with glutamine 234.[55]

The removal of the C5′ hydroxyl group from myricetin forms tricetin (CS 3.18), which is a common flavonoid in *Eucalyptus globulus* Labill. (family Myrtaceae Juss.). Tricetin induced the phosphorylation of p53 on

■ **CS 3.19** Oroxylin A

■ **CS 3.20** Eupatilin

serine 15 and serine 392 and therefore the upregulation of p21 and consequent decrease in cyclin B, cell arrest, increase in Bax and Bak, and hence apoptotis via mitochondrial insult with an IC_{50} value of 32.1 μM.[56] In HepG2 and PLC/PRF/5 cells tricetin boosted the levels of ROS, and hence DNA insult, activation of JNK, increase in DR5 levels, caspase 8 activation and apoptosis, with IC_{50} values equal to 4.8 μM and 4.2 μM, respectively.[57]

One can be reasonably certain that a B ring with a ketone at C4, a Δ_{2-3} double bond[29] and an A ring with hydroxyl substitutions at C5 and C7 are beneficial chemical features for antiproliferative activity[20] and that methoxylation of flavonoids increases their lipophilicity. One such flavonoid is the 6 methoxylated derivative of baicalein named oroxylin A (CS 3.19), which occurs in members of the genus *Scutellaria* L. (family Lamiaceae Martinov). Oroxylin A destroyed HepG2 cells, with an IC_{50} value of 230μM, via a decrease in Bcl-2 and activation of caspase 3.[58] Likewise, Hela cells (wild-type) experienced a decrease in Bcl-2, activation of caspases 3 and 8, and apoptosis, with an IC_{50} value equal to 150μM,[59] suggesting the possible activation of p53. In fact, Mu et al. (2009) showed that oroxylin A inhibited MDM2-modulated proteasome p53 degradation, hence the increase in p53 and p21 and the decrease in survivin in HepG2 cells.[60] In the same experiment, oroxylin A decreased the viability of HepG2, K562, MDA-MB-231, HUVEC and human normal liver (LO2) cells to 20%, 40%, 50%, 65% and 70%, respectively, at a dose of 200μM.[60]

Other cytotoxic methoxylated flavonoids of interest are eupatilin (CS 3.20), casticin (CS 3.21), 5-hydroxy-3,6,7,8,5′,4′-hexamethoxyflavone (CS 3.22), 5′-hydroxy-5,6,7,4′-tetramethoxyflavone (CS 3.23) and cirsilineol (CS 3.24). Eupatilin from *Artemisia asiatica* Nakai ex Pamp. (family Asteraceae Bercht. & J. Presl) reduced the viability of human

■ **CS 3.21** Casticin

■ **CS 3.22** 5-Hydroxy-3,6,7,8,5',4'-hexamethoxyflavone

■ **CS 3.23** 5'-Hydroxy-5,6,7,4'-tetramethoxyflavone

■ **CS 3.24** Cirsilineol

premalignant breast epithelial (MCF-10A) cells 'wild-type' p53 by 50%, at a dose of 100 μM, by activating p53, and hence p27 and p21, thus inhibiting cyclin D1 and cyclin-dependent kinase 2 (CDK2) independently of Akt and therefore of MDM2,[61] suggesting the possible activation of JNK or p38 MAPK by ROS. Casticin from *Achillea millefolium* L. (family Asteraceae Bercht. & J. Presl) destroyed K562 cells with a fall in Bcl-2 levels, induction of pro-apoptotic Bax, release of cytochrome c, caspase 3 activation and PARP cleavage.[62] In human breast cancer (MN-1; wild-type p53) and human breast cancer (MDD2; p53 null) cells, casticin disrupted tubulin, which resulted in p21 induction followed by CDK1 inhibition and therefore cell arrest, with an IC_{50} value equal to 2 μM.[63] In addition, casticin binds and inhibits the P-glycoprotein pump on account of the interaction of hydroxyl group at C5 and the C4 ketone with the ATP-binding site of the protein.[63] Casticin at a dose of 30 μM reduced the survival of PLC/PRF/5 and HepG2 cells to 20% and 30%, respectively, by boosting the generation of ROS, hence intracellular glutathione (GSH) depletion, endoplasmic reticulum stress, activation of CHOP and increased expression of DR5 followed by caspase 8 activation.[64] 5-Hydroxy-3,6,7,8,5′,4′-hexamethoxyflavone and 5′-hydroxy-5,6,7,4′-tetramethoxyflavone from *Citrus sinensis* (L.) Ozbeck (family Rutaceae Juss.) inhibited the proliferation of MCF-7 cells (wild-type p53), with IC_{50} values equal to 2.5 μM and 10.5 μM, respectively, via a sustained increase in free cytoplasmic Ca^{2+} of extracellular origin and endoplasmic reticulum.[65] Intriguingly, the Δ_{2-3} reduced and trimethoxylated flavonoid, cirsilineol, from *Artemisia vestita* Wall. ex Besser (family Asteraceae Bercht. & J. Presl) repressed the growth of human ovarian carcinoma (Caov-3), human hepatocellular liver carcinoma (HepG2) and human normal liver (LO2) cells by 60%, 45% and 20%, respectively, at a dose of 40 μM.[66] In Caov-3, cirsilineol induced a fall in mitochondrial membrane potential, the release of cytochrome c, caspase 3 activation and apoptosis,[66] implying that polymethoxylations enhance the cytotoxic effect of flavonoids, even with a reduced Δ_{2-3}. In addition, Wang et al. (1999) observed that 100 μM of apigenin, myricetin, quercetin and kaempferol suppressed the multiplication of HL-60 cells by 60%, 40%, 25% and 5%, respectively,[28] suggesting that the increasing occurence of multiple hydroxyl groups on ring C favors the cytotoxicity of flavonoids, whereas the presence of a hydroxy group at C3 is not a beneficial chemical feature. The deleterious effect of a hydroxyl group at C3 was further confirmed by Torkin et al. (2005) who studied the cytotoxic effect of diosmetin, apigenin, chrysin, quercetin and baicalein at a dose of 60 μM against human neuroblastoma (NUB-7) cells and found reduction in growth of 60%, 50%, 45%, 40% and 20%, respectively, whereas flavone was weakly active and naringenin inactive,[29] suggesting that the presence of C5 and C7 hydroxyl groups on ring A favors the cytotoxic activity of flavonoids. Apigenin, luteolin, naringenin, eriodictyol (CS 3.25) and hesperetin (CS 3.26) inhibited HER2/neu protein expression in MDA-MB-231

■ **CS 3.25** Eriodictyol

■ **CS 3.26** Hesperetin

cells, with IC_{50} values equal to 12.1 μM, 4.5 μM, 47.8 μM, 45.9 μM and 21.4 μM, respectively,[32] confirming that that numerous hydroxyl groups on ring C are beneficial.[32]

Several lines of evidence clearly point to the fact that the prenylation of flavonoids magnifies their cytotoxic potencies. For instance, artelastin (CS 3.27) isolated from *Artocarpus elasticus* Reinw. ex Blume (family Moraceae Gaudich.) inhibited the growth of a bewildering array of malignancies, including (IC_{50} values in parentheses): MCF-7 (6 μM), human large-cell lung cancer (NCI-H460; 5.2 μM), human glioblastoma (SF268; 10.8 μM), human renal carcinoma (TK-10; 6.2 μM) and human melanoma (UACC-62; 3.9 μM).[67] In MCF-7 cells it interfered with DNA replication and microtubules and induced death with an IC_{50} value of 5 μM.[67] Apoptosis was imposed by artonin B (CS 3.28) from *Artocarpus heterophyllus* Lam. (family Moraceae Gaudich.) on human T cell lymphoblast-like cell (CCRF-CEM) cells, with an IC_{50} value of 3.4 μmol/L, with an increase in Bax, a fall in mitochondrial membrane potential, the release of cytochrome c and caspase 3 activation.[68]

Other examples of cytotoxic prenylated flavonoids of interest are licoflavone C (CS 3.29) and isobavachin (CS 3.30), which destroyed the following cells (respective IC_{50} values in parentheses): rat hepatoma (H4IIE; 42 μmol/L, 37 μmol/L) and mouse glioma (C6; 96 μmol/L, 69 μmol/L), whereas apigenin and liquiritigenin (CS 3.31) were inactive.[69] Licoflavone C and isobavachin induced the activation of caspase 3 and 7 in H4IIE cells.[69] In the same experiment, quercetin abrogated the growth of H4IIE and C6 cells, with IC_{50} values equal to 199 μmol/L and 198 μmol/L,

■ **CS 3.27** Artelastin

■ **CS 3.28** Artonin B

■ **CS 3.29** Licoflavone C

■ **CS 3.30** Isobavachin

■ **CS 3.31** Liquiritigenin

■ **CS 3.32** Icaritin

respectively,[69] clearly implying that a prenylated moiety at C8 improves the cytotoxic potential of flavonoids. Icaritin (CS 3.32) from *Epimedium brevicornu* Maxim. (family Berberidaceae Juss.) at a dose of 50 μmol/L reduced to 20% the survival of human prostate cancer (PC-3) cells (p53 null), with increased expression of p27 and p16 resulting in the inhibition of CDK4/cyclin D and hence a decrease in phosphorylated retinoblastoma protein (pRb) followed by cycle arrest.[70] In fact, morusin (CS 3.33) from *Morus australis* Poir. (family Moraceae Gaudich.) blocked the phosphorylation of phosphatidylinositol 3-kinases (PI3K) and Akt, and activated caspases 9 and 3, resulting in apoptosis in HT-29 cells, with an IC_{50} value equal to 6.1 μM.[71] The vulnerability of wild-type p53 to prenylated flavonoids was further observed with arcommunol A (CS 3.34) from *Artocarpus communis* J.R. Forst. & G. Forst. (family Moraceae Gaudich.), which inhibited the growth of HepG2 (wild-type p53), Hep3B (p53 null),

■ **CS 3.33** Morusin –

■ **CS 3.34** Arcommunol A

human hepatocellular carcinoma (PLC5; mutant p53) and human liver adenocarcinoma (SK-HEP-1; wild-type p53) cells, with IC_{50} values equal to 9.1 μM, 20.4 μM, 30.3 μM and 2 μM, respectively. In 'wild-type' hepatocellular carcinoma cells, arcommunol A activates the pro-apoptotic protein (p53) and hence an increase in Fas and Bax, a fall in mitochondrial membrane potential, mitochondrial insult, release of AIF and activation of caspase 3.[72]

■ **CS 3.35** 3,7-Dihydroxyflavone

In addition, Neves et al. (2011) observed that the esterification of a C7 hydroxyl group by a prenyl and even a geranyl moiety enhanced the cytotoxicity of flavonoids, since 3,7-dihydroxyflavone (CS 3.35), 7-(3-methylbut-1-enyloxy)-baicalein (CS 3.36), 7-(3,7-dimethylocta-2,6-dienyloxy)-baicalein (CS 3.37), 7-(3-methylbut-1-enyloxy)-3-hydroxyflavone (CS 3.38) and 7-(3,7-dimethylocta-2,6-dienyloxy)-3-hydroxyflavone (CS 3.39) inhibited the growth of human non-small-cell lung carcinoma (H460) cells, with IC_{50} values equal to 15.4 μM, 6.8 μM, 4.1 μM, 12.6 μM and 4.9 μM, respectively.[73]

The isomerization of the C ring into a cyclohexadienone produces a series of potently cytotoxic flavonoids which occur in ferns, like protoapigenone

■ **CS 3.36** 7-(3-methylbut-1-enyloxy)-baicalein

■ **CS 3.37** 7-(3,7-Dimethylocta-2,6-dienyloxy)-baicalein

■ **CS 3.38** 7-(3-Methylbut-1-enyloxy)-3-hydroxyflavone

■ **CS 3.39** 7-(3,7-Dimethylocta-2,6-dienyloxy)-3-hydroxyflavone

(CS 3.40) from *Thelypteris torresiana* (Gaudich.) Alston (family Thelypteridaceae Ching ex Pic. Serm.). This was lethal to human prostate adenocarcinoma (LNCaP) cells, with an IC$_{50}$ value of 3.7 μM, by activation of p38 MAPK and pro-apoptotic JNK.[74] The activation of pro-apoptotic JNK by protoapigenone was further confirmed and delineated by Chen et al. (2010), who found evidence that an increase in ROS compelled the

■ **CS 3.40** Protoapigenone

■ **CS 3.41** 2'4'-Dihydroxy-6'-methoxy-3',5'-dimethylchalcone

■ **CS 3.42** Flavokawain A

■ **CS 3.43** Flavokawain B

■ **CS 3.44** Flavokawain C

dissociation of glutathione S-transferases (GSTs), which released and activated pro-apoptotic JNK. This was followed by mitochondrial insult, activation of caspase 3, cleavage of PARP, persistent activation of ERK1/2 and apoptosis with an IC_{50} value equal to 4.7 μM.[75]

The opening of the ring B results in the formation of chalcones, such as 2'4'-dihydroxy-6'-methoxy-3',5'-dimethylchalcone (CS 3.41) isolated from *Cleistocalyx operculatus* (Roxb.) Merr. & L.M. Perry (family Myrtaceae Juss.), which destroyed K562 cells, with an IC_{50} value equal to 14.2 μM.[76] The chalcones flavokawain A (CS 3.42), B (CS 3.43), and C (CS 3.44) from

Piper methysticum G. Forst. (Piperaceae Giseke) repressed the proliferation of the following cell lines (respective IC_{50} values in parentheses): human bladder cancer (T24; 16.7 μmol/L, 17.2 μmol/L, 20.8 μmol/L), human bladder carcinoma (EJ; 6.7 μmol/L, 5.7 μmol/L, 15.7 μmol/L) and human urinary bladder papilloma (RT4; 10.6 μmol/L, 14.6 μmol/L, 4.6 μmol/L.[77] Treatment of T24 cells with flavokawain A caused apoptosis with an increase in Bax, a decrease in Bcl-xL, a fall in mitochondrial membrane potential, release of cytochrome c, activation of caspases 9, 3 and PARP cleavage, and downregulation of survivin and XIAP,[76] suggesting the involment of mitochondrial insult. Indeed, flavokawain B from *Alpinia pricei* Hayata (Zingiberaceae Martinov) inhibited the multiplication of human colorectal carcinoma (HCT-116) cells at 25 μM with an increase in cytoplasmic Ca^{2+}, generation of ROS, phosphorylation of p38 MAPK and hence increase of pro-apoptotic Bim, decrease of Bcl-2, loss of mitochondrial membrane potential, release of cytochrome c and cleavage of PARP.[78] Licochalcone A (CS 3.45) from *Glycyrrhiza glabra* L. (family Fabaceae Lindl.) blocked the growth of PC-3 cells at a dose of 25 μM by lowering the levels of cyclin B1,[79] probably as a result of inhibition of STAT3 phosphorylation.[80] Sappanchalcone from *Caesalpinia sappan* L. inhibited the growth human oral cancer (HN4) and (HN12) cells, with an IC_{50} value of 25 μM, by stimulation of pro-apoptotic proteins (p53) and Bax, mitochondrial insult, release of cytochrome c and activation of caspases 9 and 3.[81] The growth of human benign prostatic hyperplasia (BPH-1) cells (wild-type p53) and human PC-3 (p53 null) cells treated with 20 μM of xanthohumol (CS 3.46) from *Humulus sativus* L. (family Cannabaceae Martinov) was decreased by 80% and 43%, respectively, with inactivation of NF-κB.[82,83]

■ **CS 3.45** Licochalcone A

■ **CS 3.46** Xanthohumol

3.1.2 *Hedyotis Biflora* (L.) Lam.

History This plant was first described by Jean Baptiste Antoine Pierre de Monnet de Lamarck, in *Encyclopédie Méthodique, Botanique*, published in 1792.

Synonyms *Hedyotis paniculata* (L.) Lam., *Hedyotis racemosa* Lam., *Oldenlandia biflora* L., *Oldenlandia crassifolia* DC., *Oldenlandia paniculata* L.

Family Rubiaceae Juss., 1789

Common Name Shuang hua er cao (Chinese)

Habitat and Description *Hedyotis biflora* is a herb which grows up to 30 cm tall in the wastelands of China, India, Southeast Asia and the Pacific Islands. The stem is quadrangular, glabrous, and somewhat articulate. The leaves are simple, subsessile, opposite and stipulate. The stipules are interpetiolar, triangular and minute.The petiole is inconspicuous. The blade is linear, 1–4 cm × 0.1 cm, glabrous, acute at base, and acute at apex. The inflorescence is a terminal or an axillary corymb. The calyx comprises four lobes which are triangular and minute. The corolla is white, hairy at throat, tubular and produces four lobes which are oblong and minute. The androecium includes four stamens. The gynaecium consists of two carpels fused into a bilocular ovary that develops a two-lobed stigma. The fruit is tiny, papery, and encloses numerous black seeds (Figure 3.2).

Rationale The cytotoxic and apoptotic potencies of coumarins depend on the functional moieties attached to their α-benzopyrone framework. This is exemplified with simple hydroxycoumarins such as 6,8-dimethoxy-7-hydroxycoumarin (isofraxidin, CS 3.47) from *Micrandra elata* (Didr.) Müll. Arg. (Euphorbiaceae Juss.) and 7-hydroxycoumarin (umbelliferone, CS 3.48) from *Angelica gigas* Nakai (family Apiaceae Lindl.), which destroyed mouse leukemia (P388) cells, with IC_{50} values equal to 1.7 µg/mL and 12 µg/mL, respectively),[84,85] suggesting that a C7 hydroxy is favorable for activity when surrounded by an ortho-congener. This was later confirmed by Finn et al. (2002), who found that 7,8-dihydroxycoumarin (CS 3.49) and 6,7-dihydroxycoumarin (esculetin) were toxic against human renal carcinoma (A498) cells, with IC_{50} values equal to 4.3 µM and 86 µM, respectively, whereas coumarin and 7-hydroxycoumarin were inactive at 500 µM.[86]

■ **CS 3.47** Isofraxidin

■ **CS 3.48** Umbelliferone

■ **CS 3.49** 7,8-Dihydroxycoumarin

■ **CS 3.50** 7,8-Dihydroxy-4-methylcoumarin

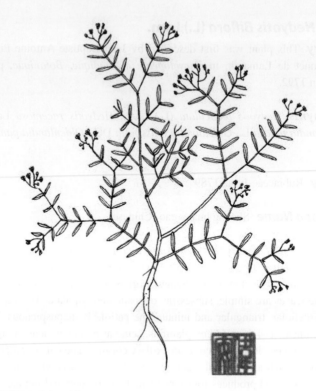

■ **FIGURE 3.2** *Hedyotis biflora* (L.) Lam.

Note that methylation of C4 reduces the cytotoxic potencies of 7,8-dihydroxycoumarin since 7,8-dihydroxy-4-methylcoumarin (CS 3.50) mitigated the viability of human lung adenocarcinoma epithelial (A549), human epitheloid cervix carcinoma (Hela) and human hepatocellular liver carcinoma (HepG2) cells, with IC_{50} values equal to 160 μg/ml, 180 μg/ml and > 200 μg/mL, respectively.[87] The methoxylation or acetoxylation of the hydroxyl moieties at C6 and C7 is deleterious: 6,7-dimethoxycoumarin (scoparone, CS 3.51) from *Artemisia capillaris* Thunb. (family Asteraceae Bercht. & J. Presl) was not toxic against human histiocytic lymphoma (U937) cells at a dose of 100 μM[88] and 5,7-dimethoxycoumarin (CS 3.52) from *Citrus limon* (L.) Osbeck (family Rutaceae Juss.) inhibited the growth of human malignant melanoma (A375) and mouse melanoma (B16) cells, with IC_{50} values equal to 300 μM and 260 μM, respectively.[89]

6-Methoxy-7-hydroxycoumarin (scopoletin), which is found in *Avena sativa* L. (family Poaceae Barnhart), weakened the multiplication and migration of human endothelial (ECV304) cells by 35% and 75%, respectively, at a dose of 500 μM.[90]

Hishihara et al. (2006) studied the structure–activity of several simple coumarins against HSC-2 cells and observed linear correlations between IC_{50}, ionization potential (eV) and absolute hardness (ή) demonstrating that the cytotoxicity of coumarins is proportional to their electron-donating properties and low absolute hardness.[91] In that study, the α-benzopyrone framework without any substituents had the highest ionization potential value and the highest absolute hardness and was therefore inactive.[91]

The cellular events responsible for the cytotoxicity of hydroxylated and methoxylated coumarins are not yet fully understood but common oncocellular features observed are a decrease in phosphorylated extracellular signal-regulated kinases (ERK1/2), mitochondrial insult and the inhibition of NF-κB, leading to apoptosis. Indeed, scoparone (from *Artemisia capillaris* Thunb.) blocked the phosphorylation of IκBα and therefore NF-κB in human histiocytic lymphoma (U937) cells.[88] 7,8-Dihydroxy-4-methylcoumarin at a dose of 160 µg/mL induced a significant decrease in phosphorylated ERK1/2, boosted the intracellular levels of c-Myc, decreased the expression of Bcl-xL and induced mitochondrial insult and hence release of cytochrome c, activation of caspases 9 and 3 and apoptosis in A549 cells.[87] Similarly, 5,7-dimethoxycoumarin in the B16 cell line reduced phosphorylated ERK1/2 at a dose of 250 µM.[89] Scopoletin (CS 3.53) inhibited the activation of ERK1/2 in ECV304 cells, at a dose of 500 µM,[90] and it induced mitochondrial membrane potential collapse, activation of caspase 3 and blocked the phosphorylation of IκBα, and therefore NF-κB, in human fibroblast-like synoviocytes (FLS) cells.[92] 7,8-Diacetoxy-4-methylcoumarin at a dose of 160 µg/mL decreased the levels of phosphorylated ERK1/2 and Bcl-xL, and abated mitochondrial membrane potential, with the consequent release of cytochrome c and activation of caspases 9 and 3 in A549 cells.[93] Note that esculetin (CS 3.54) decreased the cell growth of Hela cells, with an IC_{50} value of 37.8 µM, with generation of ROS and mitochondrial insult leading to the release of cytochrome c and apoptosis.[94] One could frame the hypothesis that hydroxylated and methoxylated coumarins (CS 3.55) act by repressing Raf, which stimulates ERK1/2 phosphorylation, NF-κB and mitochondrial insult. Note that 7-hydroxycoumarin (CS 3.56) inhibited the enzymatic activity of topoisomerase II.[95]

The importance of the C7 substitution in the cytotoxic activity of coumarins is further evidenced with alkoxy or prenyloxy C7 coumarins, which are highly lipophilic and cytotoxic. An example of a 7-prenyloxycoumarin is 7-geranyloxycoumarin (auraptene, CS 3.57) from *Poncirus trifoliata* (L.) Raf. (family Rutaceae Juss.), which inhibited the growth of mouse lymphocytic leukemia (L1210) cells, with an IC_{50} value equal to

■ **CS 3.51** Scoparone

■ **CS 3.52** 5,7-Dimethoxycoumarin

■ **CS 3.53** Scopoletin

■ **CS 3.54** Esculetin

■ **CS 3.55** Coumarin

■ **CS 3.56** 7-Hydroxycoumarin

■ **CS 3.57** Auraptene

■ **CS 3.58** Umbelliprenin

■ **CS 3.59** 7-[(*E*)-3′,7′-dimethyl-6′-oxo-2′,7′-octadienyl]oxycoumarin

■ **CS 3.60** Schinilenol

10.2 μg/mL.[96] Umbelliprenin (CS 3.58) and auraptene from *Ferula szowitsiana* DC. (family Apiaceae Lindl.) destroyed human melanoma (M4Beu) cells via apoptosis, with IC_{50} values equal to 12.4 μM and 17.2 μM, respectively, whereas the IC_{50} of coumarin was above 100 μM,[97] implying that the longer or the more unsaturated or lipophilic the isoprenyl moiety the better the activity. In fact, the hydroxylation and/or reduction of the 7-isoprenyl moiety decreases the cytotoxic activity: 7-[(*E*)-3′,7′-dimethyl-6′-oxo-2′,7′-octadienyl]oxycoumarin (CS 3.59), schinilenol (CS 3.60),

■ **CS 3.61** Schinindiol

■ **CS 3.62** 7-[(*E*)-7′-hydroxy-3′,7′-dimethylocta-2′,5′-dienyloxy]-coumarin

■ **CS 3.63** Geiparvarin

■ **CS 3.64** Murrayacoumarin B

schinindiol (CS 3.61) and 7-[(*E*)-7′-hydroxy-3′,7′-dimethylocta-2′,5′-dienyloxy]coumarin (CS 3.62) isolated from *Zanthoxylum schinifolium* Siebold & Zucc. (family Rutaceae Juss.) inhibited the growth of Jurkat human T cell lymphoblast-like cells, with IC_{50} values equal to 8.1 μM, 71.6 μM, >100 μM and 60.9 μM, respectively, whereas the control, auraptene, displayed an IC_{50} value equal to 55.3 μM.[98] Geiparvarin (CS 3.63) from *Geijera parviflora* Lindl. (family Rutaceae Juss.) killed the following cells (IC_{50} values in parentheses): human erythromyeloblastoid leukemia (K562; 2.4 μM), human promyelocytic leukemia (HL-60; 6.3 μM), human fibrosarcoma (HT1080; 9.8 μM), human lung adenocarcinoma epithelial (A549; 11.5 μM) and human neuroblastoma (SH-SY5Y; 5.7 μM).[99] Murrayacoumarin B (CS 3.64) isolated from a member of the genus *Murraya* J. König ex L. (family Rutaceae Juss.) at a dose of 30 μM abated the growth of HL-60 cells to 35% with activation of caspases 9 and 3 and a massive loss of mitochondrial membrane potential,[100] suggesting

■ **CS 3.65** 7-Demethylsuberosin

■ **CS 3.66** Psoralidin

■ **CS 3.67** Osthol

■ **CS 3.68** Ethoxy derivative of osthol

mitochondrial insult, which may very well result from a direct effect on the mitochondrial membrane.

The presence of prenyl moieties on the α-benzopyrone backbone enhance the cytotoxicity of coumarins, as evidenced with 7-demethylsuberosin (CS 3.65) from *Angelica gigas* Nakai (family Apiaceae Lindl.), which inhibited P388 cells, with an IC_{50} value equal to 19 μg/mL.[85] Psoralidin (CS 3.66) from *Psoralea corylifolia* L. (family Fabaceae Lindl.) inhibited the growth of human gastric carcinoma (SNU-1) and (SNU-16) cells, with IC_{50} values equal to 53 μg/mL and 203 μg/mL, respectively.[101] The prenylated coumarin osthol (CS 3.67) from *Cnidium monnieri* (L.) Cusson (family Apiaceae Lindl.) killed P388, Chinese hamster lung fibroblast (V79), human nasopharyngeal carcinoma (KB),) and mouse mammary tumor (MM46) cells, with IC_{50} values equal to 3.1 μg/mL, 14 μg/mL, 12 μg/mL and 8.6 μg/mL, respectively,[102] whereas ethoxy and propoxy derivatives (CS 3.68 and CS 3.69) inhibited V79 cells, with IC_{50} values equal of 5.8 μg/mL and 5.6 μg/mL, respectively.[102] A hydrogenated isoprene

■ **CS 3.69** Propoxy derivative of osthol

■ **CS 3.70** Hydrogenated derivative of osthol

■ **CS 3.71** Ferulenol

derivative (CS 3.70) was inactive,[100] implying that the double bond in the isoprene moiety and lipophilicity are crucial for cytotoxicity and confirming the beneficial effect of C7 alkoxylation. The mode of action of osthol involves cyclin B1 and p-Cdc2, a decrease in Bcl-2 and an increase in pro-apoptotic Bax levels in A549 cells (wild-type p53), with an IC_{50} value of 100 μM,[103] suggesting a DNA-damage-induced apoptosis involving ataxia-telangiectasia mutated (ATM) activation and therefore activation of Akt and pro-apoptotic p53. In fact ferulenol (CS 3.71) from *Ferula communis* L. (family Apiaceae Lindl.) abated the growth of human breast adenocarcinoma (MCF-7) cells to 50% by interfering with tubulin polymerization through binding at the colchicine site.[104]

The mammea-type coumarins are a series of potently cytotoxic prenylated compounds which are common in members of the family Calophyllaceae J. Agardh. Two examples are vismiaguianone D (CS 3.72) and E (CS 3.73) from *Vismia guianensis* (Aubl.) Choisy (family Hypericaceae Juss.), which potently inhibited the growth of KB cells, with IC_{50} values equal to 2.4 μg/mL and 3.3 μg/mL, respectively.[105] Mammeisin (CS 3.74) and mammea A/AB (CS 3.75) from *Marila tomentosa* Poepp. (family Calophyllaceae J. Agardh.) inhibited the growth of the following cells (respective IC_{50} values in parentheses): MCF-7 (0.2 μg/mL, 0.2 μg/mL),

■ CS 3.72 Vismiaguianone D

■ CS 3.73 Vismiaguianone E

■ CS 3.74 Mammeisin

■ CS 3.75 Mammea A/AB

human large-cell lung cancer (NCI-H460; 0.2 μg/mL, 0.1 μg/mL) and human glioblastoma (SF268; 0.1 μg/mL, 0.1 μg/mL),[106] confirming that Δ_{2-3} is crucial for the activity of mammea-type coumarins. Surangin D (CS 3.76) and C (CS 3.77) and theraphin B (CS 3.78) and C (CS 3.79) from a member of the genus *Mammea* L. (family Calophyllaceae J. Agardh.) inhibited the growth of Hela cells, with IC$_{50}$ values equal to

■ **CS 3.76** Surangin D

■ **CS 3.77** Surangin C

■ **CS 3.78** Theraphin B

■ **CS 3.79** Theraphin C

■ **CS 3.80** Mammea E/BB

■ **CS 3.81** 5,7,4′-Trimethoxy-8-hydroxy-4-phenylcoumarin

■ **CS 3.82** 5,7,4′-Trimethoxy-8,3′-dihydroxy-4-phenylcoumarin

■ **CS 3.83** 7,4,-Dimethoxy-5, 3′-dihydroxy-4-phenylcoumarin

13.6 μM, 4.7 μM, 10 μM and 4.7 μM, respectively,[107] indicating that the presence of a phenyl moiety at C4 enhances the cytotoxicity of prenylated coumarins. Mammea-type coumarins impose apoptosis via mitochondrial insult. Mammeisin inhibited the growth of Hela cells, with an IC_{50} value equal to 65 μM, involving translocation of pro-apoptotic Bax, downregulation of Bcl-2, mitochondrial insult, release of AIF and apoptosis.[108] Likewise, mammea E/BB (CS 3.80) from *Mesua americana* L. (family Calophyllaceae J. Agardh.) at a dose of 20 μM nullifyied the viability of human prostate cancer (PC-3) cells by 40% via a mechanism that may involve disruption of mitochondrial electron transport.[109]

4-Phenylcoumarins without isoprenes have potent cytotoxic effects: 5,7,4′-trimethoxy-8-hydroxy-4-phenylcoumarin (CS 3.81), 5,7,4′-trimethoxy-8,3′-dihydroxy-4′-phenylcoumarin (CS 3.82) and 7,4,-dimethoxy-5,3′-dihydroxy-4-phenylcoumarin (CS 3.83), isolated from *Exostema acuminatum* Urb. (family Rubiaceae Juss.), and the synthesized derivate 5,7,3′,4′-tetramethoxy-4-phenylcoumarin (CS 3.84) inhibited the growth of the following cells (respective IC_{50} values in parentheses): human lymphoma (BC-1; 0.5 μg/mL, 1.6 μg/mL, 4.3 μg/mL, 4 μg/mL), human lung adenocarcinoma (Lu-1; 1.4 μg/mL, 4.1 μg/mL, 7.9 μg/mL, 0.6 μg/mL), human colon carcinoma (Col-2; 0.1 μg/mL, 1.8 μg/mL, 4.8 μg/mL, 2.9 μg/mL), human nasopharyngeal carcinoma (KB; 0.02 μg/mL, 0.2 μg/mL, 0.3 μg/mL, 0.1 μg/mL), human neuroblastoma (SK-N-SH; 0.2 μg/mL, 0.7 μg/mL, 1.8 μg/mL, 0.8 μg/mL) and Madison lung tumor (M-109; 0.1 μg/mL, 2.1 μg/mL, 1.4 μg/mL, 1 μg/mL).[110]

■ **CS 3.84** 5,7,3′,4′-Tetramethoxy-4-phenylcoumarin

■ **CS 3.85** Decursin

■ **CS 3.86** Decursinol angelate

■ **CS 3.87** Decursinol

■ **CS 3.88** Xanthoxyletin

The fusion of a tetrahydropyran ring to C7 and C8 is not deleterious to the cytotoxic potencies of coumarins. In fact, such coumarins have been isolated from *Angelica gigas* Nakai (family Apiaceae Lindl.) and identified as decursin (CS 3.85), decursinol angelate (CS 3.86) and decursinol (CS 3.87), which destroyed P388 cells, with IC_{50} values equal to 4.5 µg/mL, 4.5 µg/mL and 36 µg/mL, respectively.[85] Decursin and decursinol angelate at a concentration of 50 µM inhibited the proliferation of K562 cells by 40 and 60%, respectively.[111] However, that the presence of a hydroxyl or methoxy group at C5 is deleterious to the cytotoxic property of 7,8-tetrahydropyran coumarins is exemplified with xanthoxyletin (CS 3.88), dentatin (CS 3.89) and nordentatin (CS 3.90) isolated from *Clausena harmandiana* (family Rutaceae Juss.), which exhibited the following, respective, IC_{50} values against the following cell lines:

■ **CS 3.89** Dentatin –

■ **CS 3.90** Nordentatin

■ **CS 3.91** Nodakenetin

human small-cell lung cancer (NCI-H187; >200 μM, 45.4 μM, 11.8 μM), human nasopharyngeal carcinoma (KB; 133.1 μM, 133.9 μM, 20.6 μM) and African green monkey kidney (Vero; >200 μM, >200 μM, 56.3 μM) cells.[112] Decursin from *Angelica gigas* Nakai (family Apiaceae Lindl.) activated protein kinase C (PKC) in K562 cells and induced the generation of ROS in U937 cells.[110,113] In addition, decursin and decursinol in human umbilical vein endothelial (HUVEC) cells inhibited the phosphorylation of ERK1/2 at a dose of 20 μM.[114] Decursin also inhibited the growth of human multiple myeloma (U266; mutant p53) by 50% at a doses of 80 μM,[115] with a fall in cyclin D1, caspase 3 activation and PARP cleavage, which suppressed the constitutive phosphorylation of JAK2 and therefore prevented STAT3 phosphorylation.[115] Note that wild-type cancer cells become apoptotic by STAT3 inhibition in a process involving pro-apoptotic protein (p53) activation followed by CD95 stimulation and hence caspase 8 activation, cleavage of Bid and mitochondrial insult.

The fusion of a tetrahydrofuran ring to C6 and C7 produces cytotoxic coumarins, such as nodakenetin (CS 3.91) from *Angelica gigas* Nakai (family Apiaceae Lindl.), which inhibited the growth of P388 cells, with an IC_{50} value equal to 2.8 μg/mL. Columbianetin (CS 3.92), from the same plant, also inhibited the growth of P388 cells but with an IC_{50} value of 50 μg/mL,[85] implying that the presence of a tetrahydrofuran at C7 and C8 is not so favorable. However, alkylation of the tetrahydrofuran ring in coumarin produces a strong cytotoxic effect, such as with hedyotiscones A, B and C (CS 3.93–95) from *Hedyotis biflora* (L.) Lam. (family Rubiaceae Juss.), which inhibited the growth of HepG2 cells, with IC_{50} values equal to 14.4 μg/mL, 17.4 μg/mL and 4.9 μg/mL, respectively.[116] Coumarin fused with a furan at C7 and C8 exhibits significant cytotoxic effects. For instance, ochrocarpins A–D (CS 3.96–99) from *Ochrocarpos punctatus* H. Perrier (family Clusiaceae, Lindl.) inhibited the growth of human ovarian cancer (A2780) cells, with IC_{50} values equal to 5.2 μg/mL, 3.8 μg/mL, 3.7 μg/mL, and 6.3 μg/mL, respectively.[117] The presence of a C6,C7 fused furan ring is a positive chemical feature: pangelin (CS 3.100) and oxypeucedanin hydrate acetonide (CS 3.101) from *Angelica dahurica*, inhibited

■ **CS 3.92** Columbianetin

■ **CS 3.93** Hedyotiscone A

■ **CS 3.94** Hedyotiscone B

■ **CS 3.95** Hedyotiscone C

■ **CS 3.96** Ochrocarpin A

■ **CS 3.97** Ochrocarpin B

■ **CS 3.98** Ochrocarpin C

■ **CS 3.99** Ochrocarpin D

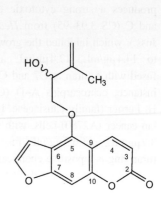

■ **CS 3.100** Pangelin

■ **CS 3.101** Oxypeucedanin hydrate acetonide

■ **CS 3.102** Oxypeucedanin

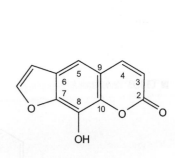

■ **CS 3.103** 8-Hydroxypsoralen

the growth of L1210 (13.3 µg/mL, 9.4 µg/mL), HL-60 (14.6 µg/mL, 9.5 µg/mL), K562 (10.1 µg/mL, 8.6 µg/mL) and murine melanoma (B16F10; 12.5 µg/mL, 9.8 µg/mL) cells,[118] confirming that alkylation of the C5 hydroxyl group is deleterious. This negative effect is confirmed with oxypeucedanin (CS 3.102), which compelled the apoptosis of human prostate carcinoma (DU-145) cells, with an IC_{50} value equal to 100 µM, accompanied by decreases in Cdc-25C, cyclins A and B1 and cdc-2, activation of caspase 3 and PARP cleavage,[119] suggesting the involvement of disrupted mitosis. 8-Hydroxypsoralen (CS 3.103) from *Clausena lansium* (Lour.) Skeels (family Rutaceae Juss.) inhibited the growth of HepG2, Hela and A549

■ **CS 3.104** (+/−)-4′-Hydroxy-3′,4′-dihydroseselin

■ **CS 3.105** (+/−)-4′-hydroxy-3′,4′-dihydroxanthyletin

■ **CS 3.106** (+/−)-4′-Acetoxy-3′,4′-dihydroxanthyletin

■ **CS 3.107** 667 Coumate

cells, with IC_{50} values equal to 0.3 µg/mL, 0.01 µg/mL and 28.2 µg/mL, respectively, probably on account of mitotic disturbances.[120]

The cytotoxic property of coumarins has compelled the synthesis of several sorts of analogue, such as the nitrocoumarins. 8-Nitro-7-hydroxycoumarin at 500 µM induced apoptosis in K562 and HL-60 cells, respectively.[121] Magiatis et al. (1998) synthesized (+/−)-4′-hydroxy-3′,4′-dihydroseselin (CS 3.104), (+/−)-4′-hydroxy-3′,4′-dihydroxanthyletin (CS 3.105) and (+/−)-4′-acetoxy-3′,4′-dihydroxanthyletin (CS 3.106), which inhibited the growth of L1210 cells, with IC_{50} values of 0.9 µM, 6.5 µM and 5 µM, respectively.[122] The 7-hydroxyl moiety mimics somewhat the 3 hydroxyl group of estradiol and series of anti-estrogenic coumarins have been developed, such as 667 coumate (CS 3.107) and SP500263 (CS 3.108). 667 Coumate is an inhibitor of steroid phosphatase which underwent phase I clinical trials in postmenopausal women with hormone-dependent breast cancer[123] but showed weak efficacy in phase II.[124] SP500263 is a selective estrogen receptor modulator (SERM).[125,126]

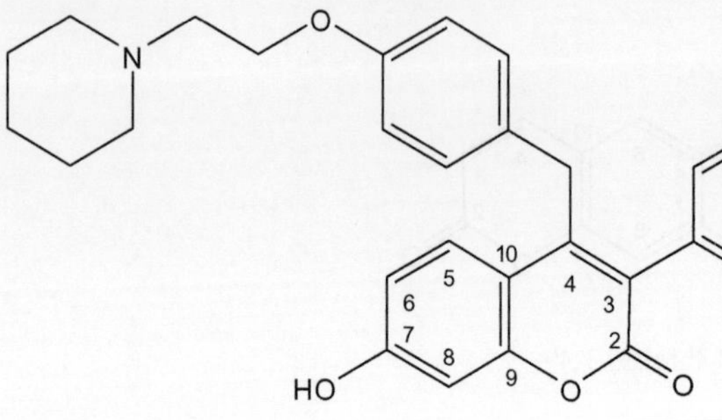

■ CS 3.108 SP500263

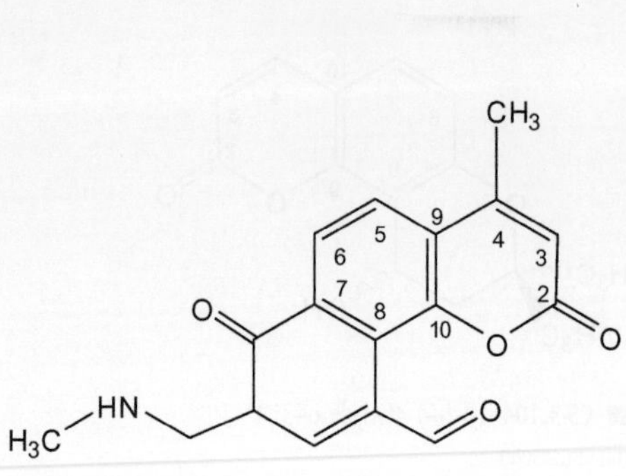

■ CS 3.109 Bc-5

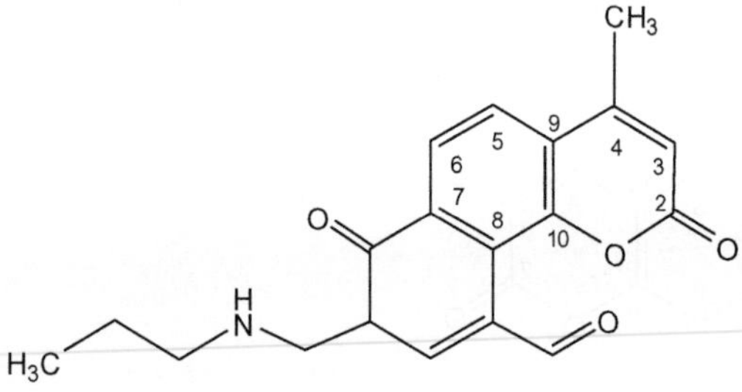

■ CS 3.110 Bc-8

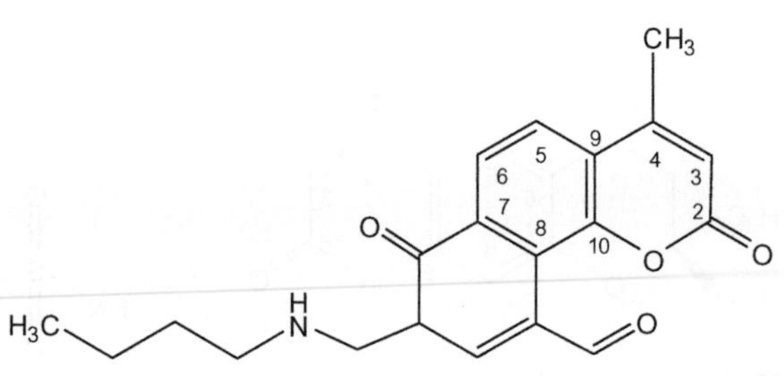

■ CS 3.111 Bc-9

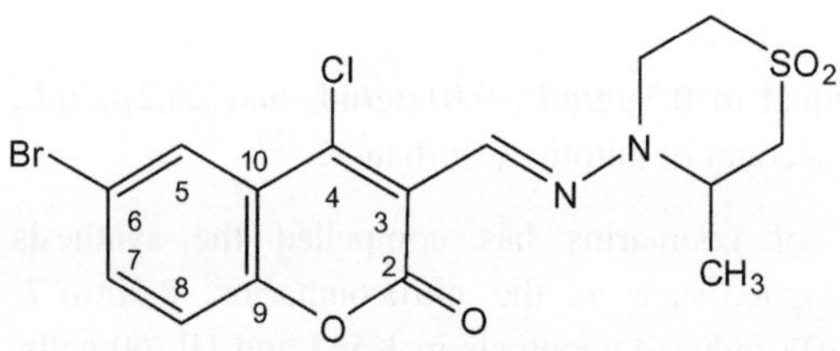

■ CS 3.112 RKS262

Likewise, the neo-tanshinlactone analogues Bc-5 (CS 3.109), Bc-8 (CS 3.110) and Bc-9 (CS 3.111) induced apoptosis in MCF-7 cells, with IC_{50} values equal to 3.8 μM, 7.9 μM and 6.5 μM, respectively, via caspase activation and apoptosis.[127] RKS262 (CS 3.112) inhibited the growth of human ovarian carcinoma (OVCAR-3) cells, with an IC_{50} value of 3 μM, associated with decreases in cyclin D1 and Ras, phosphorylation of Akt, Bcl-xl and Mcl-1, upregulation of p27, activation of Bid and Bad, a fall in mitochondrial membrane potential, activation of caspases 3 and 7, and cleavage of PARP.[128]

Combes et al. (2011) tested series of potently cytotoxic synthetic 4-phenylcoumarins against human breast epithelial (HBL100) cells. 4-(3′-Hydroxy-4′-methoxyphenyl)coumarin (CS 3.113), 4-(3′-hydroxy-4′-methoxyphenyl)-5-methoxycoumarin (CS 3.114), 4-(3′-hydroxy-4′-methoxyphenyl)-6-methoxycoumarin (CS 3.115) and 4-(3′-hydroxy-4′-methoxyphenyl)-5,7-dimethoxycoumarin (CS 3.116) inhibited the growth of these cells by 50% at doses of 84 nM, 182 nM, 39 nM, and 88 nM, respectively, via tubulin assembly inhibition,[98] implying that a C3′ hydroxyl and C4′ methoxyl benzene at C4, and a few substitutions on the A ring, are beneficial for the cytotoxic activity.[98] In addition, 4-(3′-hydroxy-4′-methoxyphenyl)-5-methoxycoumarin and 4-(3′-hydroxy-4′-methoxyphenyl)-5,7-dimethoxycoumarin at doses of 45 μM, 30 μM and 10 μM, inhibited ATP-dependent efflux pump P-glycoprotein (P-gp) and breast cancer resistance protein (BCRP) pump's activity as efficiently as 10 μM of cyclosporin A.[129] Musa et al. (2011) synthesized 4-(7-(diethylamino)-4-methyl-2-oxo-2H-chromen-3-yl)phenylacetate (CS 3.117), which inhibited the growth of A549, rat hepatoma (CRL 1548)

■ **CS 3.113** 4-(3′-Hydroxy-4′-methoxyphenyl) coumarin

■ **CS 3.114** 4-(3′-Hydroxy-4′-methoxyphenyl)-5-methoxycoumarin

■ **CS 3.115** 4-(3′-Hydroxy-4′-methoxyphenyl)-6-methoxycoumarin

■ **CS 3.116** 4-(3′-Hydroxy-4′-methoxyphenyl)-5,7-dimethoxycoumarin

■ **CS 3.117** 4-(7-(Diethylamino)-4-methyl-2-oxo-2H-chromen-3-yl)phenylacetate

and rat normal liver (CRL 1439) cells, with IC_{50} values equal to 48.1 µM, 45.1 µM and 49.6 µM, respectively.[130]

3.1.3 *Garcinia Oligantha* Merr.

History   This plant was first described by Elmer Drew Merrill, in the *Philippine Journal of Science*, published in 1923.

Family   Clusiaceae Lindl., 1836

Common Name   Dan hua shan zhu zi (Chinese)

Habitat and Description   *Garcinia oligantha* is a small tree which is found in China and grows to 4 m tall. The stems are laticiferous and striated. The leaves are simple, opposite and exstipulate. The petiole is 0.5–1 cm long. The leaf blade is broadly lanceolate, 5–10 cm × 1–5 cm, thin, wedge-shaped at base, caudate at apex, and presents five pairs of discrete secondary nerves. The female flowers are solitary and axillary. The calyx consists of four sepals of unequal size. The corolla presents four petals which are orbicular and 0.5 cm long. The gynaecium is ovoid, enveloped by 12 staminodes, and develops a peltate stigma. The fruit is fleshy, oblong, seated on the calyx, 1.5–2 cm × 1 cm, smooth, and encloses a pair of seeds (Figure 3.3).

Medicinal Uses   In China the plant is used to treat inflammation, fever and scalds.

Phytopharmacology   The plant is known to produce series of xanthones, including oliganthins A–D and gaudichaudione H.[131] The medicinal properties of the plant have not yet been validated.

Proposed Research   Pharmacological study of gaudichaudione H and derivatives for the treatment of cancer.

Rationale   Su et al. (2011) studied the structure–activity of xanthone (CS 3.118) against several types of cancer cells and observed that the γ-dibenzopyrone framework by itself was inactive.[132] In the same experiment, 1,3-dihydroxyxanthone (CS 3.119), 1,3,7-trihydroxyxanthone (CS 3.120) and 1,3,6,8-tetrahydroxyxanthone (CS 3.121) destroyed human hepatocellular liver carcinoma (HepG2) cells, with IC_{50} values equal to 71.3 µM, 15.8 µM and 9.1 µM, respectively, implying that the addition of hydroxyl groups to the γ-dibenzopyrone was crucial for the cytotoxicity

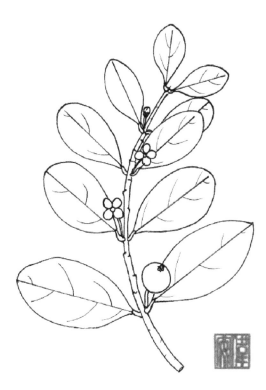

■ **FIGURE 3.3** *Garcinia oligantha* Merr.

■ **CS 3.118** Xanthone

■ **CS 3.119** 1,3-Dihydroxyxanthone

■ **CS 3.120** 1,3,7-Trihydroxyxanthone

■ **CS 3.121** 1,3,6,8-Tetrahydroxyxanthone

■ **CS 3.122** 1,5-Dihydroxyxanthone

of xanthones.[132] Examples of natural hydroxyxanthones are 1,5-dihydroxyxanthone (CS 3.122) and 1,5,6-trihydroxy-3,7-dimethoxyxanthone (CS 3.123) from *Allanblackia floribunda* Oliv. (family Clusiaceae Lindl.), which were lethal to human nasopharyngeal carcinoma (KB) cells, with IC_{50} values equal to 3.3 μg/mL and 2.5 μg/mL, respectively,[133] and gentiakochianin (CS 3.124) and gentiacaulein (CS 3.125) from *Gentiana kochiana* Perr. & Songeon (family Gentianaceae Juss.), which brought about the death of human glioblastoma (U251), with IC_{50} values of 27.3 μM and 31.4 μM, respectively, while the same compounds against mouse glioma (C6) cells gave values of 53.2 μM and 56.6 μM, respectively.[134] 1-Hydroxy-3,7,8-trimethoxyxanthone (CS 3.126) from *Gentianopsis paludosa* (Munro ex Hook. f.) Ma (family Gentianaceae Juss.) hampered the proliferation of human promyelocytic leukemia (HL-60) cells at a dose of 74.4 μM,[135] revealing that methoxylations mitigate the cytotoxic potencies of simple hydroxyxanthones.

The fusion of an alkylated tetrahydrofuran at C3,C4 boosts the cytotoxic potency of 1,5-dihydroxyxanthone as with psorospermin (CS 3.127),

■ **CS 3.123** 1,5,6-Trihydroxy-3,7-dimethoxyxanthone

■ **CS 3.124** Gentiakochianin

■ **CS 3.125** Gentiacaulein

■ **CS 3.126** 1-Hydroxy-3,7,8-trimethoxyxanthone

O-demethyl-3′,4′-deoxypsorospermin -3′,4′-diol (CS 3.128) and *O*-methyl-3′,4′-deoxypsorospermin-3′,4′-diol (CS 3.129), and *O*-ethyl-3′,4′-deoxy-psorospermin-3′,4′-diol (CS 3.130) isolated from *Psorospermum feb-rifugum* Spach (family Hypericaceae Juss.) inhibited the growth of 9PS cells, with IC$_{50}$ values equal to 10^{-4} μg/mL, 10^{-2} μg/mL, 2×10^{-4} μg/mL and 3×10^{-3} μg/mL, respectively.[136] Likewise, the fusion of an extra benzene ring to 1,3 dihydroxyxanthone magnifies its cytotoxic activity, as

■ **CS 3.127** Psorospermin

■ **CS 3.128** *O*-Demethyl-3′,4′-deoxypsorospermin -3′,4′-diol

■ **CS 3.129** *O*-Methyl-3′,4′-deoxypsorospermin -3′,4′-diol

■ **CS 3.130** *O*-Ethyl-3′,4′-deoxypsorospermin -3′,4′-diol

■ **CS 3.131** 1,3-Dihydroxy-6,7-benzoxanthone

■ **CS 3.132** 1,3-Dihydroxy-6,7-benzohydroxyxanthone

■ **CS 3.133** 1,3-Dihydroxy-7,8-benzoxanthone

■ **CS 3.134** 1-Hydroxy-3--xanthone

exemplified with 1,3-dihydroxy-6,7-benzoxanthone (CS 3.131), 1,3-dihydroxy-6,7-benzohydroxyxanthone (CS 3.132) and 1,3-dihydroxy-7,8-benzoxanthone (CS 3.133), with which IC_{50} values equal to 8.4 μM, 4 μM and 6.8 μM were obtained against HepG2 cells.[132] The attachment of an alkoxy oxirane at C3 of 1,3 dihydroxyxanthone is also beneficial whereas linear alkoxylation at C3 (CS 3.134) mitigates the activity.[132]

The cytotoxic potencies of simple hydroxyxanthones are increased by the attachmnent of isoprenyl moieties to the γ-dibenzopyrone framework. Such prenylated xanthones are found in members of the family Clusiaceae Lindl., Hypericaceae Juss. and Moraceae Gaudich. Cratoxyarborenones A, B (CS 3.135), C (CS 3.136), D (CS 3.137), E (CS 3.138) and F from *Cratoxylum sumatranum* Blume (family Hypericaceae Juss.) were cytotoxic against KB cells, with IC_{50} values equal to 4.3 μg/mL, 1 μg/mL, 1.5 μg/mL, 1.7 μg/mL, 4.3 μg/mL and 4.1 μg/mL, respectively[137] 1,4,5,6-Tetrahydroxy-7,8-diprenylxanthone (CS 3.139), 1,3,5,6-tetrahydroxy-4,7,8-triprenylxanthone (CS 3.140) from *Garcinia xanthochymus* Hook. f. (family Clusiaceae Lindl.) destroyed human breast carcinoma (MDA-MB-435) cells, with IC_{50} values equal to 1.3 μg/mL and 1.7 μg/mL, respectively. The same compounds also killed human lung adenocarcinoma epithelial (A549) cells, with IC_{50} values equal to 3.8 μg/mL and 3.3 μg/mL, respectively,[138] suggesting that a C8 isoprenyl moiety favors the cytotoxic activity.

■ **CS 3.135** Cratoxyarborenone B

■ **CS 3.136** Cratoxyarborenone C

■ **CS 3.137** Cratoxyarborenone D

■ **CS 3.138** Cratoxyarborenone E

■ **CS 3.139** 1,4,5,6-Tetrahydroxy-7,8-diprenylxanthone

■ **CS 3.140** 1,3,5,6-Tetrahydroxy-4,7,8-triprenylxanthone

■ **CS 3.141** Alvaxanthone

■ **CS 3.142** Isoalvaxanthone

■ **CS 3.143** Allanxanthone A

■ **CS 3.144** Cudratricusxanthone E

This was confirmed with alvaxanthone (CS 3.141) and isoalvaxanthone (CS 3.142) from *Cudrania cochinchinensis* (Lour.) Kudô & Masam. (family Moraceae Gaudich), which abated the survival of human squamous carcinoma (HSC-2) cells, with IC_{50} values equal to 0.02 μM.[139] A C2 isoprenyl moiety is a beneficial chemical feature since allanxanthone A (CS 3.143) from *Allanblackia floribunda* Oliv. (family Clusiaceae Lindl.) compelled the death of KB cells, with an IC_{50} value equal to 1.5 μg/mL.[133] Cudratricusxanthones E (CS 3.144) and G (CS 3.145), cudraxanthone M (CS 3.146) and toxyloxanthone C (CS 3.147) from a member of the genus *Cudrania tricuspidata* (Carrière) Bureau ex Lavalle (family Moraceae Gaudich) were lethal to the following cell lines (respective IC_{50} values in parentheses): human colorectal carcinoma (HCT-116; 4.7 μg/mL, 1.8 μg/mL, 3.4 μg/mL and 2.8 μg/mL), human hepatoma (SMMC-7721; 4,2 μg/mL, 2.7 μg/mL, 5.1 μg/mL, 8.8 μg/mL), human gastric cancer (SGC-7901;

■ **CS 3.145** Cudratricusxanthone G

■ **CS 3.146** Cudraxanthone M

■ **CS 3.147** Toxyloxanthone C

■ **CS 3.148** Isoprenylated tetrahydroxyxanthone

5.4 μg/mL, 3.4 μg/mL, 0.5 μg/mL, 11.8 μg/mL) and human gastric cancer (BGC-823; 1.6 μg/mL, 1.6 μg/mL, 2.6 μg/mL and 5.2 μg/mL).[140]

An isoprene group at C3 is profitable for the cytotoxicity of xanthones as an isoprenylated tetrahydroxyxanthone (CS 3.148), macluraxanthone B (CS 3.149) and cudraxanthone L (CS 3.150) from *Cudrania tricuspidata* (Carrière) Bureau ex Lavalle (family Moraceae Gaudich) were able to kill human lung adenocarcinoma epithelial (A549) and human ovary adeno-carcinoma (SK-OV-3) cells, with IC$_{50}$ values equal to 45.8 μM, 25.8 μM, 33.5 μM and 43.2 μM, 23.1 μM, 38 μM, respectively.[141]

Cudrafrutixanthone A (CS 3.151) from *Cudrania fruticosa* (family Moraceae Gaudich) presents a single isoprene in C7 and potently inhibited the growth

■ **CS 3.149** Macluraxanthone B

■ **CS 3.150** Cudraxanthone L

■ **CS 3.151** Cudrafrutixanthone A

■ **CS 3.152** Cudraxanthone D

of HCT-116, BGC-823, SGC-7901 and SMMC-7721, with IC_{50} values equal to 1.8 μg/mL, 2 μg/mL, 1.5 μg/mL and 2.6 μg/mL, respectively.[142] In contrast, the IC_{50} values of cudraxanthones D (CS 3.152), L (CS 3.153) and M against human gastric cancer (AGS) cells were equal to 4.7 μM, 3.9 μM and 4.1 μM, respectively,[141] inferring that an isoprene at C8 might not always be beneficial. Note that the C7 isoprenoid globulixanthone A (CS 3.154) from *Symphonia globulifera* L.f. (family Clusiaceae Lindl.) compromised the survival of KB cells, with IC_{50} values equal to 2.1 μg/mL.[143] However, *Garcinia mangostana* L. (family Clusiaceae Lindl.) produces cohorts of C8 prenylated xanthones, such as mangostenone C (CS 3.155), β-mangostin (CS 3.156), garcinone E (CS 3.157), α-mangostin (CS 3.158) and garcinone D (CS 3.159), which abolished the growth of KB cells with IC_{50} values equal to 2.8 μg/mL, 2.5 μg/mL, 2.6 μg/mL, 2 μg/mL and 3.5 μg/mL,

■ **CS 3.153** Cudraxanthone L

■ **CS 3.154** Globulixanthone A

■ **CS 3.155** Mangostenone C

■ **CS 3.156** β-Mangostin

■ **CS 3.157** Garcinone E

■ **CS 3.158** α-Mangostin

■ **CS 3.159** Garcinone D

■ **CS 3.160** γ-Mangostin

■ **CS 3.161** 1,3,7-trihydroxy-2,4-diisoprenylxanthone

■ **CS 3.162** Oliganthin A

respectively.[144] α-Mangostin from *Garcinia mangostana* L. (family Clusiaceae Lindl.) abated the proliferation of human oral squamous cell carcinoma (HN-22) by 45%.[145] The positive cytotoxic effect of an isoprene moiety at C8 was evidenced by α-mangostin and γ-mangostin (CS 3.160) from *Cratoxylum cochinchinense* (Lour.) Blume (family Hypericaceae Juss.), which blocked the proliferation of human colon cancer (HT-29) cells, with IC_{50} values of 4 μM, contrasting with 1,3,7-trihydroxy-2,4-diisoprenylxanthone (CS 3.161), which had an IC_{50} of greater than 10 μM.[146] *Pentadesma butyracea* Sabine (family Clusiaceae Lindl.) produces the 2,5 9 isoprenyl garcinone E, which annihilated the multiplication of human breast adenocarcinoma (MCF-7) cells, with an IC_{50} value equal to 1.5 μg/mL.[147] *Garcinia oligantha* Merr. (family Clusiaceae Lindl.) synthesizes oliganthins A–C (CS 3.162–164), which inhibited the growth

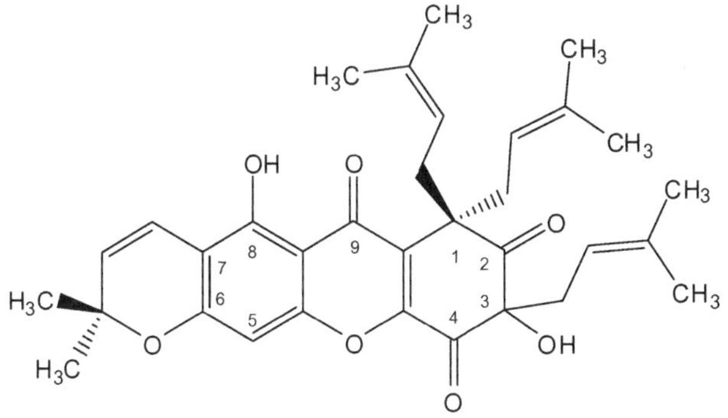

■ **CS 3.163** Oliganthin B

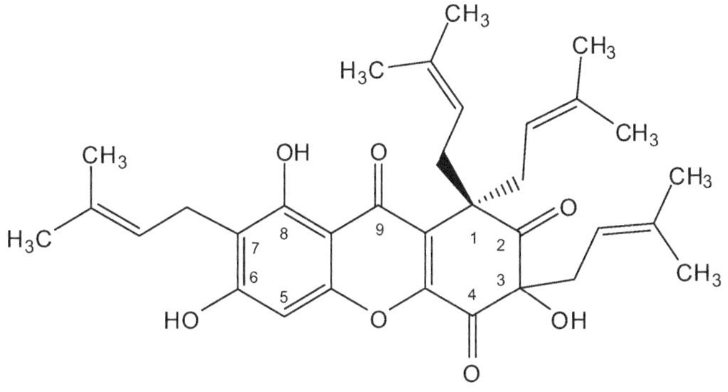

■ **CS 3.164** Oliganthin C

of human epitheloid cervix carcinoma (Hela) cells, with IC_{50} values equal to 1.5 µM, 1.5 µM and 4.1 µM, respectively.[131] The presence of geranyl moieties on the γ-dibenzopyrone framework obviously increases the lipophilicity of simple hydroxyxanthones as well as their cytotoxic potencies. Cratoxyarborenone A (CS 3.165) from *Cratoxylum sumatranum* abated the survival of KB cells, with an IC_{50} value of 4.3 µg/mL,[137] implying that a geranyl moiety at C8 is not as favorable as an isoprene moiety. This was observed in rubraxanthone (CS 3.166) from *Garcinia merguensis* Wight (family Clusiaceae Lindl.), which was mildly cytotoxic against MCF-7, human breast adenocarcinoma (MDA-MB-231), human large-cell lung cancer (NCI-H460) and human glioblastoma (SF268) cells, with IC_{50} values equal to 9 µM, 16.8 µM, 18.5 µM and 17 µM, respectively.[148] A geranyl moiety, however, is quite beneficial at C3: formoxanthone A (CS 3.167) from *Cratoxylum formosum* (Jack) Dyer (family Hypericaceae Juss.) was lethal to human KB cells, with an IC_{50} value of 3.3 µg/mL.[149] The fusion of isoprenes with adjacent hydroxyl groups yields tetrahydropyran γ-dibenzopyrones

■ **CS 3.165** Cratoxyarborenone A

■ **CS 3.166** Rubraxanthone

■ **CS 3.167** Formoxanthone A

■ **CS 3.168** Globulixanthone B

which are much less planar than simple hydroxanthones. One such compound, globulixanthone B (CS 3.168) from *Symphonia globulifera* L.f. (family Clusiaceae Lindl.), abolished the viability of KB cells, with an IC_{50} value of 1.7 μg/mL,[143] suggesting that a tetrahydropyran moiety fused at C7,C8 is not detrimental to cytotoxicity. In addition, C6,C7 tetrahydropryanxanthones such as trapezifolixanthone (CS 3.169) and manglexanthone (CS 3.170) from *Tovomita brevistaminea* Engl. (family Clusiaceae Lindl.) prevented the multiplication of KB cells, with IC_{50} values equal

■ **CS 3.169** Trapezifolixanthone

■ **CS 3.170** Manglexanthone

■ **CS 3.171** Caloxanthone A

■ **CS 3.172** Artoindonesianin P

to 4.1 µg/mL and 1.9 µg/mL, respectively.[150] Close to manglexanthone is caloxanthone A (CS 3.171) from *Calophyllum inophyllum* L. (family Calophyllaceae J. Agardh), which vitiated the ability of KB cells to cells, with IC_{50} values equal to 7.4 µg/mL.[151] This suggests that a hydroxyl group at C2 diminishes the cytotoxic potency of manglexanthone. The C5,C6 tetrahydropyranxanthones artoindonesianin P (CS 3.172), artobiloxanthon (CS 3.173), cycloartobiloxanthone (CS 3.174) and artonol B (CS 3.175) from *Artocarpus lanceifolius* Roxb. (family Moraceae Gaudich.) were cytotoxic against mouse leukemia (P388) cells, with IC_{50} values equal to 5.9 µg/mL, 1.7 µg/mL, 4.6 µg/mL and 100 µg/mL, respectively.[152] *Pentadesma butyracea* Sabine (family Clusiaceae Lindl.) produces gartanin (CS 3.176), which interrupted the growth of MCF-7 cells, with an IC_{50} value equal to 1.2 µg/mL.[147] *Garcinia oligantha* Merr. (family Clusiaceae Lindl.) contains

■ **CS 3.173** Artobiloxanthone

■ **CS 3.174** Cycloartobiloxanthone

■ **CS 3.175** Artonol B

■ **CS 3.176** Gartanin

■ **CS 3.177** Oliganthin D

■ **CS 3.178** Termicalcicolanone A

oliganthin D (CS 3.177), which prevented the proliferation of Hela cells, with an IC_{50} value of 7.8 μM.[70] *Terminalia calcicola* contains termicalcicolanones A (CS 3.178) and B (CS 3.179), which abated the proliferation of human ovarian carcinoma (A2780) cells, with IC_{50} values of 40.6 μM and 8.1 μM, respectively,[153] indicating that a tetrahydropyran at C5,C6 may be more favorable than at C6,C7.

■ **CS 3.179** Termicalcicolanone B

■ **CS 3.180** Desoxymorellin

■ **CS 3.181** Desoxygambogenin

Of tremendous oncological interest are the 'caged' xanthones which abrogate the survival of multidrug-resistant malignancies. Desoxymorellin (CS 3.180), desoxygambogenin (CS 3.181) and gambogellic acid (CS 3.182) from *Garcinia hanburyi* Hook.f. (family Clusiaceae Lindl.) destroyed Hela cells, with IC_{50} values equal to 0.3 µg/mL, 0.7 µg/mL and 1.5 µg/mL, respectively.[154] From the same plant, gaudichaudic acid (CS 3.183), gambogoic acid A (CS 3.184), isomorellic acid (CS 3.185), desoxymorellin and isomorellinol (CS 3.186) were lethal to adriamycin-resistant human erythromyeloblastoid leukemia (K562R) cells, with IC_{50} values equal to 0.6 µg/mL, 1.6 µg/mL, 1,8 µg/mL, 1.5 µg/mL and 0.6 µg/mL, respectively.[155] *Garcinia hanburyi* Hook.f. (family Clusiaceae Lindl.) also produces 30-hydroxygambogic acid (CS 3.187), 30-hydroxyepigambogic acid (CS 3.188), gambogic acid (CS 3.189) and epigambogic acid

■ **CS 3.182** Gambogellic acid

■ **CS 3.183** Gaudichaudic acid

■ **CS 3.184** Gambogoic acid A

■ **CS 3.185** Isomorellic acid

■ **CS 3.186** Isomorellinol

■ **CS 3.187** 30-Hydroxygambogic acid

■ **CS 3.188** 30-Hydroxyepigambogic acid

■ **CS 3.189** Gambogic acid

■ **CS 3.190** Epigambogic acid

■ **CS 3.191** Gambogenic acid

(CS 3.190), which inhibited the growth of K562R cells, with IC_{50} values equal to 2.8 μM, 4.4 μM, 1.3 μM and 1,1 μM.[156] Gambogic acid inhibited the growth of HCT-116, Hela, HepG2 and MCF-7 cells with IC_{50} values 1.2 μmol/L, 3.5 μmol/L, 3.8 μmol/L and 4.1 μmol/L, respectively.[157] The same compound was cytotoxic to human breast carcinoma (T-47D), human breast cancer (ZR-751) and human colon carcinoma (DLD-1) cells, with IC_{50} values equal to 0.7 μM, 1.6 μM and 0.8 μM.[158] Gambogenic acid (CS 3.191) from *Garcinia hanburyi* Hook.f. (family Clusiaceae Lindl.) prompted apoptosis in A549 cells, with an IC_{50} value of 2.5 μM.[159] Gaudichaudione A (CS 3.192) from *Garcinia gaudichaudii* Planch. & Triana (family Clusiaceae Lindl.) was cytotoxic against P388 and doxorubicin resistant mouse leukemia (P388/DOX) cells, with an IC_{50} value equal to 10 μM.[160]

Garcinia lateriflora Blume (family Clusiaceae Lindl.) produces isogaudichaudic acid (CS 3.193), isogaudichaudic acid E (CS 3.194), 11,12-dihydro-12-hydroxymorellic acid (CS 3.195), and morellic acid

■ **CS 3.192** Gaudichaudione A

■ **CS 3.193** Isogaudichaudic acid

■ **CS 3.194** Isogaudichaudic acid E

■ **CS 3.195** 11,12-Dihydro-12-hydroxymorellic acid

(CS 3.196), which were shown to be toxic to HT-29 cells, with IC_{50} values equal to 3.2 μM, 2.6 μM, 2.9 μM and 0.3 μM.[161] Furthermore, gaudichaudione H from *Garcinia oligantha* Merr. (family Clusiaceae Lindl.) inhibited the growth of Hela cells, with an IC_{50} equal to 0.9 μM.[131] Cochinchinoxanthone (CS 3.197) from *Cratoxylum cochinchinense* was cytotoxic against HT-29 cells, with an IC_{50} value equal to 5.8 μM.[146] The mode of action of simple hydroxylated xanthones is linked to DNA and microtubules. For instance, 1-hydroxy-3,7,8-trimethoxyxanthone from *Gentianopsis paludosa* (Munro ex Hook. f.) Ma (family Gentianaceae Juss.) blocked the proliferation of HL-60 cells on account of topoisomerase II inhibition.[135] In addition, gentiakochianin from *Gentiana kochiana* Perr. & Songeon (family Gentianaceae Juss.) provoked the apoptosis of mouse glioma (C6) cells by impairing microtubule homeostasis, with

■ **CS 3.196** Morellic acid

■ **CS 3.197** Cochinchinoxanthone

an IC_{50} value of 18 μM.[134] Permana et al. (1994) noted that psorospermin caused DNA damage[162] by a mechanism that was elegantly delineated by Hansen et al. (1996) and Kwok et al. (1998), whereas topoisomerase II induces DNA allosteric modification which favors the intercalation of psorospermin between the bases and electrophilic attack at the epoxide ring by N7 guanine.[163,164] Damaged DNA could very well induce apoptosis, as exemplified with 1,3-dihydroxy-6,7-benzohydroxyxanthone, where this was accompanied by a fall in mitochondrial membrane potential, activation of caspases 9 and 3, reduced levels of anti-apoptotic myeloid cell leukemia-1 (Mcl-1) and a decrease in topoisomerase II expression and enzymatic activity.[132]

Much attention has been given to the prenylated xanthone α-mangostin, the apoptotic mechanism of which implies mitochondrial insult. α-Mangostin from *Garcinia mangostana* L. (family Clusiaceae Lindl.) induced apoptosis in DLD-1 cells, with an IC_{50} value of 7.5 μM.[165] The DLD-1 cells became apoptotic when exposed to 20 μM of α-mangostin, which prompted the release of mitochondrial endonuclease-G (Endo-G) and an increase in microRNA miR-143 and hence consequent decreases in ERK1/2 and c-Myc.[165] One could reasonably frame the hypothesis that c-Myc inhibition may not result in activation of pro-apoptotic proteins (p53) and Bax since p53 is mutated in DLD-1 cells. Indeed, human oral squamous cell carcinoma (HN-22) cells (wild-type p53) exposed to 3 μg/mL of α-mangostin underwent apoptosis with increased levels of anti-apoptotic protein (p53), pro-apoptotic Bax, decrease of Bcl-2[145] and therefore mitochondrial insult. In rat pheochromocytoma (PC12) cells DNA fragmentation, a fall in mitochondrial membrane potential, release of cytochrome c, caspase 3 activation, reticulum endoplasmic stress and Ca^{2+}–ATPase inhibition were observed upon the administration of α-mangostin at a dose of 4 μM.[166] In addition, α-Mangostin at a dose of 10 μM induced apoptosis in HL-60 cells (p53 null) via a fall in mitochondrial membrane potential, release of cytochrome c and AIF, and activation of caspases 9 and 3.[167] In HT-29 cells, α-mangostin and β-mangostin inhibited p65 activation, with IC_{50} values of 15.9 μM and 12.1 μM, respectively,[168] suggesting a possible inhibition of Akt by α-mangostin. In fact, α-mangostin at a dose of 10 μg/mL decreased the levels of phosphorylated Akt, ERK1/2 and JNK, activated caspases 9, 3 and 8, decreased Bcl-2 and induced pro-apoptotic Bax and hence the release of cytochrome c in human chondrosarcoma (SW1353) cells.[169] α-Mangostin mediated apoptosis in human malignant glioma (GBM-8401) and human glioblastoma (DBTRG-05MG) cells, with IC_{50} values equal to 6.4 μM and 7.3 μM, respectively, with the formation of autophagocytic vacuoles, increased phosphorylation of the serine/threonine kinase LKB1 and its downstream target 5′ AMP-activated protein kinase (AMPK), hence inhibition of mTORC1 activity by boosting Raptor phosphorylation.[170] Human malignant glioma (GBM-8401) cells injected into rodents formed tumors, the size of which decreased upon a regimen of 2 mg/kg/day of α-mangostin.[170]

Most of what is so far undertood about the apoptotic mechanisms of caged xanthones is owed to gambogic acid, which underwent phase II clinical trials.[161] Structure–activity evidence clearly demonstrates that the Δ_1 double bond and the α,β-unsaturated ketone of caged xanthones are targeted by nucleophiles and the resulting Michael additions form covalent bonds with proteins.[158,171,172] Jurkat human T cell lymphoblast-like cells treated with 62.5 μM of gaudichaudione A and Hela exposed to 10 μM of gaudichaudione H became apoptotic via caspase 3 activation.[131,160,173] Caspase 3 activation implies mitochondrial insults which may very well occur as a result of pro-apoptotic Bax activation and translocation, as demonstrated by Zhao et al. (2004) in human gastric carcinoma (MGC-803) cells.[174] Indeed, gambogic acid induced apoptosis in SMMC-7721 and human malignant melanoma (A375) cells, with IC_{50} values of 3.2 μM and 7.5 μg/mL, the result of a decrease in Bcl-2, an increase in pro-apoptotic Bax and activation of caspase 3.[175,176] The mitochondrial involvement of gambogic acid was further confirmed in SMMC-7721 cells, where ROS increased, mitochondrial membrane potential collapsed and cytochrome c and AIF were released.[177] Likewise, K562 cells exposed to gambogic acid experienced a decrease in Bcl-2, owing to a reduction in steroid receptor coactivator-3 (SRC-3) and therefore inactivation of Akt and GSK3β and apoptosis, with an IC_{50} value of 1.9 μM.[178] A striking feature of caged xanthones is that they bring about apoptosis in both wild-type p53 and p53 null cancer cells. In HCT-116 (wild-type p53) cells, 1.2 μmol/L of gambogic acid lowered the levels of MDM2 with the consequent upregulation of pro-apoptotic p53 and apoptosis.[157,179] Furthermore, gambogic acid destroyed MCF-7 (wild-type p53) and human non-small-cell lung carcinoma (H1299) (p53 null) cells, with IC_{50} values of 3.5 μM.[180] In H1299 (p53 null) cells, 5 μM of gambogic acid induced the degradation of MDM2 and hence p21 upregulation.[180] In MCF-7 (wild-type p53) cells, gambogic acid increased the level of p53 and therefore repressed Bcl-2.[181] Note that gambogic acid provokes the degradation of mutant p53 by a chaperone-assisted ubiquitin/proteasome pathway.[182] Other compelling oncocellular features of gambogic acid are the inhibition of STAT3 and NF–κB. Gambogic acid blocked the phosphorylation of JAK and therefore the phosphorylation and activation of STAT3 followed by the reduction in IAP, survivin, Mcl-1, Bcl-2, and Bcl-xL, and cyclin D1 expression.[183] Gambogic acid inhibited the viability of A549 and U251 cells, with IC_{50} values of 1.1 μM and 2.5 μM, respectively, as a result of IKKβ inhibition and therefore dephosphorylation of IκBα.[184] In fact, the 4-oxa-tricyclo[4.3.1.0^{3,7}] dec-2-one moiety of caged xanthone is crucial for IKKβ inhibition and apoptosis.[184] Note that gambogenic acid at a dose of 1 μM in Hela cells docked into the ATP-binding domain of Hsp90 and therefore inactivated both Akt and IKK.[185] Inactivation of Akt was further observed in CNE-1 cells exposed to 1.8 μM of gambogenic acid with the consequent decrease in Bcl-2, increase in pro-apoptotic Bax, fall in mitochondrial membrane potential, release of cytochrome c, activation of caspase 9 and apoptosis.[186] Note that gambogic acid inhibits the enzymatic activity of topoisomerase II and interferes with microtubules. Gambogic acid binds to the ATPase domain of topoisomease II[187] and disrupted microtubule homeostasis in MCF-7 cells at a dose of 2.5 μM, resulting in an increase in phosphorylation levels of p38 MAPK and JNK in MCF-7 cells and therefore in apoptosis.[188]

REFERENCES

[1]   Ko HH, Yu SM, Ko FN, Teng CM, Lin CN. Bioactive constituents of *Morus australis* and *Broussonetia papyrifera*. J Nat Prod 1997;60(10):1008–11.

[2]   Nomura T, Fukai T, Shimada T, Chen IS. Components of root bark of *Morus australis*. I. Structure of a new 2-arylbenzofuran derivative, mulberrofuran D. Planta Med 1983;49(2):90–4.

[3]   Shi YQ, Fukai T, Chang WJ, Yang PQ, Wang FP, Nomura T. Phenolic constituents of the root bark of Chinese *Morus australis*. Natural Med 2001;55(3):143–6.

[4]   Zhang QJ, Tang YB, Chen RY, Yu DQ. Three new cytotoxic Diels-Alder-type adducts from *Morus australis*. Chem Biodivers 2007;4(7):1533–40.

[5]   Ferlinahayati S, Yana M, Juliawaty LD, Achmad SA, Hakim EH, Takayama H, et al. Phenolic constituents from the wood of *Morus australis* with cytotoxic activity. J Biosci 2008;63(1–2):35–9.

[6]   Ko HH, Wang JJ, Lin HC, Wang JP, Lin CN. Chemistry and biological activities of constituents from *Morus australis*. Biochim Biophys Acta 1999;1428(2–3):293–9.

[7]   Zhang QJ, Zheng ZF, Chen RY, Yu DQ. Two new dimeric stilbenes from the stem bark of *Morus australis*. J Asian Nat Prod Res 2009;11(2):138–41.

[8]   Woo KJ, Jeong YJ, Park JW, Kwon TK. Chrysin-induced apoptosis is mediated through caspase activation and protein kinase B (Akt) inactivation in U937 leukemia cells. Biochem Biophys Res Comm 2004;325(4):1215–22.

[9]   Li X, Huang Q, Ong CN, Yang XF, Shen HM. Chrysin sensitizes tumor necrosis factor-α-induced apoptosis in human tumor cells via suppression of nuclear factor-kappaB. Cancer Lett 2010;293(1):109–16.

[10] Kumar MAS, Nair M, Hema PS, Mohan J, Santhoshkumar TR. Pinocembrin triggers Bax-dependent mitochondrial apoptosis in colon cancer cells. Mol Carcinog 2007;46(3):231–41.

[11] Bestwick CS, Milne L. Influence of galangin on HL-60 cell proliferation and survival. Cancer Lett 2006;243(1):80–9.

[12] Zhang HT, Luo H, Wu J, Lan LB, Fan DH, Zhu KD, et al. Galangin induces apoptosis of hepatocellular carcinoma cells via the mitochondrial pathway. World J Gastroenterol 2010;16(27):3377–84.

[13] Lee HZ, Leung HWC, Lai MY, Wu CH. Baicalein induced cell cycle arrest and apoptosis in human lung squamous carcinoma CH27 cells. AntiCancer Res 2005;25(2 A):959–64.

[14] Liu S, Ma Z, Cai H, Li Q, Rong W, Kawano M. Inhibitory effect of baicalein on IL-6-mediated signaling cascades in human myeloma cells. European J Haematol 2010;84(2):137–44.

[15] Takahashi H, Chen MC, Pham H, Angst E, King JC, Park J, et al. Baicalein, a component of *Scutellaria baicalensis*, induces apoptosis by Mcl-1 down-regulation in human pancreatic cancer cells. Biochim Biophysi Acta 2011;1813(8):1465–74.

[16] Pan MH, Lai CS, Hsu PC, Wang YJ. Acacetin induces apoptosis in human gastric carcinoma cells accompanied by activation of caspase cascades and production of reactive oxygen species. J Agr Food Chem 2005;53(3):620–30.

[17] Lee WR, Shen SC, Lin HY, Hou WC, Yang LL, Chen YC. Wogonin and fisetin induce apoptosis in human promyeloleukemic cells, accompanied by a decrease of reactive oxygen species, and activation of caspase 3 and Ca2+-dependent endonuclease. Biochem Pharmacol 2002;63:225–36.

[18] Yu JQ, Liu HB, Tian DZ, Liu YW, Lei JC, Zou GL. Changes in mitochondrial membrane potential and reactive oxygen species during wogonin-induced cell death in human hepatoma cells. Hepatol Res 2007;37(1):68–76.

[19] Chung H, Jung YM, Shin DH, Lee JY, Oh MY, Kim HJ, et al. Anticancer effects of wogonin in both estrogen receptor-positive and -negative human breast cancer cell lines in vitro and in nude mice xenografts. Int J Cancer 2008;122(4):816–22.

[20] Chow JM, Huang GC, Shen SC, Wu CY, Lin CW, Chen YC. Differential apoptotic effect of wogonin and nor-wogonin via stimulation of ROS production in human leukemia cells. J Cell Biochem 2008;103(5):1394–404.

[21] Huang KF, Zhang GD, Huang YQ, Diao Y. Wogonin induces apoptosis and down-regulates survivin in human breast cancer MCF-7 cells by modulating PI3K-AKT pathway. Int Immunopharmacol 2012;12(2):334–41.

[22] Sung B, Pandey MK, Aggarwal BB. Fisetin, an inhibitor of cyclin-dependent kinase 6, down-regulates nuclear factor-κB-regulated cell proliferation, anti-apoptotic and metastatic gene products through the suppression of TAK-1 and receptor-interacting protein-regulated IκBα kinase activation. Mol Pharmacol 2007;71(6):1703–14.

[23] Khan N, Afaq F, Syed DN, Mukhtar H. Fisetin, a novel dietary flavonoid, causes apoptosis and cell cycle arrest in human prostate cancer LNCaP cells. Carcinogenesis 2008;29(5):1049–56.

[24] Murtaza I, Adhami VM, Hafeez BB, Saleem M, Mukhtar H. Fisetin, a natural flavonoid, targets chemoresistant human pancreatic cancer AsPC-1 cells through DR3-mediated inhibition of NF-κB. Int J Cancer 2009;125(10):2465–73.

[25] Suh Y, Afaq F, Khan N, Johnson JJ, Khusro FH, Mukhtar H. Fisetin induces autophagic cell death through suppression of mTOR signaling pathway in prostate cancer cells. Carcinogenesis 2010;31(8):1424–33.

[26] Suh Y, Afaq F, Johnson JJ, Mukhtar H. A plant flavonoid fisetin induces apoptosis in colon cancer cells by inhibition of COX2 and Wnt/EGFR/NF-κB-signaling pathways. Carcinogenesis 2009;30(2):300–7.

[27] Ying TH, Yang SF, Tsai SJ, Hsieh SC, Huang YC, Bau DT, et al. Fisetin induces apoptosis in human cervical cancer HeLa cells through ERK1/2-mediated activation of caspase-8-/caspase-3-dependent pathway. Arch Toxicol 2012;86(2):263–73.

[28] Wang IK, Lin-Shiau SY, Lin JK. Induction of apoptosis by apigenin and related flavonoids through cytochrome c release and activation of caspase-9 and caspase-3 in leukaemia HL-60 cells. Eur J Cancer 1999;35(10):1517–25.

[29] Torkin R, Lavoie JF, Kaplan DR, Yeger H. Induction of caspase-dependent, p53-mediated apoptosis by apigenin in human neuroblastoma. Mol Cancer Ther 2005;4(1):1–11.

[30] Kaur P, Shukla S, Gupta S. Plant flavonoid apigenin inactivates Akt to trigger apoptosis in human prostate cancer: an in vitro and in vivo study. Carcinogenesis 2008;29(11):2210–17.

[31] Zheng PW, Chiang LC, Lin CC. Apigenin induced apoptosis through p53-dependent pathway in human cervical carcinoma cells. Life Sci 2005;76(12):1367–79.

[32] Way TD, Kao MC, Lin JK. Degradation of HER2/neu by apigenin induces apoptosis through cytochrome c release and caspase-3 activation in HER2/neu-overexpressing breast cancer cells. FEBS Letters 2005;579(1):145–52.

[33] Chiang LC, Ng LT, Lin IC, Kuo PL, Lin CC. Anti-proliferative effect of apigenin and its apoptotic induction in human Hep G2 cells. Cancer Lett 2006;237(2):207–14.

[34] Vargo MA, Voss OH, Poustka F, Cardounel AJ, Grotewold E, Doseff AI. Apigenin-induced-apoptosis is mediated by the activation of PKCδ and caspases in leukemia cells. Biochem Pharmacol 2006;72(6):681–92.

[35] Choi SI, Jeong CS, Cho SY, Lee YS. Mechanism of apoptosis induced by apigenin in HepG2 human hepatoma cells: involvement of reactive oxygen species generated by NADPH oxidase. Arch Pharm Res 2007;30(10):1328–35.

[36] Park JH, Jin CY, Lee BK, Kim GY, Choi YH, Jeong YK. Naringenin induces apoptosis through downregulation of Akt and caspase-3 activation in human leukemia THP-1 cells. Food Chem Toxicol 2008;46(12):3684–90.

[37] Lee ER, Kang YJ, Choi HY, Kang GH, Kim JH, Kim BW, et al. Induction of apoptotic cell death by synthetic naringenin derivatives in human lung epithelial carcinoma A549 cells. Biol Pharm Bull 2007;30(12):2394–8.

[38] Sharma V, Joseph C, Ghosh S, Agarwal A, Mishra MK, Sen E. Kaempferol induces apoptosis in glioblastoma cells through oxidative stress. Mol Cancer Ther 2007;6(9):2544–53.

[39] Kim BW, Lee ER, Min H, Jeong HS, Ahn JY, Kim JH, et al. Sustained ERK activation is involved in the kaempferol-induced apoptosis of breast cancer cells and is more evident under 3-D culture condition. Cancer Biol Ther 2008;7(7):1080–9.

[40] Luo H, Rankin GO, Li Z, DePriest L, Chen YC. Kaempferol induces apoptosis in ovarian cancer cells through activating p53 in the intrinsic pathway. Food Chem 2011;128(2):513–9.

[41] Shi RX, Ong CN, Shen HM. Luteolin sensitizes tumor necrosis factor-α-induced apoptosis in human tumor cells. Oncogene 2004;23(46):7712–21.

[42] Horinaka M, Yoshida T, Shiraishi T, Nakata S, Wakada M, Nakanishi R, et al. Luteolin induces apoptosis via death receptor 5 upregulation in human malignant tumor cells. Oncogene 2005;24(48):7180–9.

[43] Selvendiran K, Koga H, Ueno T, Yoshida T, Maeyama M, Torimura T, et al. Luteolin promotes degradation in signal transducer and activator of transcription 3 in human hepatoma cells: an implication for the antitumor potential of flavonoids. Cancer Res 2006;66(9):4826–34.

[44] Choi AY, Choi JH, Yoon H, Hwang KY, Noh MH, Choe W, et al. Luteolin induces apoptosis through endoplasmic reticulum stress and mitochondrial dysfunction in neuro-2a mouse neuroblastoma cells. Eur J Pharmacol 2011;668(1–2):115–26.

[45] Yeh TC, Chiang PC, Li TK, Hsu JL, Lin CJ, Wang SW, et al. Genistein induces apoptosis in human hepatocellular carcinomas via interaction of endoplasmic reticulum stress and mitochondrial insult. Biochem Pharmacol 2007;73(6):782–92.

[46] Kuo PC, Liu HF, Chao JI. Survivin and p53 modulate quercetin-induced cell growth inhibition and apoptosis in human lung carcinoma cells. J Biol Chem 2004;279(53):55875–55885.

[47] Haghiac M, Walle T. Quercetin induces necrosis and apoptosis in SCC-9 oral cancer cells. Nutr Cancer 2005;53(2):220–31.

[48] Granado-Serrano AB, Martín MA, Bravo L, Goya L, Ramos S. Quercetin induces apoptosis via caspase activation, regulation of Bcl-2, and inhibition of PI-3-kinase/Akt and ERK pathways in a human hepatoma cell line (HepG2). J Nutr 2006;136(11):2715–21.

[49] Siegelin MD, Reuss DE, Habel A, Rami A, Von Deimling A. Quercetin promotes degradation of survivin and thereby enhances death-receptor-mediated apoptosis in glioma cells. Neuro-Oncol 2009;11(2):122–31.

[50] Jacquemin G, Granci V, Gallouet AS, Lalaoui N, Morlé A, Iessi E, et al. Quercetin-mediated Mcl-1 and survivin downregulation restores TRAIL-induced apoptosis in non-Hodgkin's lymphoma B cells. Haematologica 2012;97(1):38–46.

[51] Kim EJ, Choi CH, Park JY, Kang SK, Kim YK. Underlying mechanism of quercetin-induced cell death in human glioma cells. Neurochem Res 2008;33(6):971–9.

[52] Kuo HM, Chang LS, Lin YL, Lu HF, Yang JS, Lee JH, et al. Morin inhibits the growth of human leukemia HL-60 cells via cell cycle arrest and induction of

apoptosis through mitochondria dependent pathway. Anticancer Res 2007; 27(1 A):395–406.

[53] Ko CH, Shen SC, Hsu CS, Chen YC. Mitochondrial-dependent, reactive oxygen species-independent apoptosis by myricetin: roles of protein kinase C, cytochrome c, and caspase cascade. Biochem Pharmacol 2005;69(6):913–27.

[54] Lee KW, Kang NJ, Rogozin EA, Kim HG, Cho YY, Bode AM, et al. Myricetin is a novel natural inhibitor of neoplastic cell transformation and MEK1. Carcinogenesis 2007;28:1918–27.

[55] Kumamoto T, Fujii M, Hou DX. Akt is a direct target for myricetin to inhibit cell transformation. Mol Cell Biochem 2009;332(1–2):33–41.

[56] Hsu YAL, Uen YH, Chen Y, Liang HL, Kuo POL. Tricetin, a dietary flavonoid, inhibits proliferation of human breast adenocarcinoma MCF-7 cells by blocking cell cycle progression and inducing apoptosis. J Agr Food Chem 2009;57(18):8688–95.

[57] Hsu YL, Hou MF, Tsai EM, Kuo PL. Tricetin, a dietary flavonoid, induces apoptosis through the reactive oxygen species/c-Jun NH 2-terminal kinase pathway in human liver cancer cells. J Agr Food Chem 2010;58(23):12547–12556.

[58] Hu Y, Yang Y, You QD, Liu W, Gu HY, Zhao L, et al. Oroxylin A induced apoptosis of human hepatocellular carcinoma cell line HepG2 was involved in its antitumor activity. Biochem Biophys Res Comm 2006;351(2):521–7.

[59] Li HN, Nie FF, Liu W, Dai QS, Lu N, Qi Q, et al. Apoptosis induction of oroxylin A in human cervical cancer HeLa cell line in vitro and in vivo. Toxicol 2009;257(1–2):80–5.

[60] Mu R, Qi Q, Gu H, Wang J, Yang Y, Rong J, et al. Involvement of p53 in oroxylin A-induced apoptosis in cancer cells. Mol Carcinogen 2009;48(12):1159–69.

[61] Kim DH, Na HK, Oh TY, Kim WB, Surh YJ. Eupatilin, a pharmacologically active flavone derived from Artemisia plants, induces cell cycle arrest in ras-transformed human mammary epithelial cells. Biochem Pharmacol 2004;68(6):1081–7.

[62] Wang HY, Cai B, Cui CB, Zhang DY, Yang BF. Vitexicarpin, a flavonoid from *Vitex trifolia* L., induces apoptosis in K562 cells via mitochondria-controlled apoptotic pathway. Yao Xue Xue Bao 2005;40:27–31.

[63] Haïdara K, Zamir L, Shi QW, Batist G. The flavonoid Casticin has multiple mechanisms of tumor cytotoxicity action. Cancer Lett 2006;242(2):180–90.

[64] Yang J, Yang Y, Tian L, Sheng XF, Liu F, Cao JG. Casticin-induced apoptosis involves death receptor 5 upregulation in hepatocellular carcinoma cells. World J Gastroenterol 2011;17(38):4298–307.

[65] Sergeev IN, Li S, Colby J, Ho CT, Dushenkov S. Polymethoxylated flavones induce Ca2+-mediated apoptosis in breast cancer cells. Life Sci 2006;80(3):245–53.

[66] Sheng X, Sun Y, Yin Y, Chen T, Xu Q. Cirsilineol inhibits proliferation of cancer cells by inducing apoptosis via mitochondrial pathway. Journal of Pharm Pharmacol 2008;60(11):1523–9.

[67] Pedro M, Ferreira MM, Cidade H, Kijjoa A, Bronze-Da-Rocha E, Nascimento MSJ. Artelastin is a cytotoxic prenylated flavone that disturbs microtubules and interferes with DNA replication in MCF-7 human breast cancer cells. Life Sci 2005;77(3):293–311.

[68] Lee CC, Lin CN, Jow GM. Cytotoxic and apoptotic effects of prenylflavonoid artonin B in human acute lymphoblastic leukemia cells. Acta Pharmacol Sin 2006;27(9):1165–74.

[69] Wätjen W, Weber N, Lou Yj, Wang Zq, Chovolou Y, Kampkötter A, et al. Prenylation enhances cytotoxicity of apigenin and liquiritigenin in rat H4IIE hepatoma and C6 glioma cells. Food Chem Toxicol 2007;45(1):119–24.

[70] Huang X, Zhu D, Lou Y. A novel anticancer agent, icaritin, induced cell growth inhibition, G1 arrest and mitochondrial transmembrane potential drop in human prostate carcinoma PC-3 cells. Eur J Pharmacol 2007;564(1–3):26–36.

[71] Lee JC, Won SJ, Chao CL, Wu FL, Liu HS, Ling P, et al. Morusin induces apoptosis and suppresses NF-κB activity in human colorectal cancer HT-29 cells. Biochem Biophys Res Comm 2008;372(1):236–42.

[72] Hsu CL, Shyu MH, Lin JA, Yen GC, Fang SC. Cytotoxic effects of geranyl flavonoid derivatives from the fruit of *Artocarpus communis* in SK-Hep-1 human hepatocellular carcinoma cells. Food Chem 2011;127(1):127–34.

[73] Neves MP, Cidade H, Pinto M, Silva AMS, Gales L, Damas AM, et al. Prenylated derivatives of baicalein and 3,7-dihydroxyflavone: synthesis and study of their effects on tumor cell lines growth, cell cycle and apoptosis. Eur J Med Chem 2011;46(6):2562–74.

[74] Chang HL, Wu YC, Su JH, Yeh YT, Yuan SSF. Protoapigenone, a novel flavonoid, induces apoptosis in human prostate cancer cells through activation of p38 mitogen-activated protein kinase and c-Jun NH2-terminal kinase 1/2. J Pharmacol Exp Ther 2008;325(3):841–9.

[75] Chen WY, Hsieh YA, Tsai CI, Kang YF, Chang FR, Wu YC, et al. Protoapigenone, a natural derivative of apigenin, induces mitogen-activated protein kinase-dependent apoptosis in human breast cancer cells associated with induction of oxidative stress and inhibition of glutathione S-transferase π. Invest New Drugs 2010;29(6):1347–59.

[76] Ye CL, Qian F, Wei DZ, Lu YH, Liu JW. Induction of apoptosis in K562 human leukemia cells by 2′,4′– dihydroxy-6′-methoxy-3′,5′-dimethylchalcone. Leukemia Res 2005;29(8):887–92.

[77] Zi X, Simoneau AR. Flavokawain A, a novel chalcone from kava extract, induces apoptosis in bladder cancer cells by involvement of Bax protein-dependent and mitochondria-dependent apoptotic pathway and tumor growth in mice. Cancer Res 2005;65(8):3479–86.

[78] Kuo YF, Su YZ, Tseng YH, Wang SY, Wang HM, Chueh PJ. Flavokawain B, a novel chalcone from *Alpinia pricei* Hayata with potent apoptotic activity: involvement of ROS and GADD153 upstream of mitochondria-dependent apoptosis in HCT116 cells. Free Radical Biol Med 2010;49(2):214–26.

[79] Fu Y, Hsieh TC, Guo J, Kunicki J, Lee MYWT, Darzynkiewicz Z, et al. Licochalcone-A, a novel flavonoid isolated from licorice root (*Glycyrrhiza glabra*), causes G2 and late-G1 arrests in androgen-independent PC-3 prostate cancer cells. Biochem Biophys Res Comm 2004;322(1):263–70.

[80] Funakoshi-Tago M, Tago K, Nishizawa C, Takahashi K, Mashino T, Iwata S, et al. Licochalcone A is a potent inhibitor of TEL-Jak2-mediated transformation through the specific inhibition of Stat3 activation. Biochem Pharmacol 2008;76(12):1681–93.

[81] Lee YM, Kim YC, Choi BJ, Lee DW, Yoon JH, Kim EC. Mechanism of sappanchalcone-induced growth inhibition and apoptosis in human oral cancer cells. Toxicol in Vitro 2011;25(8):1782–8.

[82] Colgate EC, Miranda CL, Stevens JF, Bray TM, Ho E. Xanthohumol, a prenylflavonoid derived from hops induces apoptosis and inhibits NF-kappaB activation in prostate epithelial cells. Cancer Lett 2007;246(1–2):201–9.

[83] Strathmann J, Klimo K, Sauer SW, Okun JG, Prehn JHM, Gerhäuser C. Xanthohumol-induced transient superoxide anion radical formation triggers cancer cells into apoptosis via a mitochondria-mediated mechanism. FASEB J 2010;24(8):2938–50.

[84] Borris RP, Cordell GA, Farnsworth NR. Isofraxidin, a cytotoxic coumarin from *Micrandra elata* (euphorbiaceae). J. Nat Prod *(Lloydia)* 1980;43(5):641–3.

[85] Itokawa H, Yun Y, Morita H, Takeya K, Lee SR. Cytotoxic coumarins from roots of *Angelica gigas* Nakai. Nat Med 1994;48(4):334–5.

[86] Finn GJ, Kenealy E, Creaven BS, Egan DA. In vitro cytotoxic potential and mechanism of action of selected coumarins, using human renal cell lines. Cancer Lett 2002;183(1):61–8.

[87] Goel A, Prasad AK, Parmar VS, Ghosh B, Saini N. 7,8-Dihydroxy-4-methylcoumarin induces apoptosis of human lung adenocarcinoma cells by ROS-independent mitochondrial pathway through partial inhibition of ERK/MAPK signaling. FEBS Lett 2007;581(13):2447–54.

[88] Jang SI, Kim YJ, Kim HJ, Lee JC, Kim HY, Kim YC, et al. Scoparone inhibits PMA-induced IL-8 and MCP-1 production through suppression of NF-κB activation in U937 cells. Life Sci 2006;78(25):2937–43.

[89] Alesiani D, Cicconi R, Mattei M, Montesano C, Bei R, Canini A. Cell cycle arrest and differentiation induction by 5,7-dimethoxycoumarin in melanoma cell lines. Int J Oncol 2008;32(2):425–34.

[90] Pan R, Dai Y, Yang J, Li Y, Yao X, Xia Y. Anti-angiogenic potential of scopoletin is associated with the inhibition of ERK1/2 activation. Drug Development Res 2009;70(3):214–9.

[91] Ishihara M, Yokote Y, Sakagami H. Quantitative structure-cytotoxicity relationship analysis of coumarin and its derivatives by semiempirical molecular orbital method. Anticancer Res 2006;26(4 B):2883–6.

[92] Li Y, Dai Y, Liu M, Pan R, Luo Y, Xia Y, et al. Scopoletin induces apoptosis of fibroblast-like synoviocytes from adjuvant arthritis rats by a mitochondrial-dependent pathway. Drug Development Res 2009;70(5):378–85.

[93] Goel A, Prasad AK, Parmar VS, Ghosh B, Saini N. Apoptogenic effect of 7,8-diacetoxy-4-methylcoumarin and 7,8-diacetoxy-4-methylthiocoumarin in human lung adenocarcinoma cell line: role of NF-κB, Akt, ROS and MAP kinase pathway. Chem Biol Interactions 2009;179(2–3):363–74.

[94] Yang J, Xiao YL, He XR, Qiu GF, Hu XM. Aesculetin-induced apoptosis through a ROS-mediated mitochondrial dysfunction pathway in human cervical cancer cells. J Asian Nat Prod Res 2010;12(3):185–93.

[95] Finn G, Creaven B, Egan D. Investigation of intracellular signalling events mediating the mechanism of action of 7-hydroxycoumarin and 6-nitro-7-hydroxycoumarin in human renal cells. Cancer Lett 2004;205(1):69–79.

[96] Kang KS, Ryu SH, Ahn BZ. Antineopastic natural products and the analogues IV. Aurapten, the cytotoxic coumarin from *Poncirus trifoliata* against L1210 cell. Arch Pharm Res 1985;8(3):187–90.

[97]   Barthomeuf C, Lim S, Iranshahi M, Chollet P. Umbelliprenin from *Ferula szowitsiana* inhibits the growth of human M4Beu metastatic pigmented malignant melanoma cells through cell-cycle arrest in G1 and induction of caspase-dependent apoptosis. Phytomed 2008;15(1–2):103–11.

[98]   Min BK, Hyun DG, Jeong SY, Kim YH, Ma ES, Woo MH. A new cytotoxic coumarin, 7-[(E)–3′,7′-dimethyl-6′-oxo-2′,7′-octadienyl] oxy Coumarin, from the leaves of *Zanthoxylum schinifolium*. Arch Pharm Res 2011;34(5):723–6.

[99]   Viola G, Vedaldi D, dall'Acqua F, Basso G, Disarò S, Spinelli M, et al. Synthesis, cytotoxicity, and apoptosis induction in human tumor cells by geiparvarin analogues. Chem Biodivers 2004;1(9):1265–80.

[100] Murata T, Itoigawa M, Ito C, Nakao K, Tsuboi M, Kaneda N, et al. Induction of apoptosis in human leukaemia HL-60 cells by furanone-coumarins from *Murraya siamensis*. J Pharm Pharmacol 2008;60(3):385–9.

[101] Yang YM, Hyun JW, Sung MS, Chung HS, Kim BK, Paik WH, et al. The cytotoxicity of psoralidin from *Psoralea corylifolia*. Planta Med 1996;62(4):353–4.

[102] Hitotsuyanagi Y, Kojima H, Ikuta H, Takeya K, Itokawa H. Identification and structure–activity relationship studies of osthol, a cytotoxic principle from *Cnidium monnieri*. Bioorg Med Chem Lett 1996;6(15):1791–4.

[103] Xu X, Zhang Y, Qu D, Jiang T, Li S. Osthole induces G2/M arrest and apoptosis in lung cancer A549 cells by modulating PI3K/Akt pathway. J Exp Clin Cancer Res 2011;30(1):33.

[104] Bocca C, Gabriel L, Bozzo F, Miglietta A. Microtubule-interacting activity and cytotoxicity of the prenylated coumarin ferulenol. Planta Med 2002;68:1135–7.

[105] Seo EK, Wani MC, Wall ME, Navarro H, Mukherjee R, Farnsworth NR, et al. New bioactive aromatic compounds from *Vismia guianensis*. Phytochem 2000;55(1):35–42.

[106]   opez-Pérez JL, Olmedo DA, Del Olmo E, Vásquez Y, Solís PN, Gupta MP, et al. Cytotoxic 4-phenylcoumarins from the leaves of *Marila pluricostata*. J Nat Prod 2005;68(3):369–73.

[107] Ngo NTN, Nguyen VT, Van Vo H, Vang O, Duus F, Ho TDH, et al. Cytotoxic coumarins from the bark of *Mammea siamensis*. Chem Pharm Bull 2010;58(11):1487–91.

[108] Álvarez-Delgado C, Reyes-Chilpa R, Estrada-Muñiz E, Mendoza-Rodríguez CA, Quintero-Ruiz A, Solano J, et al. Coumarin A/AA induces apoptosis-like cell death in HeLa cells mediated by the release of apoptosis-inducing factor. J Biochem Mol Toxicol 2009;23(4):263–72.

[109] Du L, Mahdi F, Jekabsons MB, Nagle DG, Zhou YD. Mammea E/BB, an isoprenylated dihydroxycoumarin protonophore that potently uncouples mitochondrial electron transport, disrupts hypoxic signaling in tumor cells. J Nat Prod 2010;73(11):1868–72.

[110] Ito A, Chai HB, Shin YG, Garcia R, Mejia M, Gao Q, et al. Cytotoxic constituents of the roots of *Exostema acuminatum*. Tetrahedron 2000;56(35):6401–5.

[111] Hyeon HK, Sung SB, Jin SC, Han H, Kim IH. Involvement of PKC and ROS in the cytotoxic mechanism of anti-leukemic decursin and its derivatives and their structure–activity relationship in human K562 erythroleukemia and U937 myeloleukemia cells. Cancer Lett 2005;223(2):191–201.

[112] Songsiang U, Thongthoom T, Boonyarat C, Yenjai C. Aurailas A–D, cyto-toxic carbazole alkaloids from the roots of *Clausena harmandiana*. J. Nat Prod 2011;74(2):208–12.

[113] Ahn KS, Sim WS, Kim IH. Decursin: a cytotoxic agent and protein kinase C activator from the root of *Angelica gigas*. Planta Med 1996;62(1):7–9.

[114] Son SH, Kim MJ, Chung WY, Son JA, Kim YS, Kim YC, et al. Decursin and decursinol inhibit VEGF-induced angiogenesis by blocking the activation of extracellular signal-regulated kinase and c-Jun N-terminal kinase. Cancer Lett 2009;280(1):86–92.

[115] Kim HJ, Kim SM, Park KR, Jang HJ, Na YS, Ahn KS, et al. Decursin chemo-sensitizes human multiple myeloma cells through inhibition of STAT3 signaling pathway. Cancer Lett 2011;301(1):29–37.

[116] Chen YH, Chang FR, Wu CC, Yen MH, Liaw CC, Huang HC, et al. New cytotoxic 6-oxygenated 8,9-dihydrofurocoumarins, hedyotiscone A–C, from *Hedyotis biflora*. Planta Med 2006;72(1):75–8.

[117] Chaturvedula VSP, Schilling JK, Kingston DGI. New cytotoxic coumarins and prenylated benzophenone derivatives from the bark of *Ochrocarpos punctatus* from the Madagascar rainforest. J. Nat Prod 2002;65(7):965–72.

[118] Thanh PN, Jin W, Song G, Bae K, Kang SS. Cytotoxic coumarins from the root of *Angelica dahurica*. Arch Pharm Res 2004;27(12):1211–5.

[119] Kang TJ, Lee SY, Singh RP, Agarwal R, Yim DS. Anti-tumor activity of oxy-peucedanin from *Ostericum koreanum* against human prostate carcinoma DU145 cells. Acta Oncol 2009;48(6):895–900.

[120] Prasad KN, Xie H, Hao J, Yang B, Qiu S, Wei X, et al. Antioxidant and anti-cancer activities of 8-hydroxypsoralen isolated from wampee [*Clausena lansium* (Lour.) Skeels] peel. Food Chem 2010;118(1):62–6.

[121] Egan D, James P, Cooke D, O'Kennedy R. Studies on the cytostatic and cyto-toxic effects and mode of action of 8-nitro-7-hydroxycoumarin. Cancer Lett 1997;118(2):201–11.

[122] Magiatis P, Melliou E, Skaltsounis AL, Mitaku S, Léonce S, Renard P, et al. Synthesis and cytotoxic activity of pyranocoumarins of the seselin and xanthyle-tin series. J Nat Prod 1998;61(8):982–6.

[123] Stanway SJ, Purohit A, Woo LW, Sufi S, Vigushin D, Ward R, et al. Phase I study of STX 64 (667 Coumate) in breast cancer patients: the first study of a steroid sulfatase inhibitor. Clin Cancer Res 2006;12:1585.

[124] Purohit A, Foster PA. Steroid sulfatase inhibitors for estrogen- and androgen-dependent cancers. J Endocrinol 2012;212(2):99–110.

[125] Sutherland MK, Brady H, Gayo-Fung LM, Leisten J, Lipps SG, McKie JA, et al. Effects of SP500263, a novel selective estrogen receptor modulator, on bone, uterus, and serum cholesterol in the ovariectomized rat. Calcif Tissue Int 2003;72:710–6.

[126] Brady H, Desai S, Gayo-Fung LM, Khammungkhune S, McKie JA, O'Leary E, et al. Effects of SP500263, a novel, potent antiestrogen, on breast cancer cells and in xenograft models. Cancer Res 2002;62(5):1439–42.

[127] Sashidhara KV, Rosaiah JN, Kumar M, Gara RK, Nayak LV, Srivastava K, et al. Neo-tanshinlactone inspired synthesis, in vitro evaluation of novel substituted benzocoumarin derivatives as potent anti-breast cancer agents. Bioorg Med Chem Lett 2010;20(23):7127–31.

[128] Singh RK, Lange TS, Kim KK, Brard L. A coumarin derivative (RKS262) inhibits cell-cycle progression, causes pro-apoptotic signaling and cytotoxicity in ovarian cancer cells. Invest New Drugs 2011;29(1):63–72.

[129] Combes S, Barbier P, Douillard S, McLeer-Florin A, Bourgarel-Rey V, Pierson JT, et al. Synthesis and biological evaluation of 4-arylcoumarin analogues of combretastatins. Part 2. J Medl Chem 2011;54(9):3153–62.

[130] Musa MA, Badisa VLD, Latinwo LM, Cooperwood J, Sinclair A, Abdullah A. Cytotoxic activity of new acetoxycoumarin derivatives in cancer cell lines. Anticancer Res 2011;31(6):2017–22.

[131] Gao XM, Yu T, Cui MZ, Pu JX, Du X, Han QB, et al. Identification and evaluation of apoptotic compounds from Garcinia oligantha. Bioorg Med Chem Lett 2012;22(6):2350–3.

[132] Su QG, Liu Y, Cai YC, Sun YL, Wang B, Xian LJ. Anti-tumour effects of xanthone derivatives and the possible mechanisms of action. Invest New Drugs 2011;29:1230–40.

[133] Nkengfack AE, Azebaze GA, Vardamides JC, Fomum ZT, Van Heerden FR. A prenylated xanthone from *Allanblackia floribunda*. Phytochem 2002;60(4):381–4.

[134] Isakovic A, Jankovic T, Harhaji L, Kostic-Rajacic S, Nikolic Z, Vajs V, et al. Antiglioma action of xanthones from *Gentiana kochiana*: mechanistic and structure-activity requirements. Bioorg Med Chem 2008;16(10):5683–94.

[135] Ding L, Liu B, Qi Ll, Zhou QY, Hou Q, Li J, et al. Anti-proliferation, cell cycle arrest and apoptosis induced by a natural xanthone from *Gentianopsis paludosa* Ma, in human promyelocytic leukemia cell line HL-60 cells. Toxicol in Vitro 2009;23(3):408–17.

[136] Abou-Shoer M, Boettner FE, Chang CJ, Cassady JM. Antitumour and cytotoxic xanthones of *Psorospermum febrifugum*. Phytochem 1988;27(9):2795–800.

[137] Seo EK, Kim NC, Wani MC, Wall ME, Navarro HA, Burgess JP, et al. Cytotoxic prenylated xanthones and the unusual compounds anthraquinobenzophenones from *Cratoxylum sumatranum*. J Nat Prod 2002;65(3):299–305.

[138] Han QB, Qiao CF, Song JZ, Yang NY, Cao XW, Peng Y, et al. Cytotoxic prenylated phenolic compounds from the twig bark of *Garcinia xanthochymus*. Chem Biodivers 2007;4(5):940–6.

[139] Hou AJ, Fukai T, Shimazaki M, Sakagami H, Sun HD, Nomura T. Benzophenones and xanthones with isoprenoid groups from *Cudrania cochinchinensis*. J Nat Prod 2001;64(1):65–70.

[140] Zou YS, Hou AJ, Zhu GF, Chen YF, Sun HD, Zhao QS. Cytotoxic isoprenylated xanthones from *Cudrania tricuspidata*. Bioorg Med Chem 2004;12(8):1947–53.

[141] Lee BW, Lee JH, Lee ST, Lee HS, Lee WS, Jeong TS, et al. Antioxidant and cytotoxic activities of xanthones from *Cudrania tricuspidata*. Bioorg Med Chem Lett 2005;15(24):5548–52.

[142] Wang YH, Hou AJ, Zhu GF, Chen DF, Sun HD. Cytotoxic and antifungal isoprenylated xanthones and flavonoids from *Cudrania fruticosa*. Planta Med 2005;71(3):273–4.

[143] Nkengfack AE, Mkounga P, Fomum ZT, Meyer M, Bodo B. Globulixanthones A and B, two new cytotoxic xanthones with isoprenoid groups from the root bark of *Symphonia globulifera*. J Nat Prod 2002;65(5):734–6.

[144] Suksamrarn S, Komutiban O, Ratananukul P, Chimnoi N, Lartpornmatulee N, Suksamrarn A. Cytotoxic prenylated xanthones from the young fruit of *Garcinia mangostana*. Chem Pharml Bull 2006;54(3):301–5.

[145] Kaomongkolgit R, Chaisomboon N, Pavasant P. Apoptotic effect of alpha-mangostin on head and neck squamous carcinoma cells. Archs Oral Biol 2011;56(5):483–90.

[146] Ren Y, Matthew S, Lantvit DD, Ninh TN, Chai H, Fuchs JR, et al. Cytotoxic and NF-κB inhibitory constituents of the stems of *Cratoxylum cochinchinense* and their semisynthetic analogues. J Nat Prod 2011;74(5):1117–25.

[147] Zelefack F, Guilet D, Fabre N, Bayet C, Chevalley S, Ngouela S, et al. Cytotoxic and antiplasmodial xanthones from *Pentadesma butyracea*. J Nat Prod 2009;72(5):954–7.

[148] Kijjoa A, Gonzalez MJ, Pinto MM, Nascimento MSJ, Campos N, Mondranondra IO, et al. Cytotoxicity of prenylated xanthones and other constituents from the wood of *Garcinia merguensis*. Planta Med 2008;74(8):864–6.

[149] Boonsri S, Karalai C, Ponglimanont C, Kanjana-Opas A, Chantrapromma K. Antibacterial and cytotoxic xanthones from the roots of *Cratoxylum formosum*. Phytochem 2006;67(7):723–7.

[150] Seo EK, Wall ME, Wani MC, Navarro H, Mukherjee R, Farnsworth NR, et al. Cytotoxic constituents from the roots of *Tovomita brevistaminea*. Phytochem 1999;52(4):669–74.

[151] Yimdjo MC, Azebaze AG, Nkengfack AE, Meyer AM, Bodo B, Fomum ZT. Antimicrobial and cytotoxic agents from *Calophyllum inophyllum*. Phytochem 2004;65(20):2789–95.

[152] Hakim EH, Asnizar Y, Aimi N, Kitajima M, Takayama H. Artoindonesianin P, a new prenylated flavone with cytotoxic activity from *Artocarpus lanceifolius*. Fitoterapia 2002;73(7–8):668–73.

[153] Cao S, Brodie PJ, Miller JS, Randrianaivo R, Ratovoson F, Birkinshaw C, et al. Antiproliferative xanthones of *Terminalia calcicola* from the Madagascar rain forest. J Nat Prod 2007;70(4):679–81.

[154] Asano J, Chiba K, Tada M, Yoshii T. Cytotoxic xanthones from *Garcinia han-buryi*. Phytochem 1996;41(3):815–20.

[155] Han QB, Wang YL, Yang L, Tso TF, Qiao CF, Song JZ, et al. Cytotoxic poly-prenylated xanthones from the resin of *Garcinia hanburyi*. Chem Pharml Bull 2006;54(2):265–7.

[156] Han QB, Yang L, Wang YL, Qiao CF, Song JZ, Sun HD, et al. A pair of novel cytotoxic polyprenylated xanthone epimers from gamboges. Chem Biodivers 2006;3(1):101–5.

[157] Wang T, Wei J, Qian X, Ding Y, Yu L, Liu B. Gambogic acid, a potent inhibi-tor of survivin, reverses docetaxel resistance in gastric cancer cells. Cancer Lett 2008;262(2):214–22.

[158] Zhang HZ, Kasibhatla S, Wang Y, Herich J, Guastella J, Tseng B, et al. Discovery, characterization and SAR of gambogic acid as a potent apoptosis inducer by a HTS assay. Bioorg Med Chem 2004;12(2):309–17.

[159] Li Q, Cheng H, Zhu G, Yang L, Zhou A, Wang X, et al. Gambogenic acid inhibits proliferation of A549 cells through apoptosis-inducing and cell cycle arresting. Biol Pharm Bull 2010;33(3):415–20.

[160] Wu X, Cao S, Goh S, Hsu A, Tan BKH. Mitochondrial destabilisation and caspase-3 activation are involved in the apoptosis of Jurkat cells induced by gaudichaudione A, a cytotoxic xanthone. Planta Med 2002;68(3):198–203.

[161] Ren Y, Lantvit DD, de Blanco EJC, Kardono LBS, Riswan S, Chai H, et al. Proteasome-inhibitory and cytotoxic constituents of *Garcinia lateriflora*: absolute configuration of caged xanthones. Tetrahedron 2010;66(29):5311–20.

[162] Permana PA, Ho DK, Cassady JM, Snapka RM. Mechanism of action of the antileukemic xanthone psorospermin: DNA strand breaks, abasic sites, and protein-DNA cross-links. Cancer Res 1994;54(12):3191–5.

[163] Hansen M, Lee SJ, Cassady JM, Hurley LH. Molecular details of the structure of a psorospermin-DNA covalent/intercalation complex and associated DNA sequence selectivity. J Am Chem Soc 1996;118:5553–61.

[164] Kwok Y, Zeng Q, Hurley LH. Topoisomerase II-mediated site-directed alkylation of DNA by psorospermin and its use in mapping other topoisomerase II poison binding sites. Proc Natl Acad Sci USA 1998;95:13531–13536.

[165] Nakagawa Y, Iinuma M, Naoe T, Nozawa Y, Akao Y. Characterized mechanism of α-mangostin-induced cell death: caspase-independent apoptosis with release of endonuclease-G from mitochondria and increased miR-143 expression in human colorectal cancer DLD-1 cells. Bioorg Med Chem 2007;15(16):5620–8.

[166] Sato A, Fujiwara H, Oku H, Ishiguro K, Ohizumi Y. α-Mangostin induces Ca2+-ATPase-dependent apoptosis via mitochondrial pathway in PC12 cells. J Pharmacol Sci 2004;95(1):33–40.

[167] Matsumoto K, Akao Y, Yi H, Ohguchi K, Ito T, Tanaka T, et al. Preferential target is mitochondria in α-mangostin-induced apoptosis in human leukemia HL60 cells. Bioorg Med Chem 2004;12(22):5799–806.

[168] Han AR, Kim JA, Lantvit DD, Kardono LBS, Riswan S, Chai H, et al. Cytotoxic xanthone constituents of the stem bark of *Garcinia mangostana* (mangosteen). J Nat Prod 2009;72(11):2028–31.

[169] Krajarng A, Nakamura Y, Suksamrarn S, Watanapokasin R. α-Mangostin induces apoptosis in human chondrosarcoma cells through downregulation of ERK/JNK and Akt signaling pathway. J Agr Food Chem 2011;59(10):5746–54.

[170] Chao AC, Hsu YL, Liu CK, Kuo PL. α-Mangostin, a dietary xanthone, induces autophagic cell death by activating the AMP-activated protein kinase pathway in glioblastoma cells. J Agr Food Chem 2011;59(5):2086–96.

[171] Kuemmerle J, Jiang S, Tseng B, Kasibhatla S, Drewe J, Cai SX. Synthesis of caged 2,3,3a,7a-tetrahydro-3,6-methanobenzofuran-7(6H)-ones: evaluating the minimum structure for apoptosis induction by gambogic acid. Bioorg Med Chem 2008;16(8):4233–41.

[172] Wang X, Lu N, Yang Q, Gong D, Lin C, Zhang S, et al. Studies on chemical modification and biology of a natural product, gambogic acid (III): determination of the essential pharmacophore for biological activity. Eur J Med Chem 2011;46(4):1280–90.

[173] Wu X, Cao S, Goh S, Hsu A, Tan BKH. Mitochondrial destabilisation and caspase-3 activation are involved in the apoptosis of Jurkat cells induced by gaudichaudione A, a cytotoxic xanthone. Planta Med 2002;68(3):198–203.

[174] Zhao L, Guo QL, You QD, Wu ZQ, Gu HY. Gambogic acid induces apoptosis and regulates expressions of bax and Bcl-2 protein in human gastric carcinoma MGC-803 cells. Biol Pharm Bull 2004;27(7):998–1003.

[175] Yang Y, Yang L, You QD, Nie FF, Gu HY, Zhao L, et al. Differential apoptotic induction of gambogic acid, a novel anticancer natural product, on hepatoma cells and normal hepatocytes. Cancer Lett 2007;256(2):259–66.

[176] Xu X, Liu Y, Wang L, He J, Zhang H, Chen X, et al. Gambogic acid induces apoptosis by regulating the expression of Bax and Bcl-2 and enhancing caspase-3 activity in human malignant melanoma A375 cells. Int Dermatol 2004;48(2):186–92.

[177] Nie F, Zhang X, Qi Q, Yang L, Yang Y, Liu W, et al. Reactive oxygen species accumulation contributes to gambogic acid-induced apoptosis in human hepatoma SMMC-7721 cells. Toxicol 2009;260(1–3):60–7.

[178] Li R, Chen Y, Zeng LL, Shu WX, Zhao F, Wen L, et al. Gambogic acid induces G0/G1 arrest and apoptosis involving inhibition of SRC-3 and inactivation of Akt pathway in K562 leukemia cells. Toxicol 2009;262(2):98–105.

[179] Gu H, Wang X, Rao S, Wang J, Zhao J, Fang LR, et al. Gambogic acid mediates apoptosis as a p53 inducer through down-regulation of mdm2 in wild-type p53-expressing cancer cells. Mol Cancer Ther 2008;7(10):3298–305.

[180] Rong JJ, Hu R, Qi Q, Gu HY, Zhao Q, Wang J, et al. Gambogic acid down-regulates MDM2 oncogene and induces p21Waf1/CIP1 expression independent of p53. Cancer Lett 2009;284(1):102–12.

[181] Gu H, Rao S, Zhao J, Wang J, Mu R, Rong J, et al. Gambogic acid reduced bcl-2 expression via p53 in human breast MCF-7 cancer cells. J Cancer Res Clin Oncol 2009;135(12):1777–82.

[182] Wang J, Zhao Q, Qi Q, Gu HY, Rong JJ, Mu R, et al. Gambogic acid-induced degradation of mutant p53 is mediated by proteasome and related to CHIP. J Cell Biochem. 2011;112(2):509–19.

[183] Prasad S, Pandey MK, Yadav VR, Aggarwal BB. Gambogic acid inhibits STAT3 phosphorylation through activation of protein tyrosine phosphatase SHP-1: potential role in proliferation and apoptosis. Cancer Prev Res 2011;4(7):1084–94.

[184] Sun H, Chen F, Wang X, Liu Z, Yang Q, Zhang X, et al. Studies on gambogic acid (IV): exploring structure-activity relationship with IκB kinase-beta (IKKβ). Eur J Med Chem 2012;51:110–23.

[185] Zhang L, Yi Y, Chen J, Sun Y, Guo Q, Zheng Z, et al. Gambogic acid inhibits Hsp90 and deregulates TNF-α/NF-κB in HeLa cells. Biochem Biophys Res Comm 2010;403(3–4):282–7.

[186] Yan F, Wang M, Chen H, Su J, Wang X, Wang F, et al. Gambogenic acid mediated apoptosis through the mitochondrial oxidative stress and inactivation of Akt signaling pathway in human nasopharyngeal carcinoma CNE-1 cells. Eur J Pharmacol 2011;652(1–3):23–32.

[187] Qin Y, Meng L, Hu C, Duan W, Zuo Z, Lin L, et al. Gambogic acid inhibits the catalytic activity of human topoisomerase IIα by binding to its ATPase domain. Mol Cancer Ther 2007;6(9):2429–40.

[188] Chen J, Gu HY, Lu N, Yang Y, Liu W, Qi Q, et al. Microtubule depolymerization and phosphorylation of c-Jun N-terminal kinase-1 and p38 were involved in gambogic acid induced cell cycle arrest and apoptosis in human breast carcinoma MCF-7 cells. Life Sci 2008;83(3–4):103–9.

Topic **3.2**

Quinones

3.2.1 *Juglans Mandshurica* Maxim.

History This plant was first described by Carl Johann Maximowicz, in *Bulletin de la Classe Physico-Mathématique de l'Académie Impériale des Sciences de Saint-Pétersbourg*, published in 1856.

Synonyms *Juglans cathayensis* Dode, *Juglans cathayensis* var. *formosana* (Hayata) A.M. Lu & R.H. Chang, *Juglans collapsa* Dode, *Juglans draconis* Dode, *Juglans formosana* Hayata, *Juglans stenocarpa* Maxim.

Family Juglandaceae DC. ex Perleb, 1818

Common Names Manchurian walnut, Chinese walnut, Hu tao qiu (Chinese)

Habitat and Description *Juglans mandshurica* is a tree which grows to 20 m tall in the forests of China, Taiwan and Korea. The leaves are imparipinnate, alternate and exstipulate. The rach is 20 cm long and bears four to nine pairs of leaflets plus a terminal one. The leaflets are elliptic, 5–15 cm × 2.5–8 cm, hairy beneath, cordate at base, sessile, serrate, acute at apex and present 10–12 pairs of secondary nerves. The male spikes are cauliflorous, to 35 cm long, and the female spikes are terminal. The male flowers are minute and comprise four sepals and around 10–40 stamens. The female flowers are minute and consist of four sepals and an ovary developing a two-lobed and plumose stigma. The fruit is an ovoid drupe which is hairy, 3–5 cm across and contains a woody, ridged shell (Figure 3.4).

Medicinal Uses In Korea the plant is used to treat cancer.

Phytopharmacology The plant is known to accumulate series of the naphthalenyl glucosides, such as 1,4,8-trihydroxynaphthalene 1-*O*-β-D-[6′-*O*-(4″,5″,6″-trihydroxybenzoyl)] glucopyranoside, 1,4,8-trihydroxynaphthalene 1-*O*-β-D-[6′-*O*-(4″,6″-dimethoxy-5″-hydroxybenzoyl)] glucopyranoside, 1,5,8-trihydroxy-*R*-tetralone 5-*O*-β-D-glucopyranoside,

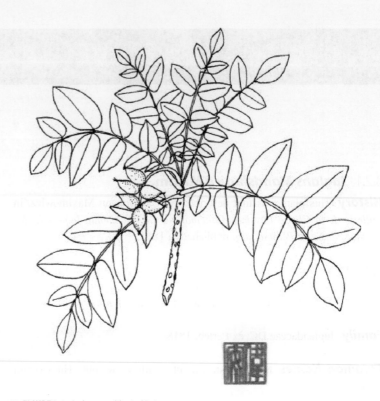

■ **FIGURE 3.4** *Juglans mandshurica* Maxim.

1,4,8-trihydroxynaphthalene-1-*O*-β-D-glucopyranoside, 1,4,8-trihydroxy-3-naphthalenecarboxylic acid 1-*O*-β-D-glucopyranoside methyl ester,[189] 1,4,8-trihydroxynaphthalene 1-*O*-[α-L-arabinofuranosyl-(1→6)-β-D-glucopyranoside] and 1,4,8-trihydroxynaphthalene 1-*O*-β-D-[6′-*O*-(3″,5″-dihydroxy-4″methoxybenzoyl)]glucopyranoside,[190,191] as well as α-tetralonyl glucopyranosides including 4,5,8-trihydroxy-*R*-tetralone 5-*O*-β-D-[6′-*O*-(3″,5″-dimethoxy-4″-hydroxybenzoyl)]glucopyranoside,[189] 4α,5,8-trihydroxy-α-tetralone 5-*O*-*b*-D-[6′-*O*-(3″,5″-dihydroxy-4″-methoxybenzoyl)]glucopyranoside and 4α,5,8-trihydroxy-α-tetralone 5-*O*-β-D-[6′-*O*-(3″,4″,5″-trihydroxybenzoyl)]glucopyranoside,[190] some diarylheptanoids,[192,193] the naphthoquinones 5-hydroxy-2-methoxy-1,4-naphthoquinone,[189] *p*-hydroxymethoxybenzobijuglone[194] and juglone,[195] the anthracene juglanthracenoside A and some anthraquinones, such as juglanthraquinones A and B.[196]

The medicinal property of the plant has not yet been validated.

Proposed Research Pharmacological study of 5-hydroxy-2-methoxy-1,4-naphthoquinone and derivatives for the treatment of lung cancer.

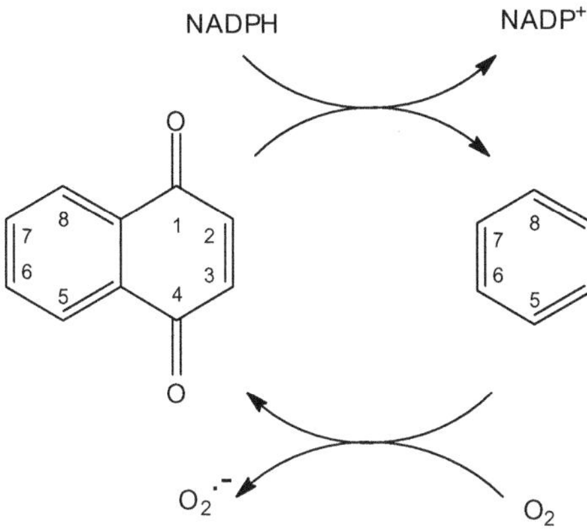

■ **CS 3.198** 'Redox cycling process'

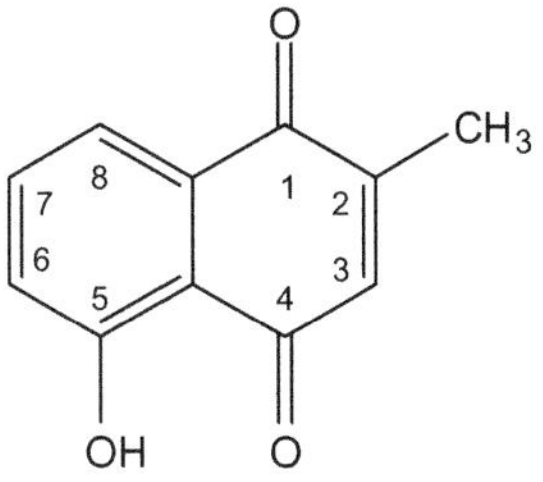

■ **CS 3.199** Plumbagin

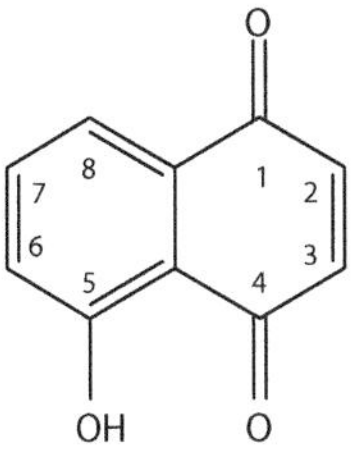

■ **CS 3.200** Juglone

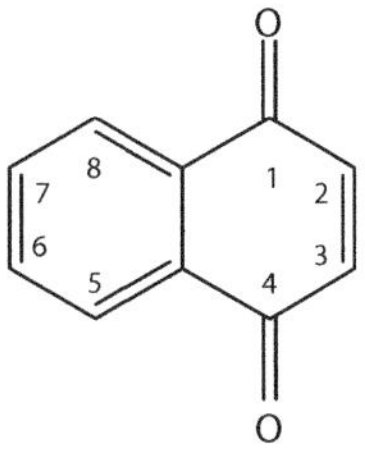

■ **CS 3.201** 1,4-Naphthoquinone

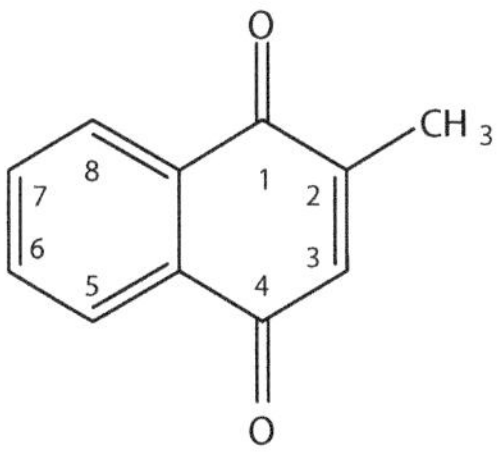

■ **CS 3.202** Menadione

Rationale A striking feature of naphthoquinones is their ability to over-whelm the cytoplasm of cancer cells with bursts of reactive oxygen species (ROS). This feature is partly as a result of the quinone moiety, which under-goes a 'redox cycling process'(CS 3.198) that starts with the formation of a semiquinone radical by NADPH cytochrome P-450 reductase, microsomal NADH-cytochrome b and mitochondrial NADH-ubiquinone oxidoreduc-tase, which reacts with oxygen to produce the highly reactive superoxide anion radical (O_2^-) in the cytoplasm of mammalian cells.[197,198,199]

In fact, Klaus et al. (2010) observed a positive correlation between the cytotoxic potency of naphthoquinones and their ability to generate ROS: plumbagin (CS 3.199), juglone (CS 3.200), 1,4-naphthoquinone (CS 3.201) and menadione (CS 3.202) abrogated the growth of human keratinocyte (HaCaT) cells, with IC_{50} values equal to 5 μM, 6 μM, 15 μM and 40 μM, respectively,[200] parallel to their ability to form ROS.[200] Note that both lawsone (CS 3.203) and lapachol (CS 3.204), which harbor C2 hydroxyl groups, were inactive in this experiment[200] whereas a hydroxyl group at C5 enhances the pro-oxidant and cytotoxic potencies of 1,4-naph-thoquinones.[200] In addition, DNA breakage was induced by menadione and juglone in HaCaT cells.[200] Juglone from *Juglans mandshurica* Maxim. inhibited the growth of human ileocecal adenocarcinoma (HCT-8) cells, with an IC_{50} value equal to 1.3 μg/mL.[201] Human promyelocytic leukemia

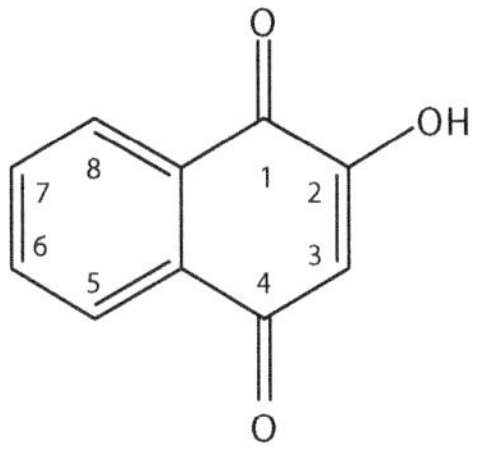

■ **CS 3.203** Lawsone

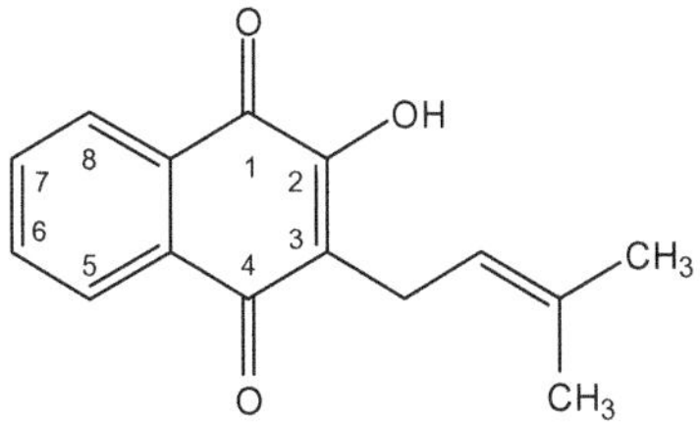

■ **CS 3.204** Lapachol

■ **CS 3.205** 5-Hydroxy-2-methoxy-1,4-naphthoquinone

■ **CS 3.206** 5-Methoxy-1,4-naphthoquinone

(HL-60) cells are sensitive to oxidative stress[202] and therefore easily destroyed by naphthoquinones.

For instance, juglone undergoes a redox cycling process[203] in human promyelocytic leukemia (HL-60) cells to form superoxide anion radicals,[204] which induced apoptosis, with an IC_{50} value of 1.4 µg/mL, associated with a decrease in anti-apoptotic protein, Bcl-2, an increase in pro-apoptotic protein (p53), pro-apoptotic Bcl-2-associated X protein (Bax), a fall in mitochondrial membrane potential, release of cytochrome c and activation of caspases 9 and 3,[195] suggesting mitochondrial insult. From the same plant, 5-hydroxy-2-methoxy-1,4-naphthoquinone (CS 3.205) was lethal to human colon cancer (HT-29) and human lung adenocarcinoma epithelial (A549) cells, with IC_{50} values equal to 2.6 µg/mL and 0.08 µg/mL, respectively.[189] Juglone and 5-methoxy-1,4-naphthoquinone (CS 3.206) inhibited the growth of the following cells (respective IC_{50} values in parentheses): HL-60 (>5 µg/mL, 0.3 µg/mL), HCT-8 (1.3 µg/mL, 0.8 µg/mL) and human breast carcinoma (MDA-MB-435; >5 µg/mL, 0.7 µg/mL),[201] suggesting that the methoxylation of C5 enhances the cytotoxicity of juglone.

Plumbagin from *Plumbago zeylanica* L. (family Plumbaginaceae Juss.) induces apoptosis via the generation of ROS.[205] In human cervical carcinoma (ME-180) cells, 3 µM of plumbagin resulted in an increase in cytoplasmic ROS and hence a fall in mitochondrial membrane potential, mitochondrial insult, release of cytochrome c, release of AIF, activation of caspases 9 and 3, DNA damage and apoptosis.[206] In A549 cells (wild-type p53), plumbagin at a dose of 11.6 µM induced the phosphorylation of c-Jun N-terminal kinase (JNK), anti-apoptotic protein (p53) and therefore the increase of pro-apoptotic Bax, decreases in the levels of Bcl-2 and Bcl-xL, mitochondrial membrane potential collapse, cytochrome c release, caspases 9 and 3 activation, poly (ADP-ribose) polymerase (PARP) cleavage and apotosis.[207] The ROS generated by plumbagin nullifies the activity of nuclear factor kappa-light-chain-enhancer of activated B cells (NF-κB) and topoisomerase II. In human chronic myeloid leukemia (KBM-5) cells, plumbagin at a dose of 5 µM inhibited the phosphorylation of IκBα by IκB kinase (IKK) and therefore blocked NF-κB transcriptional activity by a mechanism involving the oxidation of p65 Cys-38 by ROS.[208] The ROS generated by plumbagin at a dose of 9 µM oxidized thiol groups on topoisomerase II,[209–212] hence resulting in stabilization of the DNA-topo II cleavable complex[213] and DNA fragmentation. Likewise, plumbagin abated the multiplication of human breast adenocarcinoma (MDA-MB-231) and human breast adenocarcinoma (MCF-7) cells by 62% and 51%, respectively, with a fall in anti-apoptotic protein Bcl-2, activation of caspases 9 and 3, and inhibition of NF-κB.[214] Note that the plumbagin analogue ramentaceone (CS 3.207) from *Drosera aliciae* elicited the generation of ROS in HL-60 cells, resulting in pro-apoptotic Bax translocation, mitochondrial insult, a fall in mitochondrial transmembrane potential, release of cytochrome c, caspase 3 activation and apoptosis, with an IC_{50} equal to 8.7 μM.[215] Sandur et al. (2010) further observed that plumbagin induced the expression of protein tyrosine phosphatase (PTPase) in human multiple myeloma (U266) cells.[216]

Note that the phosphorylation of epidermal growth factor receptor (EGFR) by PTPase results in the activation of both extracellular-signal-regulated kinases (ERK1/2) and protein kinase B (Akt). In fact, 1,4-naphthoquinone, plumbagin and juglone at 10 μM induced the activation of the EGFR in HaCaT cells whereas menadione, lawsone and lapachol were inactive.[200] Plumbagin may inhibit PTPase by oxidation of the cysteine residue by ROS to form a cysteine sulfenate, sulfinate or sulfonate, or it could form covalent bonds with thiols to form thioethers[217] on PTPase via in Michael-type reactions at C3.[200] In fact, the 1.4-benzoquinone moiety of naphthoquinones allows the alkylation of thiols into thionylated hydroquinones,[218] except when C2 is hydroxylated because of tautomerization that impedes Michael-type addition.[200] These mechanisms of action may account for the property of 5,8-dihydroxy-1,4-naphthoquinone (CS 3.208) from members of the family Boraginaceae Juss., which caused microtubule depolymerization, release of tubulin, binding of tubulin to the p85 unit of PI3K, a consequent inhibition of Akt, an increase in the amount of pro-apoptotic Bax translocation, mitochondrial insult, caspase 3 activation, cleavage of PARP and apoptosis in A549 cells (wild-type p53), with an IC_{50} equal to 16.4 μM.[219] In rat liver epithelial (WB-F344) cells, menadione, benzoquinone and 2,3-dimethoxy-1,4-naphthoquinone (CS 3.209) at a dose of 50 μM, 100 μM and 100 μM, respectively, prompted the phosphorylation of ERK1/2.[218] Menadione alkylated PTPase, thereby stimulating EGFR and therefore the activation of mitogen-activated protein kinase (MEK) and the phosphorylation of ERK1/2.[218] Note that menadione generates ROS by both redox cycling and binding to the thiol group of glutathione.[220]

The formation of a covalent bond between the C3 of naphthoquinones and the thiol moiety of glutathione contributes to the generation of ROS, as exemplified with shikonin (CS 3.210) (β-alkannin) from *Lithospermum erythrorhizon* Siebold & Zucc. (family Boraginaceae Juss.), which, at a dose of 10 μM, induced apoptosis in HL-60 cells, with chromatin condensation and caspase 3 activation.[221,222]

In vivo, shikonin repressed the growth of mouse ascite sarcoma (S-180) cells at a dose of 10 mg/kg/day in rodents.[223] Shikonin also inhibited the enzymatic activities of both topoisomerase II (at a dose of 10^{-5} M)[224] and

■ **CS 3.207** Ramentaceone

■ **CS 3.208** 5,8-Dihydroxy-1,4-naphthoquinone

■ **CS 3.209** 2,3-Dimethoxy-1,4-naphthoquinone

■ **CS 3.210** Shikonin

■ **CS 3.211** β-Hydroxyisovalerylshikonin

protein tyrosine kinase (PTK) and therefore decreased the phosphorylation EGFR and induced the phosphorylated JNK in human epithelial carcinoma (A-431) cells, and hence mitochondrial insult and apoptosis, with an IC_{50} value equal to 2.5 μM.[225] Shikonin at a dose of 300 nM prompted the generation of ROS and hence an increase in the level of phospho-JNK and a consequent fall in mitochondrial membrane potential, mitochondrial insult, release of cytochrome c, activation of caspases 9 and 3, cleavage of PARP and apoptosis in human erythromyeloblastoid leukemia (K562) cells,[226] which are resistant to oxidative stress,[202] probably on account of their high production of glutathione. Mitochondrial-mediated apoptosis was further confirmed in human glioblastoma-astrocytoma, epithelial-like (U-MG87) cells (wild-type p53) exposed to 7.5 μM of shikonin which underwent apoptosis as a result of ROS accumulation in the cytoplasm, an increase in pro-apoptotic protein (p53) and Bax, and a decrease in Bcl-2.[198] However, shikonin induced the apoptosis of human hepatocellular carcinoma (HuH-7) and human hepatocellular carcinoma (BEL-7402) cells, with IC_{50} values equal to 4 μM and 5.3 μM, respectively, via an upsurge in ROS, reduction in the level of phosphorylated Akt, activation of caspase 8, cleavage of Bid and mitochondrial insult.[227]

β-Hydroxyisovalerylshikonin (CS 3.211) from a member of the genus *Lithospermum* L. (family Boraginaceae Juss.) abated the multiplication of HL-60 cells by 50% at a dose of 4.1×10^{-7} M, with activation of ERK1/2, JNK, p38 mitogen-activated protein kinase (p38 MAPK) and caspase 3.[228] In addition, 11-deoxyalkannin (CS 3.212), β,β-dimethylacrylalkannin (CS 3.213), 11-O-acetylalkannin (CS 3.214) and alkannin (CS 3.215) isolated from *Alkanna cappadocica* were cytotoxic against HT-29 cells, with respective IC_{50} values equal to 5.1 μM, 0.8 μM, 10.4 μM and 4 μM. From the same plant, 5-O-methyl-11-deoxyalkannin

■ **CS 3.212** 11-Deoxyalkannin

■ **CS 3.213** β,β-Dimethylacrylalkannin

■ **CS 3.214** 11-*O*-Acetylalkannin

■ **CS 3.215** Alkannin

■ **CS 3.216** 5-*O*-Methyl-11-deoxyalkannin

■ **CS 3.217** 8-*O*-Methyl-11-deoxyalkannin

(CS 3.216), 8-*O*-methyl-11-deoxyalkannin (CS 3.217), 5-*O*-methyl-11-*O*-acetylalkannin (CS 3.218), and 5-*O*-methyl-β,β-dimethylacrylalkannin (CS 3.219) yielded values against HT-29 cells of 5 μM, 5.9 μM, 0.6 μM and 0.5 μM, respectively.[229] These results suggest that an isoprenoid ether in C11 enhances the cytotoxic activity of alkannins.

■ **CS 3.218** 5-*O*-Methyl-11-*O*-acetylalkannin

■ **CS 3.219** 5-*O*-Methyl-β,β-dimethylacrylalkannin

■ **CS 3.220** β-Lapachone

■ **CS 3.221** Rhinacanthone

β-Lapachone (CS 3.220) from *Tabebuia avellanedae* Lorentz ex Griseb. (Bignoniaceae Juss.) and its congeners are potently cytotoxic. *Rhinacanthus nasutus* (L.) Kuntze (family Acanthaceae Juss.) produces rhinacanthone (CS 3.221), which is a close isomer of β-lapachone.[230] Rhinacanthone and β-lapachone abated the proliferation of human epitheloid cervix carcinoma (Hela) cells, with IC_{50} values of 1.2 μM and 2.5 μM, respectively.[230] In Hela cells rhinacanthone at a dose of 6 μM induced DNA fragmentation, increased pro-apoptotic Bax, decreased Bcl-2 proteins, induced the release of cytochrome c and activated caspases 9 and 3,[230] suggesting mitochondrial insult. *Rhinacanthus nasutus* also produces rhinacanthins C, N and Q (CS 3.222–3.224) which mitigated the multiplication of human epithelial cancer (Hela S3) cells, with IC_{50} values equal to 80 μM, 65 μM and 73 μM, respectively, via apoptosis and, in the case of rhinacanthins C and N, with caspase 3 activation.[231] The activation of caspase 3 was also observed with (+)-cordiaquinone J (CS 3.225) from

■ **CS 3.222** Rhinacanthin C

■ **CS 3.223** Rhinacanthin N

■ **CS 3.224** Rhinacanthin Q

■ **CS 3.225** (+)-Cordiaquinone J

■ **CS 3.226** Lapachol

■ **CS 3.227** α-Lapachone

Cordia leucocephala, which destroyed HL-60 cells, with the IC_{50} value of 3 μM, triggered by generation of ROS.[232] Other β-lapachone congeners are lapachol (CS 3.226) and α-lapachone (CS 3.227), from a member of the genus *Tabebuia*, which destroyed the following cells (respective IC_{50}

values in parentheses): K562 ($16\,\mu$M, $42.7\,\mu$M), vincristine-resistant human erythromyeloblastoid leukemia (Lucena-1; $20.8\,\mu$M, $47.7\,\mu$M) and human Burkitt's lymphoma (Daudi; $25.5\,\mu$M, $69.3\,\mu$M).[233] In the same experiment, lapachol was inactive against MCF-7 cells whereas α-lapachone was cytotoxic, with an IC_{50} value equal to $38\,\mu$M.[233] Note that Lucena-1 and MCF-7 cells are resistant to oxidative stress,[202,226] unlike Daudi and HL-60 cells,[229] suggesting that the generation of ROS might not be the main trigger for lapachones-induced apoptosis. In fact, β-lapachone is known to inhibit the enzymatic activity of topoisomerase II.[234,235] In HL-60 cells β-lapachone imposed apoptosis via increase in the cytoplasmic levels of ROS.[236] Strikingly, increased levels of ROS induced the apoptosis of MCF-7 (wild-type p53) and human breast carcinoma (T-47D; mutant p53) cells by β-lapachone, with IC_{75} values of $3.5\,\mu$M and $7\,\mu$M, respectively, together with an increase in cytoplasmic Ca^{2+}, activation of cysteine protease calpain, cleavage of PARP and pro-apoptotic protein (p53) activation.[237]

In addition, β-lapachone induced in human prostate epithelial cells DNA fragmentation, phosphorylation of pro-apoptotic p53, translocation of pro-apoptotic Bax, cytochrome c release and activation of caspase 3.[238] Caspase 3 activation was noted in human colorectal carcinoma (HCT-116; wild-type p53) cells, where β-lapachone decreased Bcl-2, activated caspase 3, degraded β-catenin and inhibited NF-κB.[239] Note that PARP cleavage and cytochrome c release was further observed in human multiple myeloma (ARH-77) cells treated with $4\,\mu$M β-lapachone.[240] The involvement of Ca^{2+} in the apoptotic effect of β-lapachone was confirmed in MCF-7 cells by NAD(P)H dehydrogenase (quinone 1) (NQO1) reduction, reduction of NADPH and hence release of endoplasmic reticulum (ER) Ca^{2+} into the cytoplasm followed by a fall in mitochondrial membrane potential.[241] In addition, $5\,\mu$M of β-lapachone induced apoptosis in MCF-7 (wild-type p53) cells, with PARP and pro-apoptotic p53 proteolysis and Ca^{2+}-activated protease calpain activation.[242] Increase in ROS does not always account for the apoptotic properties of β-lapachone. In fact, $3\,\mu$M of β-lapachone nullified the survival of human prostate carcinoma (DU-145) and human prostate cancer (PC-3) cells at doses of $2\,\mu$M and $5\,\mu$M, respectively, without an increase in ROS but with activation of caspase 7, topoisomerase II inhibition, hence DNA damage, induction of p21 and p27 and therefore cell arrest and apoptosis.[243] Induction of p21 was also observed in PC-3 (p53 null) cells, where β-lapachone at a dose of $3\,\mu$M inhibited the phosphorylation of retinoblastoma protein (pRB), therefore increasing the association of pRB with E2F-1 and hence resulting in the abrogation of mitosis, the induction of pro-apoptotic p53-independent binding between p21 and CDK2, leading to inhibition of CDK2, CDK4, and cyclin E.[244]

Members of the genus *Diospyros* (family Ebenaceae) produce series of mildly cytotoxic bis-naphthoquinones such as diospyrin (CS 3.228). Diospyrin from *Diospyros montana* Roxb. (family Ebenaceae) and its synthetic derivatives diospyrin dimethyl ether (CS 3.229), diospyrin dimethyl ether hydroxyderivative (CS 3.230) and diospyrin diethylether (CS 3.231) were cytotoxic against HL-60 cells, with IC_{50} values equal to $100\,\mu M$, $64\,\mu M$, $54\,\mu M$ and $30\,\mu M$, respectively,[197] via apoptosis with nuclear fragmentation and activation of caspases 3 and 8,[245] suggesting that an ethoxyl group at C8 on the diospyrin derivative is a beneficial feature for cytotoxicity.

The obvious chemotherapeutic value of natural naphthoquinones prompted the synthesis of several derivatives. One such compound is

■ **CS 3.228** Diospyrin

■ **CS 3.229** Diospyrin dimethyl ether

■ **CS 3.230** Diospyrin dimethyl ether hydroxyderivative

■ **CS 3.231** Diospyrin diethylether

2,3-dichloro-5,8-dihydroxy-1,4-naphthoquinone (CS 3.232), which, at a dose of 15 μM, abated the growth of Hela cells, with release of cytochrome c, activation of caspases 9 and 3, and cleavage of p65 by caspase 3,[246] suggesting that mitochondrial insult may very well result from pro-apoptotic Bax induction or Bid cleavage, as observed by Hae et al. (2003) in HL-60 cells.[247] The role of pro-apoptotic Bax in 2,3-dichloro-5,8-dihydroxy-1,4-naphthoquinone apoptosis was confirmed in HL-60 cells where 2,3-dichloro-5,8-dihydroxy-1,4-naphthoquinone induced apoptosis at a dose of 1 μM, with inactivation of Akt, translocation of pro-apoptotic Bax, release of cytochrome c, activation of caspases 3, 9 and 8, and subsequent cleavage of Bid.[248] The synthetic naphthoquinone atovaquone (CS 3.233) and 8-hydroxy-2-(4-phenylpiperazin-1-yl)naphthalene-1,4-dione (CS 3.234), which present long alkylated chains at C2, potently destroyed DU-145 cells, with IC$_{50}$ values equal to 29.3 μM and 2.8 μM,[249] respectively, suggesting that a piperazine at C2 is beneficial for cytotoxicity. In fact, 8-hydroxy-2-(4-phenylpiperazin-1-yl)naphthalee-1,4-dione induced the activation of caspases 9 and 3,[249] and abated the multiplication of a broad array of malignancies as follows (IC$_{50}$ values in parentheses): A549 (4.2 μM), Hela (6.9 μM), HT-29 (2.9 μM), MCF-7 (5.2 μM), human hepatocellular liver carcinoma (HepG2; 5.4 μM), human fibrosarcoma (HT1080; 8.2 μM), human nasopharyngeal carcinoma (KB; 7 μM), human lung squamous carcinoma (L-78; 3.6 μM), human renal carcinoma (OS-RC-2; 1.5 μM), human gastric cancer (SGC-7901; 3.7 μM) and human glioblastoma (U251; 6.1 μM).[249] A C3 substitution by an amino aromatic and the presence of a chlorine atom at C2 favors the cytotoxic potencies of naphthoquinones, as exemplified with TW-92 (CS 3.235), which was lethal to human histiocytic lymphoma (U937) cells[250] at a dose of 3 μmol/L via

■ **CS 3.232** 2,3-Dichloro-5,8-dihydroxy-1,4-naphthoquinone

■ **CS 3.233** Atovaquone

■ **CS 3.234** 8-Hydroxy-2-(4-phenylpiperazin-1-yl)naphthalene-1,4-dione

apoptosis, with generation of ROS, an increase in phosphorylated ERK1/2 and pro-apoptotic Bax, a fall in mitochondrial membrane potential, release of cytochrome c and activation of caspases 9 and 3.[250] Recent evidence has demonstrated that furano derivatives of naphthoquinones are potently cytotoxic.[251] One such compound is LQB-118 (CS 3.236), which suppressed the growth of K562 cells by 40% at a dose of 1.5 μM and induced the apoptosis of more than 50% of cells at a dose of 3.0 μM,[252] with caspase 3 activation, and lowered levels of survivin and X-linked inhibitor of apoptosis protein (XIAP).[252]

Another furano derivative of oncological interest is naphtho[1,2-*b*]furan-4,5-dione (CS 3.237). This might be of clinical value for the treatment of oral malignancies as it prompted the death of human gingival cancer (CA9-22) and human hepatocellular carcinoma (Hep3B) cells, with IC$_{50}$ values equal to 2.2 μM and 2.6 μM, respectively.[253,254] In Hep3B cells naphtho[1,2-b]furan-4,5-dione inhibited the activation of ERK1/2, decreased the levels of Bcl-xL and XIAP, and induced the release of cytochrome c, the activation of caspases 9 and 3, the cleavage of PARP and the dephosphorylation of IκBα, thereby blocking NF-κB nuclear translocation.[254] Likewise, in CA9-22 naphtho[1,2-*b*]furan-4,5-dione decreased the levels of phosphorylation of EGF receptors, PI3K and Akt, decreased Bcl-xL, Mcl-1 and XIAP expression and increased levels of Bad, resulting in a fall in mitochondrial membrane potential, release of cytochrome c, activation of caspases 9 and 3, and downregulation of NF-κB.[253] The inhibition of Akt by naphtho[1,2-*b*]furan-4,5-dione was further evidenced in MDA-MB-231 cells, where it inhibited JAK2, STAT3, and Akt, with the consequent upregulation of pro-apoptotic Bax and Bad, the downregulation of Bcl-2, Bcl-xL, survivin and Mcl-1, a fall in mitochondrial membrane potential, release of cytochrome c, and activation of caspases 9 and 3 and apoptosis at a dose of 8 μM.[255] Note that JAK2 is overexpressed in breast malignancies, where it enhances chemoresistance.[256]

■ **CS 3.235** TW-92

■ **CS 3.236** LQB-118

■ **CS 3.237** Naphtho[1,2-*b*] furan-4,5-dione

3.2.2 *Rubia Cordifolia* L.

History This plant was first described by Carl von Linnaeus, in *Systema Naturae*, published in 1768.

Synonyms *Rubia cordifolia* subsp. *pratensis* (Maxim.) Kitam., *Rubia cordifolia* var. *coriacea* Z.Ying Zhang, *Rubia cordifolia* var. *munjista* (Roxb.) Miq., *Rubia cordifolia* var. *pratensis* Maxim., *Rubia cordifolia* var. *rotundifolia* Franch., *Rubia manjith* Roxb. ex Fleming, *Rubia mitis* Miq., *Rubia munjista* Roxb., *Rubia pratensis* (Maxim.) Nakai, *Rubia scandens* Zoll. & Moritzi

Family Rubiaceae Juss., 1789

Common Name Qian cao (Chinese)

Habitat and Description *Rubia cordifolia* is a climber which grows in the forests of Pakistan, India, China, Korea, Japan and Mongolia. The roots have been used to dye silk and wool red since ancient times of time on account of the pigment alizarin. The stems are slender and quadrangular. The leaves are simple, opposite and stipulate, the stipule is interpetiolar. The petiole is 1.5–3 cm long. The blade is lanceolate, 1.5–5 cm×0.5–2.5 cm, cordate at base, obtuse acuminate at apex and displays three to five longitudinal nerves. The inflorescence is a terminal or axillary cyme of tiny flowers. The calyx is inconspicuous. The corolla is yellow, 0.5 cm long, salver-shaped and develops five lobes which are triangular. The fruit is a pair of black glossy berries which are globose and 0.5 cm across (Figure 3.5).

Medicinal Uses In Korea the plant is used to treat coughs, kidney stones, arthritis and gynecological bleeding and infection.

Phytopharmacology The plant produces the naphthohydroquinones furomollugin, mollugin, rubilactone and epoxymollugin;[257] naphthoquinones 2-carbamoyl-3-methoxy-1,4-naphthoquinone, 2-carbamoyl-3-hydroxy-1,4 naphthoquinone and dihydro-α-lapachone;[258] a series of anthraquinones including 1-acetoxy-3-methoxy-9,10-anthraquinone, soranjidiol, 1-hydroxy-2-methoxy-9,10 anthraquinone, alizarin 2-methyl ether,[257] 1,3,6-trihydroxy-2-hydroxymethyl-9,10-anthraquinone-3-*O*-L-rhamnopyranosyl-D-(6-*O*-acetyl)-glucopyranoside,[259] rubiacordone A,[260] cordifiol and cordifodiol,[261] and alizarin;[262] the anthracene derivatives rubiasins A–C[263] and the triterpene oleanolic acid.[257]

■ **FIGURE 3.5** Rubia cordifolia

The medicinal properties of *Rubia cordifolia* are due to anthraquinones.

Proposed Research Pharmacological study of damnacanthal and derivatives for the treatment of cancer.

Rationale Anthraquinones damage DNA by locking the topoisomerase II–DNA complex[264,265] and have been exploited to develop several anticancer drugs, such as mitoxantrone (CS 3.238), used for the treatment of acute non-lymphocytic leukemias, breast cancer and prostate cancer. Doxorubicin and daunorubicin are other successful analogues of anthraquinones used as anticancer drugs. The precise molecular mode of action of simple hydroxylated anthraquinone against topoisomerase II has been elegantly delineated by Li et al. (2010) who observed that emodin from *Rheum officinale* Baill. (family Polygonaceae) docked into and blocked the ATPase pocket of topoisomerase II, hence locking the topoisomerase II–DNA and causing DNA fragmentation.[266] However, several lines of evidence point to the fact that the inhibition of topoisomerase II does

■ **CS 3.238** Mitoxantrone

■ **CS 3.239** 1-Acetoxy-3-methoxy-9,10-anthraquinone

■ **CS 3.240** Soranjidiol

■ **CS 3.241** Alizarin 2-methyl ether

■ **CS 3.242** Danthron

not by itself account for the whole cytotoxic activity of anthraquinones *sensu lato*. For instance, Son et al. (2008) observed that 1-acetoxy-3-methoxy-9,10-anthraquinone (CS 3.239), soranjidiol (CS 3.240) and alizarin 2-methyl ether (CS 3.241) isolated from *Rubia cordifolia* inhibited the enzymatic activity of topoisomerase II by 30%, 3% and 49%, respectively, but abrogated the survival of human colon cancer (HT-29) cells, with IC_{50} values equal to 81.8 μM, >100 μM and 51.6 μM, respectively.[257] In fact, the anthraquinone framework includes a 1,4 benzoquinone moiety which undergoes a 'redox cycling process' that starts with the formation of a semiquinone radical by NADPH cytochrome P-450 reductase, microsomal NADH-cytochrome b and mitochondrial NADH-ubiquinone oxidoreductase, which reacts with oxygen to produce highly reactive superoxide anion radical (O_2^-) in the cytoplasm of mammalian cells,[197,198,199] and therefore a burst in cytoplasmic levels of ROS[267] and cellular insult. The generation of ROS by simple anthraquinones is exemplified with danthron (CS 3.242) (1,8-dihydroxyanthraquinone) from *Rheum palmatum* L. (family Polygonaceae Juss.), which imposed apoptosis in human brain glioblastoma multiforme (GBM 8401) cells at a dose of 100 μM, with increase of cytoplasmic levels of ROS and hence ER insult bringing about the release of Ca^{2+}, mitochondrial insult, a fall in mitochondrial membrane potential and activation of caspases 9, 3 and 8.[268] Caspase 8 activation suggests the involvement of CD95. This concept was exemplified in human gastric carcinoma (SNU-1) cells exposed to 150 μM of danthron, with increased levels of ROS, an increase in cytoplasmic Ca^{2+} production, activation of CD95 and consequent caspase 8 activation, cleaveage of Bid, mitochondrial insult, release of cytochrome c and activation of caspases 9 and 3.[269] The methylation of danthron at C3 produces chrysophanol (CS 3.243), obtained from the genus *Rumex* L. (family Polygonaceae Juss.), which abated the survival of human hepatocellular carcinoma (J5) cells by 50% at a dose of 120 μM, associated with an increase in ROS and Ca^{2+}.[270]

The attachment of a hydroxyl group to the methyl moiety of chrysophanol yields aloe-emodin (CS 3.244), obtained from *Aloe vera* (L.) Burm.f. (family Xanthorrhoeaceae Dumort.), which, interestingly, suppressed the growth of both human hepatocellular liver carcinoma (HepG2) (wild-type p53) and Hep3B (mutant p53) cells, with IC_{50} values equal to 11.7 μg/mL and 15.6 μg/mL, respectively.[271] The higher sensitivity of wild-type p53 to aloe-emodin was further exemplified in SJ-N-KP (wild-type p53) and SK-N-BE (mutant p53) cells, with IC_{50} values equal to 4.2 μM and 29.1 μM, respectively.[272] In SJ-N-KP (wild-type p53) cells aloe-emodin activated pro-apoptotic p53 and hence pro-apoptotic Bax translocation, mitochondrial insult, release of cytochrome c and p21 activation. In contrast, in SJ-N-KP (mutant p53) cells the toxic accumulation of mutant p53 provoked direct mitochondrial insult.[272] Aloe-emodin also prevented the survival of human lung squamous carcinoma (CH27) and human lung cancer (H460) cells at a dose of 40 μM.[273] Another striking feature of aloe-emodin is its ability to modulate the enzymatic activity of protein kinase C (PKC).[273] PKC activation by aloe-emodin was observed in H460 cells at a dose of 40 μM with decrease of ERK1/2 and p38 MAPK.[274] Note that PKC activation by aloe-emodin may very well boost the enzymatic activity of NADPH oxidase and therefore the generation of ROS,[275] as observed in HepG2 cells, where there was phosphorylation of JNK, release of cytochrome c, activation of caspases 9 and 3 and apoptosis.[276] The growth of human nasopharyngeal carcinoma (NPC-TW 039) and (NPC-TW 076) cells were repressed by a 60 μM dose of aloe-emodin by 49.3% and 43.3%, with an increase in ROS, activation of caspase 8, inhibition of Bcl-xL translocation of pro-apoptotic Bax, a fall in mitochondrial membrane potential, activation of caspase 3, cleavage of PARP and increase in Ca^{2+}.[277] Note that an increase in free cytoplasmic Ca^{2+} may very well be the direct consequence of ER stress since increase of C/EBP-homologous

■ **CS 3.243** Chrysophanol

■ **CS 3.244** Aloe-emodin

■ **CS 3.245** Rhein

protein(CHOP) and associated decrease in Bcl-2, and JNK, were observed with human renal proximal tubule (HK-2) cells exposed to 75 μM of aloe-emodin.[278]

The oxidation of aloe-emodin at C3 yields rhein (CS 3.245), obtained from *Rheum officinale* Baill. (familt Polygonaceae Juss.), which induced apoptosis in HL-60 cells (p53 null) at a dose of 100 μM by a process involving caspase 8 activation, cleavage of Bid, a fall in mitochondrial membrane potential, release of cytochrome c and activation of caspases 9 and 3,[279] suggesting CD95 activation independently of aloe-emodin-activated pro-apoptotic p53. Aloe-emodin-induced apoptosis may very well result from DNA damage caused by topoisomerase inhibition and/or ROS in wild-type p53 bearing cancer cells. In fact rhein brought about apoptosis in human epidermoid cervical carcinoma (CaSki; wild-type p53) cells at a dose of 50 μM, with an increase in ROS and consequent DNA damage, activation of pro-apoptotic p53, translocation pro-apoptotic Bax, mitochondrial membrane potential (collapse and caspase 3 activation.[274] In the same cell line, rhein induced an increase in cytoplasmic levels of Ca^{2+}, ER stress, and mitochondrial insult,[280] suggesting that the cellular homeostasis of Ca^{2+} is somewhat involved in the apoptotic potencies of anthraquinones as a result of ROS-induced ER insult. This was further exemplified with KB cells exposed to 150 μM of rhein, which resulted in increased levels of Ca^{2+}, induction of ER stress sensors ATF6 and PERK, increase in CCAAT/enhancer binding protein homologous protein (CHOP) levels, caspase 12 activation, a fall in mitochondrial membrane potential, cytochrome c release and apoptosis.[281] In addition, rhein at a dose of 50 μM abated the proliferation of human lung adenocarcinoma epithelial (A549; wild-type p53) cells via an increase in cytoplasmic ROS and hence DNA and ER damage.[282] ROS-induced DNA damage resulted in aloe-emodin-induced pro-apoptotic p53 activation and therefore cell arrest caused by p21 activation; ROS-induced ER insult provoked the release of Ca^{2+} and hence mitochondrial insult.[282] Likewise, at a dose of 50 μM, rhein provoked the death of human squamous carcinoma of the tongue (SCC-4) cells with an increase in ROS, Ca^{2+} accumulation, translocation of pro-apoptotic Bax, a fall in mitochondrial membrane potential, cytochrome c release, activation of caspases 3, 8 and 9 and, therefore, apoptosis.[283] Note that rhein at a dose of 10 μM destroyed human primary aortic smooth muscle (HASMC) cells treated by 10 ng/mL of TNF-α via dephosphorylation of IκBα[284] via probable inhibition of IKK by ROS.

The hydroxylation of chrysophanol at C8 forms emodin (CS 3.246), obtained from *Rheum palmatum* L. (family Polygonaceae Juss.), which inhibited the growth of A549, H460 and CH27 cells, with IC_{50} values of

■ **CS 3.246** Emodin

about $20\,\mu M$.[285] In A549 (wild-type p53) cells emodin increased the levels of ROS, hypohosphorylated Akt and ERK1/2 and induced the translocation of Bax, triggering a fall in mitochondrial membrane potential and activation of caspases 9 and 3.[285] The inhibition of Akt may account for the apoptosis mediated by $25\,\mu M$ of emodin in human bladder cancer (T24) cells.[286] Alternatively, in wild-type p53 cancer cells aloe-emodin-induced pro-apoptotic p53 might be activated by ATM phosphorylation and activation of ATM by ROS and therefore pro-apoptotic Bax translocation, mitochondrial insult, release of cytochrome c and activation of caspase 3, as observed in A549 cells exposed to $50\,\mu M$ of emodin.[287] Note that $30\,\mu M$ of emodin in human colon carcinoma (LS1034; mutant p53) cells induced an increase in ROS, translocation of pro-apoptotic Bax, collapse of mitochondrial membrane potential, release of cytochrome c, caspases 9 and 3 activation and an increase in free cytoplasmic Ca^{2+},[288] implying that ROS may alter the stability of the ER. Other cellular events triggered by emodin include caspase 8 activation in CH27 cells,[289] the enzymatic inhibition of mTOR in human epitheloid cervix carcinoma (Hela) cells and PKC inhibition in CH27 and H460cells at $50\,\mu M$.[273] *Hedyotis diffusa* Willd. (family Rubiaceae Juss.) produces 2-hydroxy-3-methylanthraquinone (CS 3.247) and 1-methoxy-2-hydroxyanthraquinone (CS 3.248), which inhibited the growth of HepG2 (wild-type p53) cells, with IC_{50} values equal to $51\,\mu M$ and $62\,\mu M$, respectively, and associated with reduced mitochondrial membrane potential and caspase 3 activation.[290] Methylanthraquinone (CS 3.249) from the same plant was lethal to human breast adenocarcinoma (MCF-7) cells, with an IC_{50} value equal to $18.6\,\mu M$, resulting in an increase in free cytoplasmic Ca^{2+} levels followed by calpain activation, JNK phosphorylation, decrease in Bcl-2 and translocation of pro-apoptotic Bax, mitochondrial insult, release of cytochrome c and activation of caspases 4, 9 and 7, leading to apoptosis.[291] Note that JNK is activated by ROS, hence CD95 stimulation, activation of caspase 8, cleavage of Bid and mitochondrial

■ **CS 3.247** 2-Hydroxy-3-methylanthraquinone

■ **CS 3.248** 1-Methoxy-2-hydroxyanthraquinone

■ **CS 3.249** Methylanthraquinone

■ **CS 3.250** Damnacanthal

■ **CS 3.251** 1-Hydroxy-3-[3-(dimethylamino) -propoxy]-9,10-anthraquinone

■ **CS 3.252** 1-Hydroxy-3-[3-(pyrrolidin-1-yl)
propoxy]-9,10- anthraquinone

insult. Damnacanthal (CS 3.250) from members of the genus *Morinda* L. (family Rubiaceae Juss.) at a dose of 85 µM interestingly repressed the growth of both MCF-7 (wild-type p53) cells and human hepatocellular carcinoma (SKHep-1; wild-type p53) cells by 50%.[292] In MCF-7 (wild-type p53) cells damnacanthal activated p38 MAPK and therefore resulted in induction of death receptor (DR5), caspase 8 activation, Bid cleavage, mitochondrial insult, a fall in mitochondrial membrane potential, activation of caspases 9, 3 and apoptosis.[292]

Wue et al. (2000) synthesized a series of anthraquinones and observed that 1-hydroxy-3-[3-(dimethylamino)-propoxy]-9,10-anthraquinone (CS 3.251) powerfully inhibited the growth of a range of cells as follows (IC_{50} values in parentheses): T24 (1.3 µg/mL), Hep3B (2.3 µg/mL), HepG2 (0.8 µg/mL), human cervical carcinoma (SiHa; 0.3 µg/mL), and human hepatoma (PLC/PRF/5; 1.7 µg/mL),[293] implying that an alkoxyamino chain at C3 enhances the cytotoxicity of hydroxyanthraquinones. This was further confirmed by Tu et al. (2011) with 1-hydroxy-3-[3-(pyrrolidin-1-yl)propoxy]-9,10-anthraquinone (CS 3.252), which vanquished human bladder carcinoma (NTUB1) and human prostate cancer (PC-3) cells, with IC_{50} values equal to 9.7 µM and 7.6 µM, respectively. Similarly, IC_{50} values of 8.3 µM and 9.1 µM, were obtained when testing 1-hydroxy-3-[3-(cyclohexylamino)

■ **CS 3.253** 1-Hydroxy-3-[3-(cyclohexylamino)propoxy]-9,10- anthraquinone

■ **CS 3.254** 1-Hydroxy-3-[3-(ethylamino)propoxy]-9,10-anthraquinone 7

■ **CS 3.255** Pixantrone

propoxy]-9,10-anthraquinone (CS 3.253) against the same cell lines. However, 1-hydroxy-3-[3-(ethylamino)propoxy]-9,10-anthraquinone (CS 3.254) was inactive, providing further evidence that alkylation of the nitrogen atom on the alkoxyamino chain moiety promotes cytotoxic potency.[294] Mitoxantrone, pixantrone (CS 3.255) and other derivatives, such as M2, inhibited the growth of human promyelocytic leukemia (HL-60), human ovarian cancer (A2780) and human prostate cancer (PC-3) cells, with IC_{50} values equal to 6.5 nM, 400 nM, 9.3 nM and 29 nM, 23 nM, 36 nM and 290 nM, 1800 nM and 280 nM, respectively, via apoptosis independently of ROS and as a result of the stabilization of the topoisomerase II cleavage complex.[295] Oixantrone underwent Phase III trials for the treatment of aggressive non-Hodgkins lymphoma.[296]

REFERENCES

[189] Kim SH, Lee KS, Son JK, Je GH, Lee JS, Lee CH, et al. Cytotoxic compounds from the roots of *Juglans mandshurica*. J Nat Prod 1998;61(5):643–5.

[190] Min BS, Nakamura N, Miyashiro H, Kim YH, Hattori M. Inhibition of human immunodeficiency virus type 1 reverse transcriptase and ribonuclease H activities by constituents of *Juglans mandshurica*. Chem Pharm Bull 2000;48(2):194–200.

[191] Lee SW, Lee KS, Son JK. New naphthalenyl glycosides from the roots of *Juglans mandshurica*. Planta Med 2000;66(2):184–6.

[192] Lee KS, Li G, Kim SH, Lee CS, Woo MH, Lee SH, et al. Cytotoxic diarylheptanoids from the roots of *Juglans mandshurica*. J Nat Prod 2002;65(11):1707–8.

[193] Gao LI, Ming-Lu XU, Choi HG, Lee SH, Jahng YD, Lee CS, et al. Four new diarylheptanoids from the roots of *Juglans mandshurica*. Chem Pharm Bull 2003;51(3):262–4.

[194] Li ZB, Wang JY, Jiang B, Zhang XL, An LJ, Bao YM. Benzobijuglone, a novel cytotoxic compound from *Juglans mandshurica*, induced apoptosis in HeLa cervical cancer cells. Phytomed 2007;14(12):846–52.

[195] Xu HL, Yu XF, Qu SC, Zhang R, Qu XR, Chen YP, et al. Anti-proliferative effect of juglone from *Juglans mandshurica* Maxim on human leukemia cell HL-60 by inducing apoptosis through the mitochondria-dependent pathway. Eur J Pharmacol 2010;645(1–3):14–22.

[196] Lin H, Zhang YW, Zheng LH, Meng XY, Bao YL, Wu Y, et al. Anthracene and anthraquinone derivatives from the stem bark of *Juglans mandshurica* Maxim. Helvetica Chimica Acta 2011;94(8):1488–95.

[197] O 'Brien PJ. Molecular mechanisms of quinone cytotoxicity. Chem Biol Interact 1991;80:1–41.

[198] Chen CH, Lin ML, Ong PL, Yang JT. Novel multiple apoptotic mechanism of shikonin in human glioma cells. Ann Surg Oncol 2012; in press.

[199] Forman HJ, Zhang H, Rinna A. Glutathione: overview of its protective roles, measurement, and biosynthesis. Mol Aspects Med 2009;30(1–2):1–12.

[200] Klaus V, Hartmann T, Gambini J, Graf P, Stahl W, Hartwig A, et al. 1,4-Naphthoquinones as inducers of oxidative damage and stress signaling in HaCaT human keratinocytes. Archi Biochem Biophys 2010;496(2):93–100.

[201] Montenegro RC, Araújo AJ, Molina MT, Filho JDBM, Rocha DD, Lopéz-Montero E, et al. Cytotoxic of naphthoquinones with special emphasis on juglone and its 5-*O*-methyl derivative. Chem Biological Interact 2010;184(3):439–48.

[202] Trindade GS, Capella MA, Capella LS, Affonso-Mitidieri OR, Rumjanek VM. Differences in sensitivity to UVC, UVB and UVA radiation of a multi-drug-resistant cell line overexpressing P-glycoprotein. Photochem Photobiol 1999;69:694–9.

[203] Monks TJ, Hanzlik RP, Cohen GM, Ross D, Graham DG. Quinone chemistry and toxicity. Toxicol Appl Pharmacol 1992;112:2–16.

[204] Babich H, Stern A. In vitro cytotoxicities of 1,4-naphthoquinone and hydroxylated 1,4-naphthoquinones to replicating cells. J Appl Toxicol 1993;13:353–8.

[205] Inbaraj JJ, Chignell CF. Cytotoxic action of juglone and plumbagin: a mechanistic study using HaCaT keratinocytes. Chem Res Toxicol 2004;17:55–62.

[206] Srinivas P, Gopinath G, Banerji A, Dinakar A, Srinivas G. Plumbagin induces reactive oxygen species, which mediate apoptosis in human cervical cancer cells. Mol Carcinogen 2004;40(4):201–11.

[207] Hsu YL, Cho CY, Kuo PL, Huang YT, Lin CC. Plumbagin (5-hydroxy-2-methyl-1,4-naphthoquinone) induces apoptosis and cell cycle arrest in A549 cells through p53 accumulation via c-Jun NH 2-terminal kinase-mediated phosphorylation at serine 15 in vitro and in vivo. J Pharmacol Exp Ther 2006;318(2):484–94.

[208] Sandur SK, Ichikawa H, Sethi G, Kwang SA, Aggarwal BB. Plumbagin (5-hydroxy-2-methyl-1,4-naphthoquinone) suppresses NF-κB activation and NF-κB-regulated gene products through modulation of p65 and IκBα kinase

activation, leading to potentiation of apoptosis induced by cytokine and chemo-therapeutic agents. J of Biol Chem 2006;281(25):17023–17033.

[209] Li TK, Chen AY, Yu C, Mao Y, Wang H, Liu LF. Activation of topoisomerase II-mediated excision of chromosomal DNA loops during oxidative stress. Genes Dev 1999;13:1553–60.

[210] Kawiak A, Piosik J, Stasilojc G, Gwizdek-Wisniewska A, Marczak L, Stobiecki M, et al. Induction of apoptosis by plumbagin through reactive oxygen species-mediated inhibition of topoisomerase II. Toxicol Appl Pharmacol 2007;223(3):267–76.

[211] Wang H, Mao Y, Chen AY, Zhou N, LaVoie EJ, Liu LF. Stimulation of topoisomerase II-mediated DNA damage via a mechanism involving protein thiolation. Biochem 2001;40:3316–23.

[212] Kawiak A, Piosik J, Stasilojc G, Gwizdek-Wisniewska A, Marczak L, Stobiecki M, et al. Induction of apoptosis by plumbagin through reactive oxygen species-mediated inhibition of topoisomerase II. Toxicol Appl Pharmacol 2007;223(3):267–76.

[213] Fuji N, Yamashita Y, Arima Y, Nagashima M, Nakano H. Induction of topoisomerase II-mediated DNA cleavage by the plant naphthoquinones plumbagin and shikonin. Antimicrob Agents Chemother 1992;36:2589–94.

[214] Ahmad A, Banerjee S, Wang Z, Kong D, Sarkar FH. Plumbagin-induced apoptosis of human breast cancer cells is mediated by inactivation of NF-κB and Bcl-2. J Cell Biochem 2008;105(6):1461–71.

[215] Kawiak A, Zawacka-Pankau J, Wasilewska A, Stasilojc G, Bigda J, Lojkowska E. Induction of apoptosis in HL-60 cells through the ROS-mediated mitochondrial pathway by ramentaceone from *Drosera aliciae*. J Nat Prod 2012;75(1):9–14.

[216] Sandur SK, Pandey MK, Sung B, Aggarwal BB. 5-Hydroxy-2-methyl-1,4-naphthoquinone, a vitamin K3 analogue, suppresses STAT3 activation pathway through induction of protein tyrosine phosphatase, SHP-1: potential role in chemosensitization. Mol Cancer Res 2010;8(1):107–18.

[217] Bolton JL, Trush MA, Penning TM, Dryhurst G, Monks TJ. Role of quinines in toxicology. Chem Res Toxicol 2000;13:135–60.

[218] Abdelmohsen K, Gerber PA, Von Montfort C, Sies H, Klotz LO. Epidermal growth factor receptor is a common mediator of quinone-induced signaling leading to phosphorylation of connexin-43. Role of glutathione and tyrosine phosphatases. J Biol Chem 2003;278(40):38360–38367.

[219] Acharya BR, Bhattacharyya S, Choudhury D, Chakrabarti G. The microtubule depolymerizing agent naphthazarin induces both apoptosis and autophagy in A549 lung cancer cells. Apoptosis 2011;16(9):924–39.

[220] Bellomo G, Mirabelli F, DiMonte D, Richelmi P, Thor H, Orrenius C, et al. Biochem Pharmacol 1987;36:1313–20.

[221] Gao D, Hiromura M, Yasui H, Sakurai H. Direct reaction between shikonin and thiols induces apoptosis in HL60 cells. Biol Pharm Bull 2002;25(7):827–32.

[222] Yoon Y, Kim Y, Kim N, Jeon W, Sung HJ. Shikonin, an ingredient of *Lithospermum erythrorhizon* induced apoptosis in HL60 human premyelocytic leukemia cell line. Planta Med 1999;65:532–5.

[223] Sankawa U, Ebizuka Y, Miyazaki T, Isomura Y, Otuka H, Shibata S, et al. Antitumor activity of shikonin and its derivatives. Chem Pharm Bull 1977;25:2392–5.

[224] Fujii N, Yamashita Y, Arima Y, Nagashima M, Nakano H. Induction of topoisomerase Il-mediated DNA cleavage by the plant naphthoquinones, plumbagin and shikonin. Antimicrob Agents Chemother 1992;36:2589–94.

[225] Singh F, Gao D, Lebwohl MG, Wei H. Shikonin modulates cell proliferation by inhibiting epidermal growth factor receptor signaling in human epidermoid carcinoma cells. Cancer Lett 2003;200(2):115–21.

[226] Mao X, Rong Yu, C, Hua Li, W, Xin Li, W. Induction of apoptosis by shikonin through a ROS/JNK-mediated process in Bcr/Abl-positive chronic myelogenous leukemia (CML) cells. Cell Res 2008;18(8):879–88.

[227] Gong K, Li W. Shikonin, a Chinese plant-derived naphthoquinone, induces apoptosis in hepatocellular carcinoma cells through reactive oxygen species: a potential new treatment for hepatocellular carcinoma. Free Radical Biol Med 2011;51(12):2259–71.

[228] Hashimoto S, Xu M, Masuda Y, Aiuchi T, Nakajo S, Cao J, et al. β– Hydroxyisovalerylshikonin inhibits the cell growth of various cancer cell lines and induces apoptosis in leukemia HL-60 cells through a mechanism different from those of Fas and etoposide. J Biochem 1999;125:17–23.

[229] Sevimli-Gur, C, Akgun, IH, Deliloglu-Gurhan, I, Korkmaz, KS, Bedir, E. Cytotoxic naphthoquinones from *Alkanna cappadocica*. 2010;73(5):860–864.

[230] Siripong P, Hahnvajanawong C, Yahuafai J, Piyaviriyakul S, Kanokmedhakul K, Kongkathip N, et al. Induction of apoptosis by rhinacanthone isolated from *Rhinacanthus nasutus* roots in human cervical carcinoma cells. Biol Pharm Bull 2009;32(7):1251–60.

[231] Siripong P, Yahuafai J, Shimizu K, Ichikawa K, Yonezawa S, Asai T, et al. Induction of apoptosis in tumor cells by three naphthoquinone esters isolated from Thai medicinal plant: *rhinacanthus nasutus* Kurz. Biol Pharm Bull 2006;29(10):2070–6.

[232] Marinho-Filho JDB, Bezerra DP, Araújo AJ, Montenegro RC, Pessoa C, Diniz JC, et al. Oxidative stress induction by (+)-cordiaquinone J triggers both mitochondria-dependent apoptosis and necrosis in leukemia cells. Chem Biol Interact 2010;183(3):369–79.

[233] Salustiano EJS, Netto CD, Fernandes RF, Da Silva AJM, Bacelar TS, Castro CP, et al. Comparison of the cytotoxic effect of lapachol, α-lapachone and pentacyclic 1,4-naphthoquinones on human leukemic cells. Invest New Drugs 2010;28(2):139–44.

[234] Li CJ, Averboukh L, Pardee AB. beta-Lapachone, a novel DNA topoisomerase I inhibitor with a mode of action different from camptothecin. J Biol Chem 1993;268:22463–22468.

[235] Frydman B, Marton LJ, Sun JS, Neder K, Witiak DT, Liu AA, et al. Induction of DNA topoisomerase II-mediated DNA cleavage by beta-lapachone and related naphthoquinones. Cancer Res 1997;57:620–7.

[236] Chau YP, Shiah SG, Don MJ, Kuo ML. Involvement of hydrogen peroxide in topoisomerase inhibitor beta-lapachone-induced apoptosis and differentiation in human leukemia cells. Free Radical Biol Med 1998;24:660–70.

[237] Planchon SM, Wuerzberger-Davis SM, Pink JJ, Robertson KA, Bornmann WG, Boothman DA. Bcl-2 protects against beta-lapachone-mediated caspase 3 activation and apoptosis in human myeloid leukemia (HL-60) cells. Oncol Rep 1999;6(3):485–92.

[238] Choi YH, Kim MJ, Lee SY, Lee YN, Chi GY, Eom HS, et al. Phosphorylation of p53, induction of Bax and activation of caspases during beta-lapachone-mediated apoptosis in human prostate epithelial cells. Int J Oncol 2002;21(6):1293–9.

[239] Choi BT, Cheong J, Choi YH. [β]-Lapachone-induced apoptosis is associated with activation of caspase-3 and inactivation of NF-[kappa]B in human colon cancer HCT-116 cells. Anti-Cancer Drugs 2003;14(10):845–50.

[240] Li Y, Li CJ, Yu D, Pardee AB. Potent induction of apoptosis by beta-lapachone in human multiple myeloma cell lines and patient cells. Mol Med 2000;6(12):1008–15.

[241] Tagliarino C, Pink JJ, Dubyak GR, Nieminenll AL, Boothman DA. Calcium is a key signaling molecule in β-lapachone-mediated cell death. J Biol Chem 2001;276(22):19150–19159.

[242] Tagliarino C, Pink JJ, Reinicke KE, Simmers SM, Wuerzberger-Davis SM, Boothman DA. μ-Calpain activation in β-lapachone-mediated apoptosis. Cancer Biol Ther 2003;2(2):141–52.

[243] Don JD, Chang YH, Chen KK, Ho LK, Chau YP. Induction of CDK inhibitors (p21WAF1 and p27Kip1) and bak in the b-lapachone-induced apoptosis of human prostate cancer cells. Mol Pharmacol 2001;59:784–94.

[244] Choi YH, Ho SK, Yoo MA. Suppression of human prostate cancer cell growth by β-lapachone via down-regulation of pRB phosphorylation and induction of Cdk inhibitor P21 WAF1/CIP1. J Biochem Mol Biol 2003;36(2):223–9.

[245] Chakrabarty S, Roy M, Hazra B, Bhattacharya RK. Induction of apoptosis in human cancer cell lines by diospyrin, a plant-derived bisnaphthoquinonoid, and its synthetic derivatives. Cancer Lett 2002;188(1–2):85–93.

[246] Kang KH, Lee KH, Kim MY, Choi KH. Caspase-3-mediated cleavage of the NF-κB subunit p65 at the NH 2 terminus potentiates naphthoquinone analog-induced apoptosis. J Biol Chem 2001;276(27):24638–24644.

[247] Hae JK, Jung YM, Young JC, Kyung HC, Sung WH, Mie YK. Effects of a naphthoquinone analog on tumor growth and apoptosis induction. Arch Pharm Res 2003;26(5):405–10.

[248] Kim HJ, Kang SK, Mun JY, Chun YJ, Choi KH, Kim MY. Involvement of Akt in mitochondria-dependent apoptosis induced by a cdc25 phosphatase inhibitor naphthoquinone analog. FEBS Lett 2003;555(2):217–22.

[249] Zhou J, Duan L, Chen H, Ren X, Zhang Z, Zhou F, et al. Atovaquone derivatives as potent cytotoxic and apoptosis inducing agents. Mutagenesis 2009;24(5):413–8.

[250] Hallak M, Win T, Shpilberg O, Bittner S, Granot Y, Levy I, et al. The anti-leukaemic activity of novel synthetic naphthoquinones against acute myeloid leukaemia: induction of cell death via the triggering of multiple signalling pathways. Br J Haematol 2009;147(4):459–70.

[251] Netto CD, Silva AJ, Salustiano EJ, Rica IG, Cavalcante MC, Rumjanek VM, et al. New pterocarpanquinones: synthesis, antineoplasic activity on cultured human malignant cell lines and TNF-α modulation in human PBMC cells. Bioorg Med Chem 2010;18:1610–6.

[252] Maia RC, Vasconcelos FC, de Sá Bacelar T, Salustiano EJ, da Silva LFR, Pereira DL, et al. LQB-118, a pterocarpanquinone structurally related to lapachol [2-hydroxy-3-(3-methyl-2-butenyl)-1,4-naphthoquinone]: a novel class of agent with high apoptotic effect in chronic myeloid leukemia cells. Invest New Drugs 2011;29:1143–55.

[253] Chien CM, Lin KL, Su JC, Chuang PW, Tseng CH, Chen YL, et al. Naphtho[1,2-b]furan-4,5-dione induces apoptosis of oral squamous cell carcinoma: involvement of EGF receptor/PI3K/Akt signaling pathway. Eur J Pharmacol 2010;636(1–3):52–8.

[254] Chiu CC, Chen JYF, Lin KL, Huang CJ, Lee JC, Chen BH, et al. P38 MAPK and NF-κB pathways are involved in naphtho[1,2-b] furan-4,5-dione induced anti-proliferation and apoptosis of human hepatoma cells. Cancer Lett 2010;295(1):92–9.

[255] Lin KL, Su JC, Chien CM, Tseng CH, Chen YL, Chang LS, et al. Naphtho[1,2-b]furan-4,5-dione disrupts Janus kinase-2 and induces apoptosis in breast cancer MDA-MB-231 cells. Toxicol in Vitro 2010;24(4):1158–67.

[256] Cance WG, Craven RJ, Weiner TM, Liu ET. Novel protein kinases expressed in human breast cancer. Int J Cancer 1993;54:571–7.

[257] Son JK, Jung SJ, Jung JH, Fang Z, Lee CS, Seo CS, et al. Anticancer constituents from the roots of *Rubia cordifolia* L. Chem Pharm Bull 2008;56(2):213–6.

[258] Koyama J, Ogura T, Tagahara K, Konoshima T, Kozuka M. Two naphthoquinones from *Rubia cordifolia*. Phytochem 1992;31(8):2907–8.

[259] Jeong SY, Zhao BT, Lee CS, Son JK, Min BS, Woo MH. Constituents with DNA topoisomerases I and II inhibitory activity and cytotoxicity from the roots of *Rubia cordifolia*. Planta Med 2012;78(2):177–81.

[260] Li X, Liu Z, Chen Y, Wang LJ, Zheng YN, Sun GZ, et al. Rubiacordone A: a new anthraquinone glycoside from the roots of *Rubia cordifolia*. Molecules 2009;14(1):566–72.

[261] Abdullah ST, Ali A, Hamid H, Ali M, Ansari SH, Alam MS. Two new anthraquinones from the roots of *Rubia cordifolia* Linn. Pharmazie 2003;58(3):216–7.

[262] Angelini LG, Pistelli L, Belloni P, Bertoli A, Panconesi S. *Rubia tinctorum* a source of natural dyes: agronomic evaluation, quantitative analysis of alizarin and industrial assays. Ind Crop Prod 1997;6:303–11.

[263] Chang LC, Chavez D, Gills JJ, Fong HHS, Pezzuto JM, Kinghorn AD. Rubiasins A–C, new anthracene derivatives from the roots and stems of *Rubia cordifolia*. Tetrahedron Lett 2000;41(37):7157–62.

[264] Zwelling LA, Mayes J, Altschuler E, Satitpunwaycha P, Tritton TR, Hacker MP. Activity of two novel anthracene-9,10-diones against human leukemia cells containing intercalator-sensitive or -resistant forms of topoisomerase II. Biochem Pharmacol 1993;46:265–71.

[265] Müller SO, Eckert I, Lutz WK, Stopper H. Genotoxicity of the laxative drug components emodin, aloe-emodin and danthron in mammalian cells: topoisomerase II mediated? Mutat. Res. 1996;371:165–73.

[266] Li Y, Luan Y, Qi X, Li M, Gong L, Xue X, et al. Emodin triggers DNA double-strand breaks by stabilizing topoisomerase II-DNA cleavage complexes and by inhibiting ATP hydrolysis of topoisomerase II. Toxicological Sci 2010;118(2):435–43.

[267] Huang HC, Chang JH, Tung SF, Wu RT, Foegh ML, Chu SH. Immunosuppressive effect of emodin, a free radical generator. Eur J Pharmacol 1992;211:359–64.

[268] Lu HF, Wang HL, Chuang YY, Tang YJ, Yang JS, Ma YS, et al. Danthron induced apoptosis through mitochondria- and caspase-3-dependent pathways in human brain glioblastoma multiforms GBM 8401 cells. Neurochem Res 2010;35(3):390–8.

[269] Chiang JH, Yang JS, Ma CY, Yang MD, Huang HY, Hsia TC, et al. Danthron, an anthraquinone derivative, induces DNA damage and caspase cascades-mediated apoptosis in SNU-1 human gastric cancer cells through mitochondrial permeability transition pores and Bax-triggered pathways. Chem Res Toxicol 2011;24(1):20–9.

[270] Lu CC, Yang JS, Huang AC, Hsia TC, Chou ST, Kuo CL, et al. Chrysophanol induces necrosis through the production of ROS and alteration of ATP levels in J5 human liver cancer cells. Mol Nutr Food Res 2010;54(7):967–76.

[271] Kuo YC, Sun CM, Ou JC, Tsai WJ. Antitumor cell growth inhibitor from *Polygonum hypoleucum* Ohwi. Life Sci 1997;61:2335.

[272] Pecere T, Sarinella F, Salata C, Gatto B, Bet A, Dalla Vecchia F, et al. Involvement of p53 in specific anti-neuroectodermal tumor activity of aloe-emodin. Int J Cancer 2003;106:836–47.

[273] Lee HZ. Protein kinase C involvement in aloe-emodin- and emodin-induced apoptosis in lung carcinoma cell. Br J Pharmacol 2001;134:1093.

[274] Yeh FT, Wu CH, Lee HZ. Signaling pathway for aloe-emodin-induced apoptosis in human H460 lung nonsmall carcinoma cell. Int J Cancer 2003;106(1):26–33.

[275] Chen KC, Zhou Y, Zhang W, Lou MF. Control of PDGF-induced reactive oxygen species (ROS) generation and signal transduction in human lens epithelial cells. Mol Vis 2007;14(13):374–87.

[276] Lu GD, Shen HM, Chung MCM, Ong CN. Critical role of oxidative stress and sustained JNK activation in aloe-emodin-mediated apoptotic cell death in human hepatoma cells. Carcinogenesis 2007;28(9):1937–45.

[277] Lin ML, Lu YC, Chung JG, Li YC, Wang SG, Ng SH, et al. Aloe-emodin induces apoptosis of human nasopharyngeal carcinoma cells via caspase-8-mediated activation of the mitochondrial death pathway. Cancer Lett 2010;291(1):46–58.

[278] Zhu S, Jin J, Wang Y, Ouyang Z, Xi C, Li J, et al. The endoplasmic reticulum stress response is involved in apoptosis induced by aloe-emodin in HK-2 cells. Food Chem Toxicol 2012;50(3–4):1149–58.

[279] Lin S, Fujii M, Hou DX. Rhein induces apoptosis in HL-60 cells via reactive oxygen species-independent mitochondrial death pathway. Arch Biochem Biophys 2003;418:99–107.

[280] Ip S-W, Weng Y-S, Lin S-Y, Yang M-D, Tang N-Y, Su C-C, et al. The role of Ca^{+2} on rhein-induced apoptosis in human cervical cancer Ca Ski cells. Anticancer Res 2007;27(1A):379–90.

[281] Lin ML, Chen SS, Lu YC, Liang RY, Ho YT, Yang CY, et al. Rhein induces apoptosis through induction of endoplasmic reticulum stress and Ca^{2+}-dependent mitochondrial death pathway in human nasopharyngeal carcinoma cells. Anticancer Res 2007;27(5A):3313–22.

[282] Hsia TC, Yang JS, Chen GW, Chiu TH, Lu HF, Yang MD, et al. The roles of endoplasmic reticulum stress and Ca^{2+} on rhein-induced apoptosis in A-549 human lung cancer cells. Anticancer Res 2009;2(1):309–18.

[283] Lai WW, Yang JS, Lai KC, Kuo CL, Hsu CK, Wang CK, et al. Rhein induced apoptosis through the endoplasmic reticulum stress, caspase- and mitochondria-dependent pathways in SCC-4 human tongue squamous cancer cells. In Vivo 2009;23(2):309–16.

[284] Heo SK, Yun HJ, Park WH, Park SD. Rhein inhibits TNF-α-induced human aortic smooth muscle cell proliferation via mitochondrial-dependent apoptosis. J Vascular Res 2009;46(4):375–86.

[285] Su YT, Chang HL, Shyue SK, Hsu SL. Emodin induces apoptosis in human lung adenocarcinoma cells through a reactive oxygen species-dependent mitochondrial signaling pathway. Biochem Pharmacol 2005;70:229–41.

[286] Lin JG, Chen GW, Li TM, Chouh ST, Tan TW, Chung JG. Aloe-emodin induces apoptosis in T24 human bladder cancer cells through the p53 dependent apoptotic pathway. J Urol 2006;175:343–7.

[287] Lai JM, Chang JT, Wen CL, Hsu SL. Emodin induces a reactive oxygen species-dependent and ATM-p53-Bax mediated cytotoxicity in lung cancer cells. Eur J Pharmacol 2009;623(1–3):1–9.

[288] Ma YS, Weng SW, Lin MW, Lu CC, Chiang JH, Yang JS, et al. Antitumor effects of emodin on LS1034 human colon cancer cells in vitro and in vivo: roles of apoptotic cell death and LS1034 tumor xenografts model. Food ChemToxicol 2012;50(5):1271–8.

[289] Lee HZ. Effects and mechanisms of emodin on cell death in human lung squamous cell carcinoma. Br J Pharmacol 2001;134:11.

[290] Shi Y, Wang CH, Gong XG. Apoptosis-inducing effects of two anthraquinones from *Hedyotis diffusa* Willd. Biol Pharm Bull 2008;31(6):1075–8.

[291] Liu Z, Liu M, Liu M, Li J. Methylanthraquinone from Hedyotis diffusa WILLD induces Ca2+-mediated apoptosis in human breast cancer cells. Toxicol Vitro 2010;24(1):142–7.

[292] Lin FL, Hsu JL, Chou CH, Wu WJ, Chang CI, Liu HJ. Activation of p38 MAPK by damnacanthal mediates apoptosis in SKHep 1 cells through the DR5/TRAIL and TNFR1/TNF-α and p53 pathways. Eur J Pharmacol 2011;650(1):120–9.

[293] Wei BL, Wu SH, Chung MI, Won SJ, Lin CN. Synthesis and cytotoxic effect of 1,3-dihydroxy-9,10-anthraquinone derivatives. Eur J Med Chem 2000;35(12):1089–98.

[294] Tu HY, Huang AM, Teng CH, Hour TC, Yang SC, Pu YS, et al. Anthraquinone derivatives induce G2/M cell cycle arrest and apoptosis in NTUB1 cells. Bioorg Med Chem 2011;19(18):5670–8.

[295] Evison BJ, Pastuovic M, Bilardi RA, Forrest RA, Pumuye PP, Sleebs BE, et al. M2, a novel anthracenedione, elicits a potent DNA damage response that can be subverted through checkpoint kinase inhibition to generate mitotic catastrophe. Biochem Pharmacol 2011;82(11):1604–18.

[296] Mukherji D, Pettengell R. Pixantrone for the treatment of aggressive non-Hodgkin lymphoma. Expert Opin Pharmacother 2010;11:1915–23.

Lignans

3.3.1 *Houpoea Obovata* (Thunb.) N.H. Xia & C.Y. Wu

History This plant was first described by Xia in *Flora of China*, published in 2008

Synonyms *Liriodendron liliiflorum* Steud., *Magnolia glauca* Thunb., *Magnolia honogi* P. Parm., *Magnolia hoonokii* Siebold, *Magnolia hypoleuca* Siebold & Zucc., *Magnolia hypoleuca* var. *concolo*r Siebold & Zucc., *Magnolia obovata* Thunb.,*Yulania japonica* var. *obovata* (Thunb.) P. Parm.

Family Magnoliaceae Juss., 1789

Common Name Ri ben hou po (Chinese)

Habitat and Description *Houpoea obovata* is a tree which grows in China, Korea and Japan. The bole reaches 30 m high. The leaves are simple, spiral and exstipulate. The petiole is 2.5–5 cm long. The blade is obovate, 20–35 cm × 10–20 cm, glaucous beneath, cuneate at base, acute at apex, and presents 20–25 pairs of secondary nerves. The flowers are terminal, massive, up to 15 cm in diameter and fragrant. The perianth includes 10 tepals which are pure white, fleshy, spoon-shaped and 5–12 cm × 1.5–5 cm. The androecium includes numerous spirally arranged stamens which are 2 cm long, and reddish and flattened at base. The gynecium consist of numerous free carpels arranged in a column. The fruit is a woody, red, 15-cm-long syncarp bearing numerous carpels which are beaked (Figure 3.6).

Medicinal Uses In Japan the plant is used to heal stomach ulcers and to treat allergies.

Phytopharmacology The plant contains: some lignans, such as honokiol, magnolol, obovatol, obovatal,[297] clovanemagnolol,[298] magnolianin,[299] 4-methoxymagnaldehyde B, coumanolignan[300] and 4-methoxy-honokiol;[301] the flavonoids quercetin 3-*O*-rhamnoside and rutin;[302] and

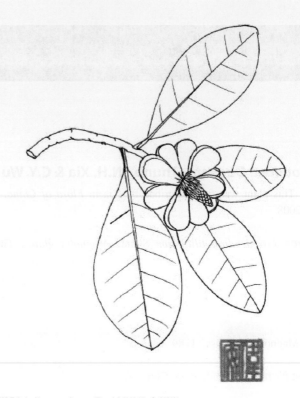

■ **FIGURE 3.6** *Houpoea obovata* (Thunb.) N.H. Xia & C.Y. Wu

aporphine alkaloids *N*-acetylanonaine, *N*-acetylxylopine, *N*-formylanonaine, liriodenine and lanuginosine.[303]

The medicinal properties of *Houpoea obovata* (Thunb.) N.H. Xia & C.Y. Wu may be due to lignans such as honokiol and/or magnolol.

Proposed Research Pharmacological study of justicidin A and derivatives for the treatment of cancer.

Rationale Accumulated evidence points to the fact that lignans are cytotoxic. One such lignan is the aryltetralin lactone podophyllotoxin from *Podophyllum peltatum* L. (family Berberidaceae Juss.), which at a dose of 1 μM arrests cancer cell division because it binds to tubulin and therefore inhibits tubulin polymerization into microtubules.[304]

Indeed, podophyllotoxin (CS 3.256) potently inhibits the growth of a broad array of malignancies, including mouse leukemia (P388), human lung adenocarcinoma epithelial (A549) and human colon cancer (HT-29) cells,

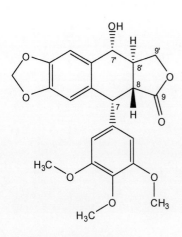

■ **CS 3.256** Podophyllotoxin

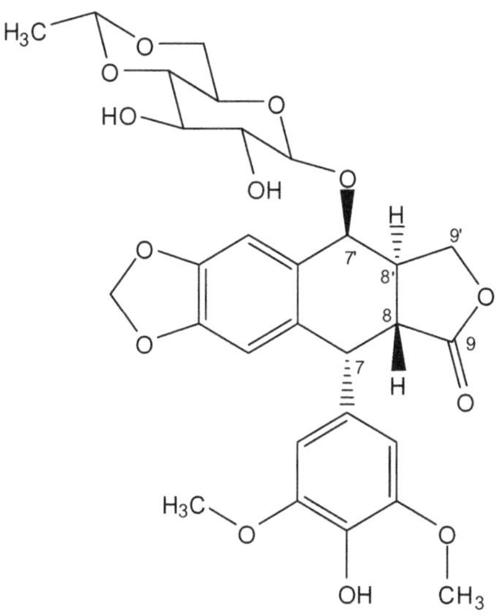

■ **CS 3.257** Etoposide

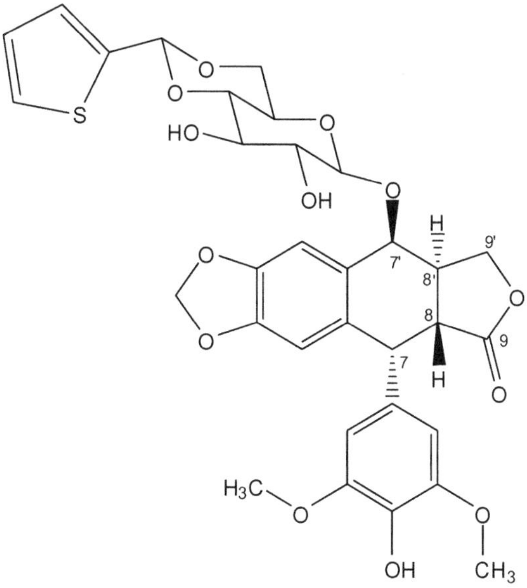

■ **CS 3.258** Teniposide

with IC_{50} values equal to 0.01 μM, 0.01 μM and 0.02 μM, respectively,[305] but in clinical trials it showed mild activity and heavy side effects. Subsequently the podophyllotoxin analogues etoposide (VP-16-213) (CS 3.257) and teniposide (VM-26) (CS 3.258) were developed into drugs for the treatment of cancers.[306]

Although etoposide and podophyllotoxin share an aryltetralin lactone lignan framework, just differing by a β glucoside moiety at C4 and a hydroxyl at the C4′ position, their modes of action are completely different: etoposide imposes the stabilization of topoisomerase II with DNA.[305,306] Therefore, one can reasonably draw the exciting inference that particular cytotoxic mechanisms are associated with particular lignan structures, and since lignans are extremely diverse, studies on the apoptotic mechanisms of lignans will contribute to the discovery of new apoptotic pathways.

The simple dibenzylbutane lignan 7′-hydroxy-3′,4′,5,9,9′-pentamethoxy-3,4-methylenedioxylignan (CS 3.259) from *Phyllathus urinaria* imposed apoptosis in HEp2 epthelial cells, with an IC_{50} value equal to 4.4 μM, following inhibition of telomerase activity and stimulation of microtubule assembly, an increase in c-Myc and caspases 3 and 8 activation (2002). Caspase 3 activation was further observed with macelignan (CS 3.260), oleiferin C (CS 3.261) and *meso*dihydroguaiaretic acid (CS 3.262) from

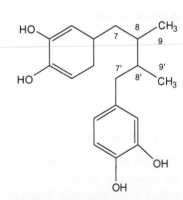

■ **CS 3.259** 7′-Hydroxy-3′,4′,5,9,9′-pentamethoxy-3,4-methylenedioxylignan

■ **CS 3.260** Macelignan

■ **CS 3.261** Oleiferin C

■ **CS 3.262** *Meso*dihydroguaiaretic acid

■ **CS 3.263** Nordihydroguaiaretic acid

Machilus thunbergii Siebold & Zucc. (family Lauraceae Juss.) at a dose of 100 μM in human promyelocytic leukemia (HL-60) cells (p53 null).[305] In addition, nordihydroguaiaretic acid (CS 3.263) from Larrea tridentata (Sessé & Moc. ex DC.) Coville (family Zygophyllaceae R. Br.) brought about the death of human colon adenocarcinoma (SW480) cells, with an IC$_{50}$ value equal to 1.9 μM, via Bcl-xL decrease and collapse of mitochondrial membrane potential and apoptosis.[307]

The attachment of a lactone at C8 and C8′ to simple dibenzylbutane lignan produces arctigenin (CS 3.264), matairesinol (CS 3.265) and 7,8-didehydroarctigenin (CS 3.266) from *Arctium lappa* L. (family Asteraceae

■ **CS 3.264** Arctigenin

■ **CS 3.265** Matairesinol

■ **CS 3.266** 7,8-Didehydroarctigenin

Bercht. & J. Presl), which abrogated the survival of mouse hybridoma (MH60) cells, with IC_{50} values equal to 1 μM, 8.4 μM and 13.5 μM, respectively, via apoptosis.[308] Arctigenin inhibited the growth of SW480 cells, with an IC_{50} value equal to 42.5 μM, via the phosphorylation of β-catenin and therefore downregulation of cyclin D1 and survivin.[309] In addition, arctigenin inflicted lethal effects on human gastric carcinoma (SNU-1) cells by 50% at a dose of 3 μM by inhibiting the phosphorylation of retinoblastoma protein (Rb), and therefore cell division via activation of p21.[310]

The presence of a covalent bond between C2′ and C7 dibenzylbutane lactone lignans forms aryltetralin lactone lignans, such as justicidin A (CS 3.267) and diphyllin (CS 3.268) from *Justicia procumbens* L. (family Acanthaceae Juss.), which were potently cytotoxic against human nasopharyngeal carcinoma (KB) cells, with IC_{50} values inferior to 1 μg/mL. In contrast, justicidin D (CS 3.269) resulted in an IC_{50} value equal to 9 μg/mL, implying that the γ-lactone ring carbonyl α at C8 instead of C8′ enhances the cytotoxicity.[311] Justicidin A destroyed the following cells (IC_{50} values in parentheses): HT-29 (0.1 μM), human colorectal carcinoma (HCT-116; 0.4 μM), human cervical carcinoma (SiHa; 0.02 μM), human breast adenocarcinoma (MCF-7; 1.5 μM) and human bladder cancer (T24; 0.004 μM), via apoptosis involving pro-apoptotic Bcl-2-associated X protein (Bax) translocation, mitochondrial insult, collapse of mitochondrial membrane potential, release of cytochrome c and activation of caspases 9 and 3.[312] Furthermore, justicidin A induced the death of human hepatocellular carcinoma (Hep3B), human hepatocellular liver carcinoma (HepG2), human epitheloid cervix carcinoma (Hela) and peripheral blood mononuclear (PBMC) cells, with IC_{50} values equal to 0.04 μM, 0.05 μM, 0.9 μM and 23 μM, respectively,[313] implying that this lignan targets pro-apoptotic

■ **CS 3.267** Justicidin A

■ **CS 3.268** Diphyllin

■ **CS 3.269** Justicidin D

■ **CS 3.270** Taiwanin E

protein (p53)-compromised malignancies. In hepatoma cells justicidin A induced the activation of caspase 8, cleavage of Bid, mitochondrial insult, activation of caspases 9, 3 and 8, and apoptosis.[313] In addition, diphyllin from *Cleistanthus collinus* (Roxb.) Benth. (family Phyllanthaceae Martinov) arrested the proliferation of human gastric cancer (SGC-7901) cells by 50% at a dose of 7.8 μM via inhibition of V-ATPase and hence hypophosphorylation of low-density lipoprotein receptor-related protein 6 (LRP6), inhibition of β-catenin and a decrease in c-Myc and cyclin D1.[314] Taiwanin E (CS 3.270) from *Taiwania cryptomerioides* Hayata (family Taxodiaceae Saporta) repressed the growth of MCF-7 cells, with an IC_{50} value equal to 1.4 μg/mL.[315]

The presence of an ether moiety between C7 and C7′ on the dibenzylbutane framework of lignans forms furanolignans such as manassantin A (CS 3.271) from *Saururus chinensis* (Lour.) Baill. (family Saururaceae Rich. ex T. Lestib.). This inhibited nuclear factor kappa-light-chain-enhancer of activated B cells (NF-κB) in Hela cells, with an IC_{50} value equal to 2.5 μM,[316] and at a dose of 200 ng/mL destroyed human prostate cancer (PC-3; p53 null) cells via hypophosphorylation of Rb, a reduction in Bcl-2, an increase in pro-apoptotic Bax and activation of caspases 9, 3, and 8, and apoptosis.[317] Phosphorylation of STAT3 was not induced by interleukin-6 (IL-6) in Hep3B cells exposed to 60 μM of manassantin A.[318] *Bupleurum scorzonerifolium* Willd. (family Apiaceae Lindl.) contains the furanolignan isochaihulactone (CS 3.272), which abated the survival of the following cells (IC_{50} values in parentheses): A549 (18.5 μM), HT-29 (30.4 μM), MCF-7 (39.6 μM), HepG2 (12.1 μM), KB (23.2 μM), human non-small-cell lung carcinoma (H1299; 25.4 μM) and human ovarian carcinoma (OVCAR-3; 17.2 μM).[319] At a dose 60 μM isochaihulactone activated pro-apoptotic p53, and hence also activated caspases 9 and 3

■ **CS 3.271** Manassantin A

■ **CS 3.272** Isochaihulactone

■ **CS 3.273** Epiashantin

■ **CS 3.274** Syringaresinol

and the p21 protein, which was followed by a decrease in cyclin B1/cdc2 kinases and inhibition of tubulin polymerization in A549 (wild-type p53) cells.[319] *Artemisia absinthium* L. elaborates the difuranolignan epiashantin (CS 3.273) which was cytotoxic against SW480 cells, with an IC_{50} value equal to 9.8 μM.[307] Another example of a difuranolignan is syringaresinol (CS 3.274) from *Daphne genkwa* Siebold & Zucc. (family Thymelaeaceae Juss.), which induced both apoptosis and cell cycle arrest in HL-60 (p53 null) cells, with an IC_{50} equal to 5.8 μM, via translocation of pro-apoptotic Bax, release of cytochrome c, activation of caspases 9 and 3, cleavage of poly (ADP-ribose) polymerase (PARP) and induction of p21.[320] Sesamin

■ **CS 3.275** Sesamin

■ **CS 3.276** Taiwanin A

(CS 3.275) from *Sesamum indicum* L. (family Pedaliaceae R. Br.) inhibited the proliferation of MCF-7 cells to 45% at 100 μM, via the dephosphorylation of Rb protein and a fall in cyclin D1, and stopped multiplication of the following cells by 50% (dose levels in parentheses): human chronic myeloid leukemia (KBM-5; 42.7 μmol/L), human erythromyeloblastoid leukemia (K562; 48.3 μmol/L), human multiple myeloma (U266; 51.7 μmol/L), human prostate carcinoma (DU-145; 60.2 μmol/L), human colorectal carcinoma (HCT-116; 57.2 μmol/L), human pancreatic carcinoma (MIA PaCa-2; 58.3 μmol/L), human non-small-cell lung carcinoma (H1299; 40.1 μmol/L), and human breast adenocarcinoma (MDA-MB-231; 51.1 μmol/L).[321] At a dose of 100 μmol/L sesamin blocked the activation of NF-κB in human myeloma (RPMI-8226), human non-small-cell lung carcinoma (H1299) and human colorectal carcinoma (HCT-116) cells by inhibiting the enzymatic activity of IκB kinase (IKK).[321]

The presence of a covalent bond between C2 and C2′ yields a series of dibenzocyclooctadiene lignans such as taiwanin A (CS 3.276) from *Taiwania cryptomerioides* Hayata (family Taxodiaceae Saporta), which repressed the growth of MCF-7 cells, with an IC_{50} value equal to 0.5 μg/mL, via apoptosis.[315] In fact taiwanin A has a broad array of cytotoxic action as it inhibited the growth of the following cells (IC_{50} values in parentheses): HL-60 (2.4 μM), PC-3 (3.1 μM), OVCAR-3 (2.4 μM), human colon adenocarcinoma (CoLo 205; 2 μM) and human breast carcinoma (T-47D; 4 μM).[322] In addition, taiwanin A was lethal to HepG2 (wild-type p53) and Hep3B cells, with IC_{50} values equal to 0.6 μM and 28.5 μM, respectively,

■ **CS 3.277** Deoxyschizandrin –

■ **CS 3.278** Gomisin N –

■ **CS 3.279** Gomisin A

suggesting the implication of pro-apoptotic p53, and hence activation of caspases 9 and 3 and p21,[322] in hepatomas. The dibenzocyclooctadiene lignans deoxyschizandrin (CS 3.277) and gomisin N from a member of the genus *Schisandra* Michx (family Schisandraceae Blume) nullified the growth of HL-60 (p53 null) cells, with IC_{50} values of 60 μM and 56 μM, respectively, and associated with Bid cleavage, a fall in mitochondrial membrane potential, release of cytochrome c, activation of caspase 9 and 3, and apoptosis[323] independently of pro-apoptotic protein (p53). Indeed, gomisin N (CS 3.278) from *Schisandra chinensis* (Turcz.) Baill. (family Schisandraceae Blume) induced apoptosis in human histiocytic lymphoma (U937; mutant p53) cells at a dose of 150 μM, via reduction of Bcl-2, release of cytochrome c and activation of caspases 9 and 3,[324] whereas gomisin A was inert. In fact, gomisin A (CS 3.279) at a dose of 40 μM inhibited the proliferation of HCT-116 (wild-type p53) cells via caspase 7 activation and PARP cleavage[325] implying that gomisin A is cytotoxic against pro-apoptotic p53-competent cells. In multidrug-resistant human hepatocellular liver carcinoma (HepG2-DR) cells, gomisin A inhibited the enzymatic activity of Pgp-ATPase and hampered substrate binding to Pgp by compelling conformational changes at a dose of 200 μM.[326] Note that 100 μM of gomisin N inhibited the phosphorylation of IKK and therefore the degradation of IκBα, and therefore blockade of NF-κB and apoptosis in Hela cells.[327]

The blockage of NF-κB has been observed in simple biphenyl lignans such as magnoliol and hinokiol. Magnolol (CS 3.280) produced by *Houpoea obovata* (Thunb.) N.H. Xia & C.Y. Wu (family Magnoliaceae Juss.) imposed the

■ **CS 3.280** Magnolol

■ **CS 3.281** Honokiol

apoptosis of human fibrosarcoma (HT1080) at a dose of 100 μM via the activation of caspases 3 and 8.[328] Magnolol at a dose of 80 μM inhibited TNF-α activation of NF-κB in Hela, MCF-7, HL-60 and U937 cells by inhibition of IKK.[329] Magnolol was furthermore cytotoxic against PC-3, DU-145 and human prostate adenocarcinoma (LNCaP) cells, with IC_{50} values equal to 60 μM, 70 μM and 53 μM, respectively.[330] In PC-3 cells (p53 null) 60 μM of magnolol inhibited both PI3K and protein kinase B (Akt), resulting in translocation of pro-apoptotic Bax, mitochondrial insult, release of cytochrome c, activation of caspases 9, 3 and 8 and cleavage of PARP.[330] This mechanism was equally observed with honokiol (CS 3.281) from *Houpoea officinalis* (Rehder & E.H. Wilson) N.H. Xia & C.Y. Wu (family Magnoliaceae Juss.), which inhibited NF-κB activation by TNF-α in Hela, MCF-7, HL-60 and U937 cells via inhibition of IKK.[331] Honokiol also induced apoptosis in B-cell chronic lymphocytic leukemia (BCLL), with an IC_{50} value equal to 49 μM, with translocation of pro-apoptotic Bax, activation of caspases 9, 3 and 8, and cleavage of the DNA repair enzyme PARP.[332]

REFERENCES

[297] Ito K, Iida T, Ichino K, Tsunezuka M, Hattori M, Namba T. Obovatol and obovatal, novel biphenyl ether lignans from the leaves of *Magnolia obovata* Thunb. Chem Pharm Bull 1982;30(9):3347–53.

[298] Fukuyama Y, Otoshi Y, Kodama M. Structure of clovanemagnolol, a novel neurotrophic sesquiterpene-neolignan from *Magnolia obovata*. Tetrahedron Lett 1990;31(31):4477–80.

[299] Fukuyama Y, Otoshi Y, Miyoshi K, Hasegawa N, Kan Y, Kodama M. Structure of magnolianin, a novel trilignan possessing potent 5-lipoxygenase inhibitory activity. Tetrahedron Lett 1993;34(6):1051–4.

[300] Youn U, Chen QC, Lee IS, Kim H, Yoo JK, Lee J, et al. Two new lignans from the stem bark of *Magnolia obovata* and their cytotoxic activity. Chem Pharm Bull 2008;56(1):115–7.

[301] Zhou HY, Shin EM, Guo LY, Youn UJ, Bae K, Kang SS, et al. Anti-inflammatory activity of 4-methoxyhonokiol is a function of the inhibition of iNOS and

COX-2 expression in RAW 264.7 macrophages via NF-κB, JNK and p38 MAPK inactivation. Eur J Pharmacol 2008;586:340–9.

[302] Pyo MK, Koo YK, Yun-Choi HS. Anti-platelet effect of the phenolic constituents isolated from the leaves of *Magnolia obovata*. Nat Prod Sci 2002;8(4):147–51.

[303] Pyo MK, Yun-Choi HS, Hong Y-J. Antiplatelet activities of aporphine alkaloids isolated from leaves of *Magnolia obovata*. Planta Med 2003;69(3):267–9.

[304] Loike JD, Horwitz SB. Effects of podophyllotoxin and VP-16-213 on micro-tubule assembly in vitro and nucleoside transport in HeLa cells. Biochem 1976;15(25):5435–43.

[305] Park BY, Min BS, Kwon OK, Oh SR, Ahn KS, Kim TJ, et al. Increase of cas-pase-3 activity by lignans from *Machilus thunbergii* in HL-60 cells. Biol Pharm Bull 2004;27(8):1305–7.

[306] Berger NA, Chatterjee S, Schmotzer JA, Helms SR. Etoposide (VP-16-213)-induced gene alterations: potential contribution to cell death. Proc Natl Acad Sci USA 1991;88(19):8740–3.

[307] Hausott B, Greger H, Marian B. Naturally occurring lignans efficiently induce apoptosis in colorectal tumor cells. J Cancer Res Clin Oncol 2003;129(10):569–76.

[308] Matsumoto T, Hosono-Nishiyama K, Yamada H. Antiproliferative and apoptotic effects of butyrolactone lignans from *Arctium lappa* on leukemic cells. Planta Med 2006;72(3):276–8.

[309] Yoo J-H, Lee HJ, Kang K, Jho EH, Kim CY, Baturen D, et al. Lignans inhibit cell growth via regulation of Wnt/β-catenin signaling. Food Chem Toxicol 2010;48:2247–52.

[310] Jeong JB, Hong SC, Jeong HJ, Koo JS. Arctigenin induces cell cycle arrest by blocking the phosphorylation of Rb via the modulation of cell cycle regulatory proteins in human gastric cancer cells. Int Immunopharmacol 2011;11(10):1573–7.

[311] Fukamiya N, Lee K-H. Antitumor agents, 81. Justicidin-A and diphyllin, two cytotoxic principles from *Justicia procumbens*. J Nat Prod 1986;49(2):348–50.

[312] Lee J-C, Lee C-H, Su C-L, Huang C-W, Lin C-N, Won S-J. Justicidin A decreases the level of cytosolic Ku70 leading to apoptosis in human colorectal cancer cells. Carcinogenesis 2005;26(10):1716–30.

[313] Su C-L, Huang LLH, Huang L-M, Lee J-C, Lin C-N, Won S-J. Caspase-8 acts as a key upstream executor of mitochondria during justicidin A-induced apoptosis in human hepatoma cells. FEBS Lett 2006;580(13):3185–91.

[314] Shen W, Zou X, Chen M, Liu P, Shen Y, Huang S, et al. Effects of diphyl-lin as a novel V-ATPase inhibitor on gastric adenocarcinoma. Eur J Pharmacol 2011;667:330–8.

[315] Chang S-T, Wang DS-Y, Wu C-L, Shiah S-G, Kuo Y-H, Chang C-J. Cytotoxicity of extractives from *Taiwania cryptomerioides* heartwood. Phytochem 2000;55(3):227–32.

[316] Hwang BY, Lee J-H, Nam JB, Hong Y-S, Lee JJ. Lignans from *Saururus chinensis* inhibiting the transcription factor NF-κB. Phytochem 2003;64(3):765–71.

[317] Song SY, Lee I, Park C, Lee H, Hahm JC, Kang WK. Neolignans from *Saururus chinensis* inhibit PC-3 prostate cancer cell growth via apoptosis and senescence-like mechanisms. Int J Mol Med 2005;16(4):517–23.

[318] Chang JS, Lee SW, Kim MS, Yun BR, Park MH, Lee SG, et al. Manassantin A and B from *Saururus chinensis* inhibit interleukin-6-induced signal transducer and activator of transcription 3 activation in Hep3B cells. J Pharmacol Sci 2011;115(1):84–8.

[319] Chen YL, Lin SZ, Chang JY, Cheng YL, Tsai NM, Chen SP, et al. In vitro and in vivo studies of a novel potential anticancer agent of isochaihulactone on human lung cancer A549 cells. Biochem Pharmacol 2006;72(3):308–19.

[320] Park B-Y, Oh S-R, Ahn K-S, Kwon O-K, Lee H-K. (–)-Syringaresinol inhibits proliferation of human promyelocytic HL-60 leukemia cells via G1 arrest and apoptosis. Int Immunopharmacol 2008;8(7):967–73.

[321] Harikumar KB, Sung B, Tharakan ST, Pandey MK, Joy B, Guha S, et al. Sesamin manifests chemopreventive effects through the suppression of NF-κB-regulated cell survival, proliferation, invasion, and angiogenic gene products. Mol Cancer Res 2010;8(5):751–61.

[322] Ho P-J, Chou C-K, Kuo Y-H, Tu L-C, Yeh S-F. Taiwanin A induced cell cycle arrest and p53-dependent apoptosis in human hepatocellular carcinoma HepG2 cells. Life Sci 2007;80(5):493–503.

[323] Lin S, Fujii M, Hou D-X. Molecular mechanism of apoptosis induced by schizandrae-derived lignans in human leukemia HL-60 cells. Food Chem Toxicol 2008;46(2):590–7.

[324] Kim JH, Choi YW, Park C, Jin CY, Lee YJ, Park DJ, et al. Apoptosis induction of human leukemia U937 cells by gomisin N, a dibenzocyclooctadiene lignan, isolated from *Schizandra chinensis* Baill. Food Chem Toxicol 2010;48(3):807–13.

[325] Hwang D, Shin SY, Lee Y, Hyun J, Yong Y, Park JC, et al. A compound isolated from Schisandra chinensis induces apoptosis. Bioorg Med Chem Lett 2011;21(20):6054–7.

[326] Wan C-K, Zhu G-Y, Shen X-L, Chattopadhyay A, Dey S, Fong W-F. Gomisin A alters substrate interaction and reverses P-glycoprotein-mediated multidrug resistance in HepG2-DR cells. Biochem Pharmacol 2006;72(7):824–37.

[327] Waiwut P, Shin M-S, Inujima A, Zhou Y, Koizumi K, Saiki I, et al. Gomisin N enhances TNF-α-induced apoptosis via inhibition of the NF-κB and EGFR survival pathways. Mol Cell Biochem 2011;350(1–2):169–75.

[328] Ikeda K, Nagase H. Magnolol has the ability to induce apoptosis in tumor cells. Biol Pharml Bull 2002;25(12):1546–9.

[329] Tse AK-W, Wan C-K, Zhu G-Y, Shen X-L, Cheung H-Y, Yang M, et al. Magnolol suppresses NF-[kappa] B activation and NF-[kappa] B regulated gene expression through inhibition of IkappaB kinase activation. Mol Immunol 2007;44(10):2647–58.

[330] Lee DH, Szczepanski MJ, Lee YJ. Magnolol induces apoptosis via inhibiting the EGFR/PI3K/Akt signaling pathway in human prostate cancer cells. J Cell Biochem 2009;106(6):1113–22.

[331] Tse AK-W, Wan C-K, Shen X-L, Yang M, Fong W-F. Honokiol inhibits TNF-α-stimulated NF-κB activation and NF-κB-regulated gene expression through suppression of IKK activation. Biochem Pharmacol 2005;70(10):1443–57.

[332] Battle TE, Arbiser J, Frank DA. The natural product honokiol induces caspase-dependent apoptosis in B-cell chronic lymphocytic leukemia (B-CLL) cells. Blood 2005;106(2):690–7.

Index of Natural Products

A

Abietatrien-3β-ol, 136
Acacetin, 267–268
Aceriphyllic acid A, 225–244
12-Acetoxyamoorastatin, 193–208
3β-Acetoxycycloartan-24-ene-1α,2α-diol, 162–165, 163f
8-Acetoxy-4,10-dihydroxy-2,11(13)-guaiadiene-12,6-olide, 115
6α-Acetoxy-14,15β-dihydroxyklaineanone, 184–186, 187f
1β-Acetoxyeudesman-4(15), 7(11)-dien-8α,12-olide, 101–102, 102f
1β-Acetoxy, 8β-hydroxyeudesman-4(15), 7(11)-dien-8α,12-olide, 101–102, 102f
(+)-8α-Acetoxy-4α-hydroxyguaia-1(10), 2,11(13)-trien-12,6α-olide, 112–118, 115f
6α-Acetoxy-15β-hydroxyklaineanone, 184–186, 187f
17-Acetoxyjolkinolides A and B, 99, 133–134
19-Acetoxyl-ent-3β,17-dihydroxykaur-15-ene, 118–119
19-Acetoxyl-ent-3β-hydroxykaur-15-en-17-al, 118–119
1-Acetoxy-3-methoxy-9,10-anthraquinone, 354–361
3β-Acetoxy-olean-12-en-27-oic acid, 225–244, 226f
11-O-Acetylalkannin, 346–347
N-Acetylanonaine, 369–370
3-O-Acetyl-16-O-p-bromobenzoylpachysandiol B, 73
25-Acetylcucurbitacin F, 169–170
1-Acetyl-2-deacetyltrichilin H, 193–208
1-Acetyl-3-deacetyltrichilin H, 193–208
12-Acetyl-13,18-dihydroeurycomanone, 184–186
7α-Acetyldihydronomilin, 177
19-Acetyl-ent-3-β,17-dihydroxykaur-15-ene, 118–119
17-O-Acetylechitamine, 28–29
15-Acetyl-13α(18)-epoxyeurycomanone, 184–186
2′-(R)-O-Acetylglaucarubinone, 187–189
19-Acetyl-ent-3 β-hydroxykaur-15-en-17-al, 118–119
15β-Acetyl-14-hydroxy-klaineanone, 184–186, 187f

19-Acetyl-ent-3β-hydroxyl-kaur-16-ene, 118–119
Acetyl-11-keto-β-boswellic acid, 98, 243–244
7α-Acetylobacunol, 192–193
1-Acetyltrichilin H, 193–208
N-Acetylxylopine, 369–370
Actinodaphnine, 54–55
Acutaxylines A and B, 81
Agnuside, 136
Ajmalicine, 26–27
19-epi-Ajmalicine, 16–17
Akuammicine, 22–23, 28–29
Alamaridine, 35–37
Alangimarckine, 35–37
Alangimaridine, 35–37
Alangimarine, 35–37
Alangine, 35–37
Alangiside, 35–37
Alangiumkaloid A, 2, 37–38, 38f
Alismaketone 23-acetate, 147–149
Alismorientols A and B, 147–149
Alismoxide 10-O-methyl ether, 147–149
Alisols A–C, F–I, O and P, 147–149
Alisol A acetate, 147–149
Alisol B acetate, 97–98, 149–151
Alisol C acetate, 147–149
Alisol E acetate, 147–149
Alisol E 23 acetate, 147–149
Alisol J acetate, 147–149
Alisol K acetate, 147–149
Alisol L 23 acetate, 147–149
Alisol M 23 acetate, 147–149
Alisol N 23 acetate, 147–149
Alismol, 147–149
Alismoxide, 147–149
Alisolide, 147–149
Alizarin, 354–355
Alizarin 2-methyl ether, 268–269, 354–361
Alkannin, 346–347
Allanxanthone A, 310–313
Allocryptopin, 57
Alloyohimbine, 16–17
Aloe-emodin, 268–269, 357–360
Alstogustine, 28–29
Alstoyunines A–H, 16–17
Alvaxanthone, 310–313
Amazoquinone, 235–242
Amyrin, 28–29
Amyrin acetate, 28–29

14-Angeloyloxy-2α,3α-epoxy-1-oxo-O-methylcacalol, 120
3β-(Angeloyloxy)eremophil-7(11)-en-12,8β-olid-14-oic acid, 120–121, 120f
Angustifolin, 138–139
Angustilobine B, 28–29
6,7-seco-Angustilobine B, 28–29
25-Anhydroalisol, 148
25-Anhydroalisol A, 148
25-Anhydroalisol A 24 acetate, 147–149
(−)-Anolobine, 52–53
Anonaine, 1
Anthothecol, 197–200
Antofine, 86–92
(−)-10α-Antofine N-oxide, 86
(+)-Anwulignan, 152–153
Anwuweizic acid, 152–153
Apigenin, 100–101, 108, 268–269, 275–278, 280–285
Arcommunol A, 268–269, 283–285
Arctigenin, 372–373
Argentatin A, 162–165
Argenteanones A and E, 162–165
Aristofolins A–D, 38–39
Aristolactams I, II and IIIa, 38–39
Aristolactams AII, AIIIa, BII and FII, 38–40, 40f, 42–43, 52–53
Aristolic acid, 38–39
Aristolic acid II, 42
Aristolic acid methyl ester, 38–39
Aristolochic acid, 38–40, 42
Aristolochic acids I, II, IIIa, IV, IVa and VIIa, 38–39, 42
Aristolochic acids C and D, 38–39
Aristolochic acid I methyl ester, 42
Aristolochic acid II methyl ester, 42
Aristolochic acid III methyl ester, 42
Aristolochic acid IV methyl ester, 42
Aristolochic acid IVa methyl ester, 42
Aristolochic acid VII methyl ester, 42
Aristoloside, 42–43
Artelastin, 283
Artemetin, 136
Artobiloxanthon, 313–320
Artoindonesianin P, 313–320
Artonin B, 283
Artonol B, 313–320
Arucanolide, 109–110
Astragalin, 168
Atractylenolides I and II, 97–98

381

Index of Oncological and Related Terms

Human epidermoid carcinoma (9-KB) cells, 19–20, 40

Human epidermoid cervical carcinoma (CaSki) cells, 217–219, 358

Human epithelial cancer (Hela S3) cells, 173, 202–203, 348–350

Human epithelial carcinoma (A-431) cells, 40, 42–43, 139–141, 235–242

Human epithelial ovarian cancer (HO-8910) cells, 102, 225–244

Human epithelial tumor (HONE-1) cells, 86–92

Human epitheloid cervix carcinoma (Hela) cells, 4, 26–27, 54–55, 57–59, 81–84, 101, 106, 109–110, 131, 139–141, 161–162, 176–177, 202–203, 346–347, 358–360

Human erythromyeloblastoid leukemia (K562) cells, 9–10, 50–51, 54–55, 109–110, 120, 134–135, 162–165, 321–324, 345–346, 348–350, 374–376

Human fibrosarcoma (HT1080) cells, 86–92, 149–151, 175–176, 216, 235–242, 351–353, 377–378

Human foreskin fibroblast (HS68) cells, 242–243

Human gastric cancer (AZ521) cells, 153–166

Human gastric cancer (BGC-823) cells, 50–51, 57–59, 200–202, 310–320

Human gastric cancer (MKN-45) cells, 217–219

Human gastric cancer (SGC-7901) cells, 103, 150–151, 310–320, 351–353, 373–374

Human gastric cancer (SNU-638) cells, 86–92

Human gastric carcinoma (AGS) cells, 272–273, 313–320

Human gastric carcinoma (MGC803) cells, 105–107

Human gastric carcinoma (SNU-5) cells, 37–38

Human gastric carcinoma (TMK-1) cells, 86–92

Human gingival cancer (CA9-22) cells, 221–223, 353

Human glioblastoma (LN229) cells, 275–277

Human glioblastoma (SF268) cells, 6, 37–38, 169–172, 235, 283, 295–298, 313–320

Human glioblastoma (T98G 7) cells, 275–277

Human glioblastoma (U251) cells, 120, 162–165, 193–208, 306–328, 351–353

Human glioblastoma (U-343) cells, 215

Human glioblastoma (U373) cells, 216, 235–242

Human glioblastoma (U87) cells, 12–13, 231–235

Human glioblastoma-astrocytoma, epithelial-like (U-MG87) cells, 105–107, 275–279, 345–346

Human hepatic (L02) cells, 102

Human hepatic (WRL-68) cells, 242–243

Human hepatocellular carcinoma (BEL-7402) cells, 50–51, 57–59, 177, 273–274, 345–346

Human hepatocellular carcinoma (BEL-7404) cells, 193–208

Human hepatocellular carcinoma (HuH-7) cells, 18–19, 157–158, 345–346

Human hepatocellular carcinoma (PLC5), 283–285

Human hepatocellular carcinoma cell lines Mahlavu cells, 54–55

Human hepatocellular liver carcinoma (HepG2) cells, 7, 23–24, 37–38, 45–46, 57–59, 89–92, 101–103, 112–118, 120, 131, 134–135, 139–141, 161–162, 175, 214, 217–219, 231–235, 272–273, 280–283, 290, 351–353, 357–358, 373–374, 376–377

Human hepatoma (Hep3B) cells, 45–46, 176–177, 221–223, 272–273, 275–277, 283–285, 353, 357–358, 373–374, 376–377

Human hepatoma (PLC/PRF/5) cells, 272–273, 275–277, 279–283, 360–361

Human hepatoma (SH-J1) cells, 109–110

Human hepatoma (SMMC-7721) cells, 29–30, 102, 310–320, 327–328

Human histiocytic lymphoma (U937) cells, 71–72, 103, 175–176, 193–208, 217–219, 221–223, 228–229, 272–273, 275–277, 290–291, 299–300, 351–353, 377–378

Human ileocecal adenocarcinoma (HCT-8) cells, 50–51, 115, 343–344

Human keratinocyte (HaCaT) cells, 7, 242–243, 343–344

Human kidney carcinoma (Caki-1) cells, 171–172

Human large-cell immunoblastic lymphoma (SR) cells, 175–176

Human large-cell lung cancer (NCI-H460) cells, 6, 105–107, 169–170, 283, 295–298, 313–320

Human laryngeal carcinoma (Hep-2) cells, 140, 216, 231–235

Human leukemia (CEM) cells, 150

Human leukemia (CEM-C7) cells, 45–46

Human leukemia (HPB-ALL) cells, 140–141

Human leukemic lymphoblast (CCRF-CEM) cells, 165–166, 171–172, 175–176, 243–244, 283

Human liver adenocarcinoma (SK-HEP-1) cells, 54–55, 283–285

Human liver cancer (BEL-7402) cells, 50–51, 57–59, 177, 273–274, 345–346

Human lung adenocarcinoma (Lu-1) cells, 159–161, 298

Human lung adenocarcinoma (SPC-A-1) cells, 139–141

Human lung adenocarcinoma epithelial (A549) cells, 24, 26–27, 29–30, 37–38, 40, 48, 50–51, 57–59, 67–69, 86–92, 101, 105–107, 115, 120, 130–132, 150–151, 153–166, 173, 181–182, 193–208, 217–219, 278–279, 290, 298, 310–313, 344, 358, 370–371

Human lung cancer (95-D) cells, 153–166

Human lung cancer (H522) cells, 197–200

Human lung carcinoma (CH27) cells, 272–273, 357–360

Human lung fibroblast (HLF) cells, 277–278

Human lung giant-cell carcinoma (Lu-65) cells, 217–219

Human lung large-cell carcinoma (Lu-99) cells, 217–219, 221–223

Human lung tumor (CHAGO) cells, 161–165, 203–205

Human lung tumor (VA-13) cells, 112–118

Human lymphoblast (RPMI 8402) cells, 37–38

Human lymphoblastoid (BM-13674) cells, 45–46

Human lymphoblastoid leukemia (REH) cells, 130–132

Human lymphocyte (MT1) cells, 140–141

Human lymphocyte (MT2) cells, 140–141

Human lymphoma (BC-1) cells, 228–229, 235–242, 298

Index of Plants

Subject Index

A

Acronychia esquirolii H. Lév.. *See* Alstonia
Yunnanensis Diels
akar kalong. *See* Piper Caninum Blume
Alangium lamarckii Thwaites.
See Alangium Salviifolium (L.f.)
Wangerin
Alangium Salviifolium (L.f.) Wangerin
common names, 35
family, 35
habitat and description, 35
history, 35
illustration, *36*
medicinal uses, 35
phytopharmacology, 35–37
proposed research, 37
rationale, 37–38
synonyms, 35
Alisma Orientale (Sam.) Juz.
common names, 147
family, 147
habitat and description, 147
history, 147
illustration, *148*
medicinal uses, 147
phytopharmacology, 147–149
proposed research, 149
rationale, 149–151
synonyms, 147
Alisma plantago-aquatica L. var.
orientale Sam. *See Alisma Orientale*
(Sam.) Juz.
Alisma plantago-aquatica subsp. *orientale*
(Sam.) Sam. *See Alisma Orientale*
(Sam.) Juz.
alkaloids. *See also specific compounds*
overview, 1–2
amide, 2–15
indole, 16–34
isoquinoline, 35–66
terpenoid, 67–80
other alkaloids, 81–95
Alstonia esquirolii H. Lév.. *See* Alstonia
Yunnanensis Diels
Alstonia glaucescens (Wallich ex G. Don)
Monachino. *See* Winchia Calophylla
A. DC.
Alstonia pachycarpa Merrill & Chun. *See*
Winchia Calophylla A. DC.

Alstonia rostrata C.E.C. Fisch. *See*
Winchia Calophylla A. DC.
Alstonia Yunnanensis Diels
common names, 16
family, 16
habitat and description, 16
history, 16
illustration, *17*
medicinal uses, 16
phytopharmacology, 16–17
proposed research, 17
rationale, 17–19
synonyms, 16
Alyxia glaucescens Wallich ex G. Don. *See*
Winchia Calophylla A. DC.
amide alkaloids, 2–15. *See also specific
compounds*
amuyon. *See Goniothalamus Amuyon*
(Blco.) Merr.
Anemone chinensis Bunge. *See* Pulsatilla
Chinensis (Bunge) Regel
Anemone pulsatilla var. chinensis (Bunge)
Finet & Gagnep. *See* Pulsatilla
Chinensis (Bunge) Regel
ankol. *See Alangium Salviifolium* (L.f.)
Wangerin
Aquilegia fauriei H. Lév.. *See* Dictamnus
Dasycarpus Turcz.
argadi. *See* Piper longum L.
Aristolochia Cucurbitifolia Hayata
common name, 38
family, 38
habitat and description, 38
history, 38
illustration, *39*
medicinal uses, 38
phytopharmacology, 38–39
proposed research, 39
rationale, 40
Aristolochia manshuriensis Kom.
common names, 41
family, 41
habitat and description, 41
history, 40
illustration, *41*
medicinal uses, 42
phytopharmacology, 42
proposed research, 42
rationale, 42–43

synonyms, 41
Atractylis lancea var. *chinensis*
(Bunge) Kitam. *See* Atractylodes
Macrocephala Koidz.
Atractylis macrocephala (Koidz.)
Hand.-Mazz. *See* Atractylodes
Macrocephala Koidz.
Atractylis macrocephala (Koidz.) Nemoto.
See Atractylodes Macrocephala Koidz.
Atractylis macrocephala var. *hunanensis* Y.
Ling. *See* Atractylodes Macrocephala
Koidz.
Atractylodes Macrocephala Koidz
common names, 100
family, 100
habitat and description, 100
history, 99
illustration, *100*
medicinal uses, 100
phytopharmacology, 100–101
proposed research, 101
rationale, 101–103
synonyms, 99

B

bag sun. *See* Dictamnus Dasycarpus Turcz.
bai shu. *See* Atractylodes Macrocephala
Koidz.
bai tou weng. *See* Pulsatilla Chinensis
(Bunge) Regel
bai xian (bi xian pi). *See* Dictamnus
Dasycarpus Turcz.
bangkal. *See* Nauclea Orientalis Willd
beach vitex. *See* Vitex Rotundifola L.F
benzopyrones, 269–340. *See also specific
compounds*
bi ba. *See* Piper longum L.
Blume, Carl Ludwig von, 5, 67
Bocconia microcarpa Maxim..
See Macleaya Microcarpa (Maxim.)
Fedde
Brucea Javanica (L.) Merr
common name, 179
family, 179
habitat and description, 179
history, 178
illustration, *179*
medicinal uses, 179
phytopharmacology, 180

403

Printed and bound by CPI Group (UK) Ltd, Croydon, CR0 4YY

08/05/2025

01864979-0002